Computational Techniques for Smart Manufacturing in Industry 5.0

Methods and Applications

Editors

Kim Phuc Tran

Ph.D., Senior Associate Professor
University of Life, GEMTEX, Lille, France

Zhenglei He

Ph.D., Assistant Professor
South China University of Technology
Guangzhou, China

CRC Press
Taylor & Francis Group
Boca Raton London New York

CRC Press is an imprint of the
Taylor & Francis Group, an **informa** business
A SCIENCE PUBLISHERS BOOK

First edition published 2025
by CRC Press
2385 NW Executive Center Drive, Suite 320, Boca Raton FL 33431

and by CRC Press
4 Park Square, Milton Park, Abingdon, Oxon, OXl 4 4RN

CRC Press is an imprint of Taylor & Francis Group, LLC

Library of Congress Cataloging-in-Publication Data (applied for)

ISBN: 978-1-032-50620-3 (hbk)
ISBN: 978-1-032-50621-0 (pbk)
ISBN: 978-1-003-39932-2 (ebk)

DOI: 10.1201/9781003399322

Typeset in Times New Roman
by Prime Publishing Services

Preface

We are witnessing the rapid development and powerful application of advanced computational technologies, leading to the 5th industrial revolution. Industry 5.0 will be characterized by the cooperation between machines and human beings, with the ultimate aim to give added value to production, by creating personalized products able to meet customers' requirements. In the context of Industry 5.0, intelligent manufacturing systems have attracted various industrial sectors (e.g., automobiles, power supply, chemistry) in order to realize data-driven innovations for delivering highly customizable products and services faster, cheaper, better, and greener. In an intelligent manufacturing system, computational techniques constitute a key issue for providing efficient decision support tools based on data measured through the Internet of Things and ubiquitous sensing mechanisms. In complex industrial production scenarios, these computational techniques enable the enhancement of the quality of products and their manufacturing processes as well as sustainability (i.e., minimization of environmental and social impacts), and reduce uncertainties related to human factors. This book, consisting of 14 chapters, presents recent advancements in research, new methods and techniques, and applications of advanced computational technologies in intelligent manufacturing for modeling, simulating, optimization, decision-making, etc. which are very typical problems in the manufacturing processes. The book undertakes to stimulate scientific exchange, ideas, and experiences in the field of intelligent manufacturing applications. Researchers and practitioners alike will benefit from this book to enhance their understanding of Industry 5.0 which focuses on combining human creativity and craftsmanship with the speed, productivity, and consistency of AI systems. The real-world case studies in various fields with guidance and recommendations for the practical applications of these studies are introduced in each chapter. Status, trends, future directions, opportunities, etc. will be discussed, making it friendly for beginners and young researchers.

Finally, I would like to thank all the chapter authors and reviewers for their selfless contribution and support to this work.

Zhenglei He
Guangzhou, China
April 2024

Contents

1. Introduction to Computational Techniques for Smart Manufacturing in Industry 5.0: Methods and Applications

Zhenglei He[1,*] and Kim Phuc Tran[2,3,*]

1 Scope of the Research Domain

We are in the midst of rapid development and an era of using powerful ap-plications of advanced technologies, resulting in the 5th industrial revolution. The wide use of cyber-physicl systems and the Internet ofThings have led to the era of Big Data, Yau et al.[1] . Advanced computational technolo-gies and their broad applications have attracted various industrial sectors (e.g. automobile, power supplying, chemistry) in order to realize data-driven innovations for delivering highly customizable products and services faster, cheaper, better, and greener [2], for more details, see Javaid and Haleem [3]; Xu et al. [4]; Nahavandi [5]; George and George [6]; Breque et al. [7]; Ha et al. [8]. In a smart manufacturing system, computational techniques constitute a key issue for providing efficient decision support tools based on data measured through the Internet of

[1] State Key Laboratory of Pulp and Paper Engineering, South China University of Technology, 510640, China.

[2] University of Lille, ENSAIT, GEMTEX, F-59000 Lille, France.

[3] International Chair in DS & XAI, International Research Institute for Artificial Intelligence and Data Science, Dong A University, Danang, Vietnam.

* Corresponding author: hezhenglei@scut.edu.cn; kim-phuc.tran@ensait.fr

Things and ubiquitous sensing mechanisms. In complex industrial production scenarios, these computational techniques enable enhancement of the quality of products and their manufacturing processes as well as the sustainability (i.e. minimization of environmental and social impacts on the plant), and reduce uncertainties related to human factors. However, there is currently no reference book presenting systematically and in detail the methods and applications of computational techniques in the context of smart manufacturing. This book presents recent advancements in research, new methods and techniques, and applications of advanced computational technologies in smart manufacturing for modeling, simulating, optimization, and decision-making. which are very typical problems in the manufacturing processes. Computational techniques extract knowledge from big data effectively and interpret decisions. It explores deep learning, cyber-physical systems, reinforcement learning for multi-variable modeling[10;11;12], multi-objective optimization[13;14;15], multi-criteria decision-making[16], and multi-agent systems[9] in smart manufacturing with big data. The book undertakes to stimulate scientific exchange, ideas, and experiences in the field of smart manufacturing applications. Researchers and practitioners alike will benefit from this book to enhance their understanding of computational technologies, and use in smart manufacturing scenarios. Real-world case studies in various fields with guidance and recommendations for the practical applications of these studies are introduced in each chapter.

2 Structure of the Book

This book presents recent advancements in research, new methods, and techniques with applications of computational techniques for smart manufacturing. It explores prediction models for papermaking process monitoring[10], surrogate models for greenhouse gas emission prediction[17], optimization of the hybrid energy system[18], a lossy compression of noisy images, recommendation and design of personalized garments, evaluating garment fit levels, control charts, multi-agent systems, raw material selection optimization, process fault diagnosis, scheduling of flexible flow shops and, industrial digital twins. The book contains 14 chapters.

In the Introductory chapter "Introduction to Computational Techniques for smart manufacturing: Methods and Applications", the book editors Zhenglei He and Kim Phuc Tran elaborate the peculiarities of computational techniques in smart manufacturing. They determine recent research streams and summarize the structure and contribution of the book.

Jiwei Qian, Zhenglei He, Yi Man, Jigeng Li, and Mengna Hong develop in their chapter data-driven soft sensors to monitor key paper properties and apply the models to optimize the papermaking process. Four different data-driven methods, including random forest, gradient boosting regression, ridge regression, and K-nearest neighbor, were employed to establish soft sensor models for monitoring folding endurance, bursting strength, smoothness, and transverse ring compressive strength. The developed soft sensors can make a significant contribution to the papermaking industry.

Hongze Zheng, Zhenglei He, Mengna Hong, and Yi Man released a Kriging surrogate model with high precision in their chapter to calculate the greenhouse gas emissions generated in the papermaking wastewater treatment process. Based on the Kriging model of GHG emission, it also analyzes the impact of different operating parameters on GHG emissions. The model supports that reasonable optimization and control of parameters could significantly reduce direct GHG emissions in the biochemical reaction process.

Vladimir Lukin, Bogdan Kovalenko, Sergii Kryvenko, Victoriya Naumenko, Benoit Vozel consider the task of lossy compression of noisy images by employing Better Portable Graphics (BPG). It is demonstrated that an Optimal Operation Point (OOP) is possible for this noise type as well. Finally, it is shown that the proposed prediction method is valid for images of different fields.

Zhujun Wang, Xuyuan Tao, Xianyi Zeng, Yingmei Xing, Zhenzhen Xu, and Pascal Bruniaux put forth a new interactive personalized garment design recommendation approach based on a series of intelligent techniques, including fuzzy logic, genetic algorithms, and support vector regression. This system can be easily combined with classical 3D clothing design software (e.g. Lectra Modaris 3D, Clo 3D, Vstitcher) to form a new interactive personalized garment design process, enhancing the level of mass customization for various fashion brand companies. It can effectively characterize relations between consumer profiles and garment profiles to find the most relevant garment design solution. It can also be used for the estimation of the garment design technical parameters.

Zhujun Wang, Yingmei Xing, Xianyi Zeng, Pascal Bruniaux, and Xuyuan Tao put forward a new garment fit level prediction approach using Probabilistic Neural Network (PNN) technology. The PNN model performs better in the prediction of garment fit levels for both loose and tight garments. It can predict not only the comprehensive fit levels, but also the local fit levels of the garments fast, precisely, and automatically. It provides a feasible and efficient solution for perceiving the garment fit levels without real try-ons, which can be utilized in both online and in-store shopping to enhance the consumers' satisfaction. Additionally, the model can be enhanced continuously by learning new datasets.

Aamir Saghir, Kim Duc Tran, and Kim Phuc Tran have discussed Machine learning-based control charts for anomaly detection, specifically for smart manufacturing in the new digital era in the future generation of the industry. Machine learning has demonstrated its effectiveness in various applications;

however, several challenges should be addressed, including stationarity problems, hyperparameter optimizations, resource restrictions, and interpretability. In addition, several state-of-the-art approaches and methods to address the limitations of ML-based anomaly detection are also presented. After that, perspectives of combining solutions toward efficient and explainable ML-based control charts for anomaly detection systems based on deep learning are shown through an end-to-end framework.

Thi Thuy Van Nguyen, Cédric Heuchenne, Kim Duc Tran, and Kim Phuc Tran discuss in their chapter, a review of several research studies, providing a comprehensive overview of control charts for monitoring the ratio of two normal random variables and Compositional Data.

Zhiqiang Zeng, Le Gao, Zhiyong Hong, and Wenyi Liang study the problem of energy-aware scheduling of a two-stage flexible flow shop, considering the inconsistency of the process differences of jobs. Group technology is used to reduce setup energy consumption, and dynamic split strategy is used to solve the problem of unbalanced machine load caused by group technology. The experimental results show that the group technology can not only reduce the energy consumption, but also reduce the make-span, and the dynamic split strategy can effectively balance the load of the machine. By comparing with the existing excellent algorithms, it shows that the improved multi-objective scatter search algorithm performs better.

Chang Liu, Zhenglei He, Yi Man, Jigeng Li, and Mengna Hong launched a representative day acquisition model and applied kernel density estimation of wind and solar power generation, establishing a multi-energy synergy dispatching model considering the combination of economy and low-carbon objectives. The model operation benefits are demonstrated in four typical scenarios.

Xiangyao Liang, Ningkui Wang, Mengna Hong, Yi Man, and Zhenglei He propose a trust estimation model in multi-agent systems, and it is concluded that direct trust resolves uncertainties caused by many factors, and it enables agents to react quickly to unexpected changes in performance. Rational confidence is introduced in the accepted testimony in three aspects: consistency, which reflects the quality of the evidence; credibility, indicating its reliability in the light of past experience; and certainty of the testimony estimated from the perspective of entropy.

Xiuli He, Rongyuan Wang, Mengna Hong, and Zhenglei He propose a batch combination model with a formula calculation model based on the logical combination algorithm and improved Sequential Quadradic Programming (SQP) algorithm. These models were verified in a case study of glass production, outperforming manual calculations. The batch combination model, reduce the deviation of quality indicators from standard values by 0.89 percent, and the standard deviation among combinations by 3 percent.

Qingyuan Zhang, Zhenglei He, Yi Man, Jigeng Li, and Mengna Hong present the construction of a knowledge graph for the paper drying process, as well as a knowledge graph for identifying and diagnosing faults. The utilization of such knowledge graphs has resulted in significant progress toward the identification and resolution of certain drying process faults.

Yishui Zhang, Rongyuan Wang, Mengna Hong, and Zhenglei He take the drying section of a paper machine as the research object. Through the analysis of the paper-making process, based on industrial data, process mechanism, and intelligent algorithms, the paper-making process parameters prediction and solution are conducted based on digital twin to strengthen the sustainable development of the pulp and paper industry.

3 Conclusion

This book is dedicated to promoting scientific exchange, ideas, and experiences in the field of computational techniques in smart manufacturing. Researchers, graduate students, and practitioners alike will benefit from this book to enhance their understanding of machine learning, optimization models, and their application to industrial practices such as decision support or industrial control. Real-world case studies in various fields are presented with guidance and advice for the practical application of these studies in each chapter.

References

[1] Kok-Lim Alvin Yau, Heejeong Jasmine Lee, Yung-Wey Chong, Mee Hong Ling, Aqeel Raza Syed, Celimuge Wu, and Hock Guan Goh. Augmented intelligence: Surveys of literature and expert opinion to understand relations between human intelligence and artificial intelligence. *IEEE Access*, 9:136744–136761, 2021.

[2] K.P. Tran S. Thomassey X. Zeng C. Yi Z. He, J. Xu. Modeling of textile manufacturing processes using intelligent techniques: a review. *The International Journal of Advanced Manufacturing Technology*, pages 1–29, 2021.

[3] Mohd Javaid and Abid Haleem. Critical components of industry 5.0 towards a successful adoption in the field of manufacturing. *Journal of Industrial Integration and Management*, 5(03):327–348, 2020.

[4] Xun Xu, Yuqian Lu, Birgit Vogel-Heuser, and Lihui Wang. Industry 4.0 and industry 5.0—inception, conception and perception. *Journal of Manufacturing Systems*, 61:530–535, 2021.

[5] Saeid Nahavandi. Industry 5.0—a human-centric solution. *Sustainability*, 11(16):4371, 2019.

[6] A SHAJI George and AS HOVAN George. Industrial revolution 5.0: the transformation of the modern manufacturing process to enable man and machine to work hand in hand. *Journal of Seybold Report ISSN NO*, 1533:9211, 2020.

[7] Maija Breque, Lars De Nul, and Athanasios Petridis. Industry 5.0: towards a sustainable, human-centric and resilient european industry. *Luxembourg, LU: European Commission, Directorate-General for Research and Innovation*, 2021.

[8] Do Thu Ha, Nguyen Xuan Hoang, Nguyen Viet Hoang, Nguyen Huu Du, Truong Thu Huong, and Kim Phuc Tran. Explainable anomaly detection for industrial control system cybersecurity. *arXiv preprint arXiv:2205.01930*, 2022.

[9] Zhenglei He, Kim Phuc Tran, Sebastien Thomassey, Xianyi Zeng, Jie Xu, and Changhai Yi. Multi-objective optimization of the textile manufacturing process using deep-q-network based multi-agent reinforcement learning. *Journal of Manufacturing Systems*, 62:939–949, 2022.

[10] Zhenglei He, Jiwei Qian, Jigeng Li, Mengna Hong, and Yi Man. Data-driven soft sensors of papermaking process and its application to cleaner production with multi-objective optimization. *Journal of Cleaner Production*, 372:133803, 2022.

[11] Jie Xu, Zhenglei He, Sheng Li, and Wenbo Ke. Production cost optimization of enzyme washing for indigo dyed cotton denim by combining kriging surrogate with differential evolution algorithm. *Textile Research Journal*, 90(15-16):1860–1871, 2020.

[12] Zhenglei He, Kim-Phuc Tran, Sébastien Thomassey, Xianyi Zeng, Jie Xu, and Yi Changhai. Modeling color fading ozonation of reactive-dyed cotton using the extreme learning machine, support vector regression and random forest. *Textile Research Journal*, 90(7-8):896–908, 2020.

[13] Huanhuan Zhang, Jigeng Li, Mengna Hong, Yi Man, and Zhenglei He. Cost optimal production-scheduling model based on vns-nsga-ii hybrid algorithm—study on tissue paper mill. *Processes*, 10(10):2072, 2022.

[14] Zhenya Zhang, Xiuli He, Yi Man, and Zhenglei He. Multi-objective scheduling in dynamic of household paper workshop considering energy consumption in production process. *Journal of Smart Environments and Green Computing*, 3:87–105, 2023.

[15] Jie Xu, Feng Liu, Zhenglei He, Zongao Zhang, and Sheng Li. Cost optimization of sodium hypochlorite bleaching washing for denim by combining ensemble of surrogates with particle swarm optimization. *Journal of Engineered Fibers and Fabrics*, 16:15589250211022331, 2021.

[16] Zhenglei He, Guojian Chen, Mengna Hong, Qingang Xiong, Xianyi Zeng, and Yi Man. Process monitoring and fault prediction of papermaking by learning from imperfect data. *IEEE Transactions on Automation Science and Engineering*, 2023.

[17] Zhenglei He, Mengna Hong, Hongze Zheng, Jinfeng Wang, Qingang Xiong, and Yi Man. Towards low-carbon papermaking wastewater treatment process based on kriging surrogate predictive model. *Journal of Cleaner Production*, page 139039, 2023.

[18] Zhenglei He, Chang Liu, Yutao Wang, Xu Wang, and Yi Man. Optimal operation of wind-solar-thermal collaborative power system considering carbon trading and energy storage. *Applied Energy*, 352:121993, 2023.

2. Research and Application of Raw Paper Quality Prediction Model for Cardboard Papermaking Process

Jiwei Qian,[1] Zhenglei He,[1,*] Yi Man,[1,2] Jigeng Li[1] and Mengna Hong[1]

1 Introduction

The papermaking industry is an important pillar industry for China's national economy and social development[1]. As China continues to move towards the goal of building a modern and powerful country, the development potential of the papermaking industry will be further unleashed[2].Cardboard is the main product of China's papermaking industry. Due to the rapid development of industries such as e-commerce, logistics, and foreign trade, the demand for cardboard has grown rapidly in recent years. Additionally, the supply and demand gap of cardboard will further expand due to factors such as relevant tax policies and the diversified demand for end products.

Ensuring product quality is the foundation of production. The quality of cardboard needs to be strictly controlled to ensure that the end products have good compressibility, moisture resistance, cushioning and other capabilities.

[1] State Key Laboratory of Pulp and Paper Engineering, South China University of Technology, Guangzhou.
[2] Guangdong Artificial intelligence and Digital Economy Laboratory(Guangzhou) (Pazhou Lab), Guangzhou.
* Corresponding author: hezhenglei@scut.edu.cn

Therefore, the quality inspection of raw paper is one of the critical production processes for cardboard manufacturing enterprises.

However, the manufacturing process of cardboard is lengthy and involves a series of complex procedures such as thermal dispersion, forming, pressing, drying, and calendering[3]. Although enterprises have improved the accessibility and traceability of production process variables through the extensive installation of physical sensors, distributed control systems, manufacturing execution systems and associated production management systems[4;5], the monitoring methods for key indicators of raw paper still heavily rely on offline manual detection. This method suffers from long feedback cycle, unstable test results, and insufficient representativeness of test paper samples, which makes it difficult to grasp the dynamic relationship between the process parameters and the quality of the raw paper in real-time. It makes it difficult to effectively analyze the complex production process and the functioning of its parameters, which ultimately leads to uncertainty in the process of cardboard quality control[6].

Though some online inspection methods based on optical and acoustic principles have been applied to achieve online monitoring of general indicators such as moisture, thickness, and quantification, online monitoring methods are still lacking key cardboard indicators, such as folding endurance, bursting strength, smoothness, and transverse ring compressive strength. Building a prediction model based on the mapping relationship between easily measurable and difficult-to-measure variables, known as soft measurement, is an effective solution to the aforementioned problems. This approach has been widely used and validated in related fields[7].

In addition, during the actual manufacturing process, adjustments are often made to the amount of additives to improve the quality of the raw paper. As a result, the quality control process can impact costs, the environment, and other aspects. Due to changes in policies and market demand, the number of major pulping and papermaking enterprises has rapidly decreased from around 2800 to around 2500 in the five years from 2016 to 2021[1;8]. Existing paper companies not only need to focus on product quality, but also urgently need to pay attention to their own production costs[9], resource utilization[10], energy consumption[11], and environmental impact[12], and accelerate the promotion of intelligent, clean, and high-quality development.

Therefore, conducting research on quality prediction modeling to achieve real-time and online monitoring of products and developing production process optimization models to guide process optimization is of great significance for the development of the papermaking industry.

2 Literature review and motivation

Paper plays an important role in people's daily lives, and paper and pulp manufacturing is crucial to the development of the process industry. The scope of the papermaking process considered in this study is shown in Fig.??, and it consists of a long series of production processes, including cleaning, screening, refining, washing, dewatering and dispersion. These processes involve a wealth of variables, and it is essential to monitor them dynamically. However, the high nonlinearity, time-delay, and coupling underlying such a long chain of processes make it complicated to describe the relationships between variables accurately [13]. In the real application, the manipulators rely heavily on over-inputted power, resources and other operations to avoid potential uncertainty, which has not only raised unnecessary costs, and wastes immensely. Additionally, it has also increased hazardous emissions to the environment [14].

The variables of papermaking include parameters of materials with regard to wood chips, fiber resources, recycled fibers, and chemicals, as well as the control and state parameters in terms of jet angle, net speed ratio, dewatering ratio, speed difference, among others. Each of these parameters determines the quality of the final paper product to some degree, and different scenarios may result in significant differences in production cost, energy consumption and emissions [15;16]. For instance, the wastage rate of fiber and chemicals like retention aid, dry strength agent conditions the production properties and the cost directly [9]. The press section dehydrates wet paper under pressure, which impacts the quality of the fiber net and the energy consumption in following drying sections as a whole, and the states of the cylinder and dryer felt (where the pressure comes from) also perform similar functions in the process [5]. Sizing is settled after the drying process. The water contained in the sizing agent is not desired from an energy-saving point of view in drying, but it is necessary for maintaining the sizing quality.

It is illustrated that monitoring the dynamic variables in the papermaking process is complicated and crucial. However, certain key variables, such as bursting strength and folding endurance, can only be tested offline through physical experiments on paper scrolls [17]. Obviously, such an approach not only causes a severe delay in providing process information but also increases the risk of standard deviation in the products, making it challenging for decision-makers to make sense of the data before the scrolls are unloaded from the line. Besides the direct effects, the possible issues may also include the gap of testing frequency among variables, the time-lag of testing results due to different locations of sensors installed, and the difficulty in mapping real-time collected process data to off-line tested data.[18].

In addressing the aforesaid issues, soft-sensor techniques have been widely proposed in the process industry. For example, in the food industry, soft sensors have been investigated for real-time monitoring of milk powder moisture content during drying in a spouted bed dryer[19], online estimation of air temperature in mobile small-scale cooling chambers[20], monitoring of multiple process stages in industrial sugar production[21], and evaluating crystal size distribution parameters in cane sugar crystallization process[22]. In the chemical industry, applications include assessing the quality properties of acrylonitrile-butadiene-styrene in the production process by fusing near-infrared and process sensor data[23], predicting critical diesel properties in a wide variety of diesel fuel fractions and blends collected from an industrial refinery[24], observing the production processes of Tennessee Eastman and the purified terephthalic acid[25], which enable chemical industrial plants to operate more effectively and stably[26]. Furthermore, the related models have also been constructed to test the online moisture content information and temperature of lentil seeds in a microwave fluidized bed dryer in agriculture area[27], and employed to optimize wastewater treatment process of aerobic granular sludge[28] and estimate the total Kjeldahl nitrogen[29].

In the pulp and paper industry, one of the most typical process industries, many studies have been conducted on soft sensors in the production process. Zhang et al.[30] estimated the beating degree and wet weight of pulp after refining, aiming to assess the performance of the process, evaluate the operational conditions, and optimize the process of stock preparation. They have also attempted to estimate the Canadian standard pulp freeness and outlet consistency after the high consistency refining stage of the alkaline peroxide mechanical pulping process[31]. Liu et al.[32] tried to monitor online the crystal size distribution in crystallization processes, the molecular weight distribution in polymerization processes, the particle size distribution in powder industries, and the pulp fiber length distribution in paper industries. Raffaele and Ondruch[33] presented the exploitation of a data-driven soft sensor for paper strength parameters in a production environment, and it is confirmed in an experiment-based validation that the resulting soft sensor for tensile

Table 1: Sensor variables

Symbol	Meaning	Dimension
P_{w1}	Line pressure of the first wet press	kN/m
P_{w2}	Line pressure of the second wet press	kN/m
P_{cal}	Line pressure of calendering machine	kN/m
BW	Paper basis weight	g/m^2
ABW	Absolute dry paper basis weight	g/m^2
W_{us}	Amount of sizing agent on the upper surface of paper	g/m^2
W_{ls}	Amount of sizing agent on the lower surface of paper	g/m^2
M_{t1}	Amount of dry strength agent on top layer	kg/min
M_{b1}	Amount of dry strength agent on bottom layer	kg/min
M_{t2}	Amount of retention aid on top layer	kg/min
M_{b2}	Amount of retention aid on bottom layer	kg/min
M_{t3}	Amount of filler on top layer	kg/min
M_{b3}	Amount of filler on bottom layer	kg/min
F_{cal}	Tension of paper during calendaring	N/m
F_{cur}	Tension of paper during curling	N/m
v_{gap}	The gap between pulp speed and net speed	m/min
v_{ratio}	Ratio of pulp speed to net speed	%
T_{cal}	Temperature of calendering roller surface	$^{\circ}C$
T_{wind}	Temperature of the airflow in the sizing machine	$^{\circ}C$
R	Moisture content of paper	%
h	Thickness of paper	mm

index deployed at a paper machine shows an exemplary performance. Li et al.[34] measured pulp quality of the refining process with respect to the fiber length and the Canadian standard freeness by means of soft sensors, and the effectiveness of such a method was validated in industrial experiments.

The wastewater treatment process of papermaking, as another important aspect of the pulp and paper industry, has been paid attention to develop soft-sensors. Liu et al.[35] published their study outcomes about the prediction of effluent quality in papermaking wastewater treatment processes in regard to the key quality indices of effluent chemical oxygen demand (COD), and in another paper[36], they uncovered a more complicated case investigating soft sensors with multiple target variables of effluent suspended solids (SS), effluent chemical oxygen demand (COD), effluent ammonia concentration (SNH), and effluent nitrate concentration (SNO).

Table 2: Generated variables

Symbol	Meaning	Dimension
a_{ts}	Total amount of sizing agent	g/m^2
a_{t1}	Total amount of dry strength agent	kg/min
a_{t2}	Total amount of retention aid	kg/min
a_{t3}	Total amount of filler	kg/min
x_1	M_{t1}/BW	$10^3 \cdot m^2/min$
x_2	M_{t2}/BW	$10^3 \cdot m^2/min$
x_3	M_{t3}/BW	$10^3 \cdot m^2/min$

Table 3: Target variables

Symbol	Meaning	Dimension
R_f	Folding endurance	$times$
F_b	Bursting strength	kPa
S	Smoothness	s
F_t	Transverse ring compressive strength	kN/m

On the basis of reviewing literatures about soft sensor, particularly those for the papermaking industry, several trends of this research topic have been summarized. Data-driven soft sensors are alternating the mechanism-based and knowledge-based methods to deal with the growing complicated engineering troubles in the process[7]. It is universally acknowledged that the two kinds of conventional techniques have made significant contributions to industrial development[37], but their performance leans on the detailed and accurate description of the mechanism of the process, or requires broadly available experience and knowledge about process. As these preconditions are increasingly hard to be realized in today's industrial processes, data is readily accessible in the environment of internet of things (IoT). In contrast data-driven soft sensors are therefore having a more promising prospect[28].

There are a growing number of variables being considered in the soft sensor construction. Taking literatures 35 and 36 as an instance, it is worth noting that not only were the adopted factors enhanced, but also the predicted targets boosted in the models. In recent years, the application of computational intelligent techniques is blooming taking advantage of its aptitudes in tackling problems in the complex system of highly nonlinear design, large data sets, data drift, missing data, and more[38]. Resulting from the rapid development of computational power and intelligent techniques, as well as the accessibility of data owing to IoT, the dimension of research questions about soft sensors applied to industrial processes has become very diverse nowadays[39;40].

The increment of variables in published soft sensor studies involves the attribution of people's growing concern about the industry's impact on the ecology and environment. It is easy to infer from the literature review that, researchers are taking more environmental impact related variables into account, rather than caring only about quality in traditional investigations[41]. More precisely, studies tend to devote their efforts to building soft sensors for energy intensive processes intensive processes[42], water-intensive[43] manufacturing processes, and monitoring variables highly related to environmental impacts in addition to quality indices[44;45].

Following the analysis of the papermaking process and related literature, in order to effectively address the aforementioned issues of soft sensors for papermaking this paper studied and developed a series of models for predicting the variables of folding endurance, bursting strength, smoothness, and transverse ring compressive strength in paper manufacturing dynamic. These models were coupled with production cost, energy-consumption, and greenhouse gas (GHG) emission of the papermaking process to optimize it with regard to multiple objectives. This paper is organized as follow: Section 3 presents the problem formulation and system model. It is followed by results and discussion of the system application in Section 4. Finally, conclusions and future works are summarized in Section 5.

3 System model

3.1 Problem formulation

Folding endurance (R_f), bursting strength (F_b), smoothness (S), and transverse ring compressive strength (F_t) provide the most important information about the paper product. However, their variation cannot be estimated online as the testing methods destroy. Taking such measurements involves stopping the process which is unpractical in the process industry. In engineering applications, factories prefer to overfeed the resource, energy and other inputs to guarantee the paper product quality in terms of these indexes. The present approach has caused serious wastes in resource and power, and arising GHG emissions causing environmental problems. Therefore, there is a dire need for building soft sensors of these indicies, and integrating them with the production cost, energy consumption and GHG emissions to construct a multi-optimization system aimed at promoting the papermaking process to be more intelligent, efficient, and cleaner.

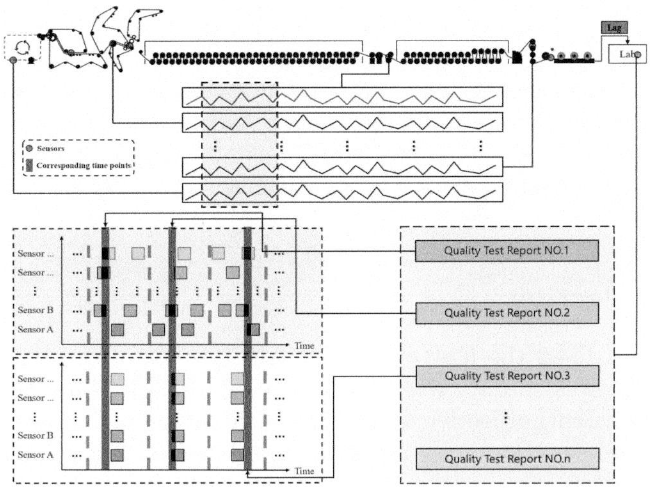

Figure 1: Representation of data collection in the papermaking production processes

3.2 Data processing

As shown in Table1 to Table3, depending on papermaking mechanism and availability of data in the cooperated papermaking factories, a number of the most related variables are taken as the potential models' inputs for predicting the target variables in this research. It is worth mentioning that a couple of generated variables are taken into account as well by processing the initial tested variables following experts' suggestions. a_{ts}, a_{t1}, a_{t2} and a_{t3} were generated based on the characteristics of the study subjects, i.e., negligible differences in the amount of chemicals used on the top and bottom layers and on the upper and lower surfaces. Further, x_1, x_2, and x_3 were generated by dividing BW on the basis of the breadth measures a_{t1}, a_{t2} and a_{t3}. These generated variables can additionally reveal the accumulative or distributed situation of certain initial variables in the dynamic process.

Figure1 illustrates how factories collect data from the manufacturing processes. There are many sensors monitoring the state variables throughout the process, these data would be collected with manipulating variables as well as the features about inputted resources in quality inspection reports. However, the collection frequency of sensors varies dramatically in different environment or with different device as the production environment and devices are not that uniform and normative. On the other hand, recording manipulating variables and resources' features relies on manual operations, which would furthermore increase the chaos of collected data. As the reports cannot be strictly conducted with the sensor time in line, it is necessary to narrow the gap due to their collection frequency and denoise the collected data at the very beginning.

3.3 Feature selection

It is demonstrated in Fig.2 the flow chart of following steps of constructing the system model. Processed data will support the feature selection and soft sensor modeling and multi-objective optimization in the following sections. Among which, feature selection is an important process reducing the redundancy and analyze correlation of the extracted features. It one of the key approaches on data before modeling that facilitate the model training to efficiently exploit more samples in the limited space, enhance the computation speed by decreasing calculation dimension, and improve the models' generalization capability as a whole[?] . Specifically, the feature selection is based on analysis of Person's coefficients and papermaking mechanism of variables.

3.4 Soft sensor modeling

In this work, four data-driven models are proposed to develop the soft sensors of R_f, F_b, S and F_t in the papermaking process, namely random forest (RF), gradient boosting regression (GBR), ridge regression and K-nearest neighbor (KNN). Taking the features of papermaking process and collected data, the proposal of these methods involves that, RF and GBR are typical ensemble learning algorithms of bagging and boosting respectively which are assumed good at generalization and small data sets[?] [?] [?] [?] , ridge regression is relative more robotic than other regressors to avoid over-fitting[?] , KNN is a supervised learning algorithm usually applied to classification problems so that is supposed can reveal the clustering tendency of process data[?] . In

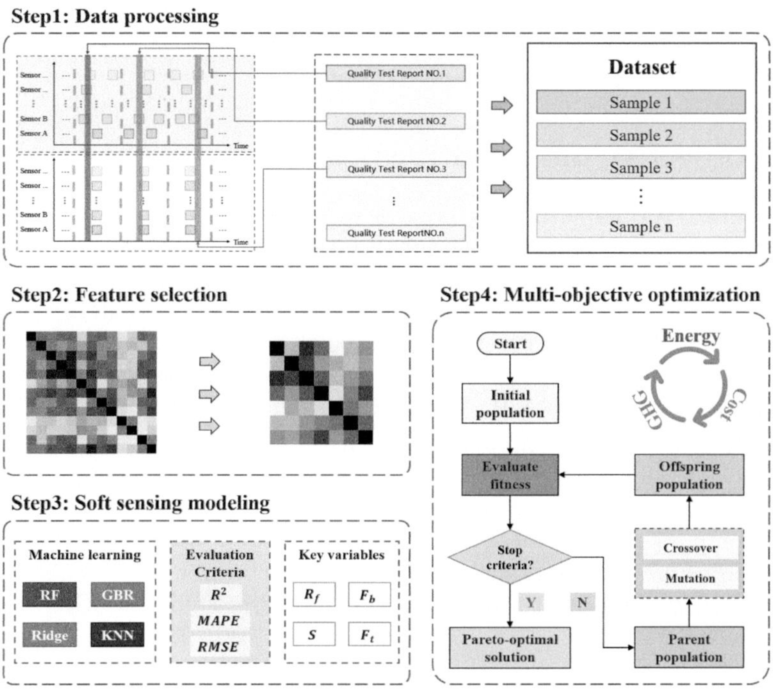

Figure 2: The framework of building soft sensors and formulating multi-objective optimization system of papermaking process

Eq.1-3, three criteria are used in this study for assessing the performance of obtained soft sensor models, which include R-square (R^2), root-mean-square error ($RMSE$), mean absolute percentage error ($MAPE$).

$$R^2 = 1 - \frac{\sum_{i=1}^{n} (y_i - \hat{y}_i)^2}{\sum_{i=1}^{n} (y_i - \bar{y})^2} \tag{1}$$

$$RMSE = \sqrt{\frac{1}{n} \cdot \sum_{i=1}^{n} (y_i - \hat{y}_i)^2} \tag{2}$$

$$MAPE = \frac{\sum_{i=1}^{n} |(y_i - \hat{y}_i) / y_i|}{n} \cdot 100\% \tag{3}$$

where n is number of testing samples, y_i indicates the value of real tested sample i, $\hat{y}_i$ and $\bar{y}$ are the predicted value and the corresponding average value respectively.

3.5 Multi-objective optimization

In order to optimize the papermaking process in terms of production cost, energy consumption, and GHG emissions simultaneously subjected to quality standard, this work developed a multi-objective optimization model for the production process aimed at minimizing material consumption costs, indirect energy consumption, and indirect greenhouse gas emissions, combining practical production constraints and product quality constraints. The specific solving algorithms used include NSGA-II, R-NSGA-II[46], and NSGA-III[47;48;49].

The NSGA-II algorithm is proposed on the basis of the Non dominated Sorting Genetic Algorithm (NSGA)[50], which further improves computational performance by introducing fast non dominated sorting, elite retention strategy, crowding comparison operator, etc. It has been widely used in solving complex multi-objective problems such as production scheduling[51], formula optimization[52], and waste heat utilization[53]. The basic process of the NSGA-II algorithm is as follows:

1. Randomly initialize the first generation parent population, perform basic genetic (selection, crossover, mutation) operations after non dominated sorting, and generate a first generation offspring population of N (i.e. the second generation parent population);

2. Starting from the second generation, after completing genetic operations, the parent population needs to merge with the generated offspring population (with a quantity of 2N), and priority is given to selecting individuals based on fast non dominated sorting to form a new parent population. For the last non dominated layer (i.e. critical layer) included in the selection, individuals are selected based on the size of crowding distance, and individuals with larger crowding distance are preferred to be retained until the population of the new parent reaches N;

3. Repeat step 2 until the termination condition is met, usually when the population iterations reach the set upper limit.

R-NSGA-II solves the preference setting problem of NSGA-II, using reference points to guide population evolution direction, and can find a large number of feasible solutions around the reference points, thereby supporting subsequent decision-making with preferences. In addition, R-NSGA-II allows for the simultaneous setting of multiple reference points, guiding the population to evolve towards multiple regions of interest to decision-makers. Compared to the crowding comparison operator of NSGA-II, R-NSGA-II forms a new parent population by comparing hierarchical relationships, normalized Euclidean distances from individuals to reference points, and so on. Specifically, the selection of critical layer individuals for R-NSGA-II mainly includes the following three aspects:

1. Calculate the normalized Euclidean distance between different individuals and each reference point, and sort the individuals in ascending order according to the calculation results;
2. After completing all reference point calculations, for different individuals, the minimum value of the calculation result is used as the corrected crowding distance for the corresponding individual. Subsequently, individuals with smaller corrected crowding distances are prioritized and retained;
3. To maintain population diversity, the sum of normalized differences for all targets $\leq \epsilon$ are divided into groups. Randomly select one individual from each group and delete other individuals.

NSGA-III is an optimization algorithm based on reference direction, which is used to solve problems such as poor algorithm convergence and poor population diversity that may occur when optimizing three or more objectives with the crowding degree concept proposed by NSGA-II. After merging parent and offspring populations and completing non dominated sorting, NSGA-III first maps the individual normalization to the hyperplane where the reference point is located. Then, associate the individual with the corresponding reference point. Finally, niche selection strategies are applied to each reference point, pruned and merged to form a new parent population.

Unlike R-NSGA-II reference points that need to be set in advance by decision-makers, NSGA-III reference points can be predefined in a structured manner. In this study, Das and Dennis method[54] is used to generate structured reference points. Its generated points are on a $(M-1)$ dimensional hyperplane, and M is the target space dimension, that is, the number of optimization targets. If each target is divided into p parts, the number of reference points H is calculated as follows:

$$H = C^{p}_{M+p-1} \tag{4}$$

The hyperplane of NSGA-III is the ideal point and extreme point obtained through calculation according to the combined population, and finally generated by normalization, while the individual is mapped to the hyperplane in the way of adaptive normalization, and then linked to the reference point according to the distance from the reference line. Afterwards, for each reference point, the number of ρ_j is used to evaluate the density and apply niche selection strategies to select excellent individuals, thereby forming a group of evenly distributed new parent populations. Among them, the number of small living environments ρ_j represents the number of individuals associated with the reference point j (other non-dominated layer individuals need to be considered). The main logic of niche selection strategy is: priority selection ρ_j which with the smallest j (if there are multiple, randomly select one of them), and if there are individuals associated with it, the ρ of this reference point plus one. At the same time, randomly select one of the associated individuals to join the new parent population until the number of individuals in the new parent population reaches N.

NSGA-II, R-NSGA-II, and NSGA-III have similar process frameworks. Fast non-dominated sorting is the main individual selection method when applying elite retention strategies, while for critical layer individuals, crowding distance between individuals, modified crowding distance based on normalized Euclidean distance, and niche strategy based on reference points are used to complete the selection.

4 Results and discussion

4.1 Data process

As presented in Fig.1, the raw data collected from different physical sensors may have gaps regarding to time and granularity, a basic transformation process was conducted on these data before feeding them to the models. The mean function is applied to the data in every 30 seconds to average collected data and comb the granularity. In this process, 30 seconds is determined by the lost information raised in comparison with 60 seconds and 5 minutes, as well as the considerations of matching effect of sensor data with quality inspection reports as the time discharging scrolls from machine is stochastic and the quality indexes mainly reflect the variations of paper quality around discharging time. The processed data can be simplified illustrated in Fig.3, where the time with same granularity can be unified to the quality reports, and a sample with processed data is given at the right of it.

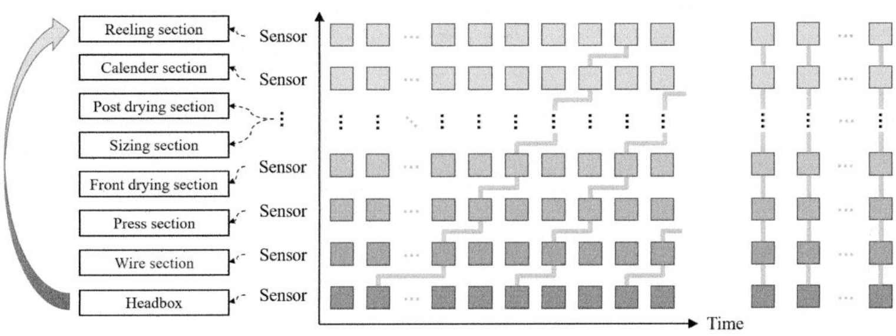

Figure 3: Simplified illustration of the method processing raw data in this study

4.2 Feature selection

As claimed there are 21 original variables collected from a papermaking factory, which consist of line pressure of the first wet press (P_{w1}), line pressure of the second wet press (P_{w2}), line pressure of calendering machine(P_{cal}), paper basis weight (BW), absolute dry paper basis weight (ABW), amount of sizing agent on the upper surface of paper (W_{us}), amount of sizing agent on the lower surface of paper (W_{ls}), amount of dry strength agent on top layer (M_{t1}), amount of dry strength agent on bottom layer (M_{b1}), amount of retention aid on top layer (M_{t2}), amount of retention aid on bottom layer (M_{b1}), amount of filler on top layer (M_{t3}), amount of filler on bottom layer (M_{b3}), tension of paper during calendaring (F_{cal}), tension of paper during curling (F_{cur}), gap between pulp speed and net speed (v_{gap}), ratio between pulp speed and net speed (v_{ratio}), temperature of calendering roller surface (P_{cal}), temperature of airflow in sizing machine (T_{wind}), moisture content of paper, thickness of paper (R), thickness of paper (h).

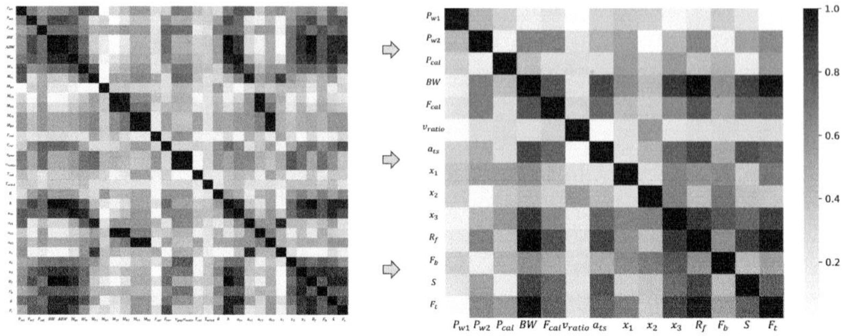

Figure 4: Analysis of relative importance in comparison of Pearson's coefficient of variables

Table 4: Testing results of developed soft sensor models

	RF			GBR			Ridge			KNN		
	R^2	$MAPE$	$RMSE$	R^2	$MAPE$	$RMSE$	R^2	MAPE	RMSE	R^2	MAPE	RMSE
R_f	0.839	2.969	0.333	0.831	2.971	0.340	0.700	4.137	0.454	0.831	3.034	0.341
F_b	0.994	1.510	3.526	0.994	1.544	3.572	0.987	2.406	5.086	0.993	1.735	3.679
S	0.849	4.602	0.608	0.853	4.592	0.600	0.795	5.583	0.708	0.809	5.387	0.685
F_t	0.956	4.593	36.117	0.970	3.777	29.534	0.934	6.593	44.289	0.880	8.124	59.514

In the papermaking process, operations on agents, filler, and aids may be conducted at different time, and on different position of different surface. In order to reinforce the variables' relationship to the targets, following factory's suggestion, the distributions of them are additionally drawn in by generating 7 variables summing up agents, filler, and aids or further dividing them into paper basis weight (BW). The generated variables can better characterize the paper composition as the latter is more closely related to material quality in a physical sense. Upon the initial variables and generated variables (28 in total), a comparison of their relative importance is analyzed by Pearson's coefficient showing in Fig.4. It is worth noting that the four target variables are quite relevant to each other, and overall, the top 7 variables can be filtered basically from the list for first, which are P_{w2}, P_{cal}, BW, F_{cal}, a_{ts}, x_1, and x_2. However, taking results of several trials on different combinations and the papermaking mechanism into account, it is suggested to put P_{w1}, v_{ratio}, and x_3 into selected variables group additionally[33]. Therefore, a 10-set of features (right one shown in Fig.4) were selected to model the papermaking variables of R_f, F_b, S, and F_t as a whole.

4.3 Soft sensor models

For training models of RF, GBR, Ridge regression and KNN, a total number of 1037 sets of data are used by 8:2 distribution for training and testing respectively. An optimization process of models' hyperparameters are conducted referring to the method of literature[55]. The performance of four kinds of models is evaluated by R_2, $MAPE$, and $RMSE$, respectively, and the results can be observed in Table 4. The visual demonstration of results is provided by an area comparison in triangle composed by three evaluation indexes posited at the angles(shown in Fig.5), it indicates clearly that the smaller the area in a triangle, the better the represented model performs in all dimensions. Where the RF models of R_f, and F_b have the most accurate results overall in all the three dimensions in terms of R^2, $MAPE$, and $RMSE$ equaling to 0.839, 2.97, 0.333, and 0.994, 1.51, 3.526 respectively. For S, and F_t soft sensors, the GBR models dominate all the competitors with R^2, $MAPE$, and $RMSE$ of 0.839, 2.97, 0.333, and 0.994, 1.51, 3.526 respectively. As a result, these models are preliminarily chosen as the optimal ones of soft sensor for papermaking process in this work.

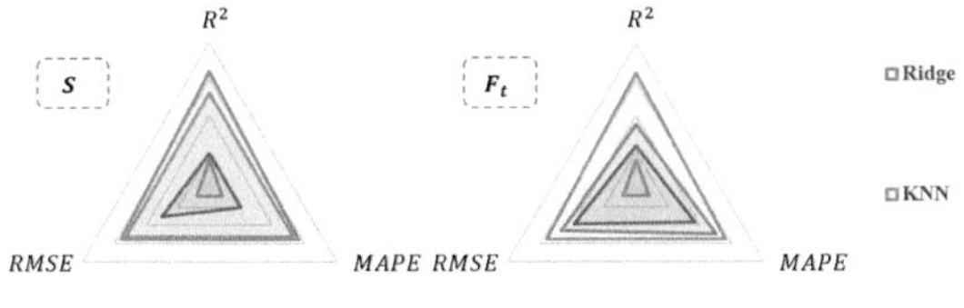

Figure 5: Triangle diagram of model performance in terms of R^2, $MAPE$, and $RMSE$

In order to further verified the optimality of the chosen models, a comparison is given in Fig.6 of detailed performance of every soft sensor in the real application with regard to tested results of developed models on 20 samples with real values. It is found that, in general, all the proposed models have well predicted the target variables, especially for the soft sensors of bursting strength. However, there are a diversity of samples disclose models' disparity on outputting different targets. For example, for the 5th, 6th, 8th, 11th, and 19th samples in folding endurance prediction and the 2nd, 6th, 7th, 8th, 10th, 15th, 18th, and 19th samples in smoothness prediction. Such phenomenon can easily be observed in the bottom right corner of Fig.6 as well. Although the disparity for bursting strength is not as clear as others, the relatively poorer coverage effect of ridge regression and KNN models can still be found in comparison.

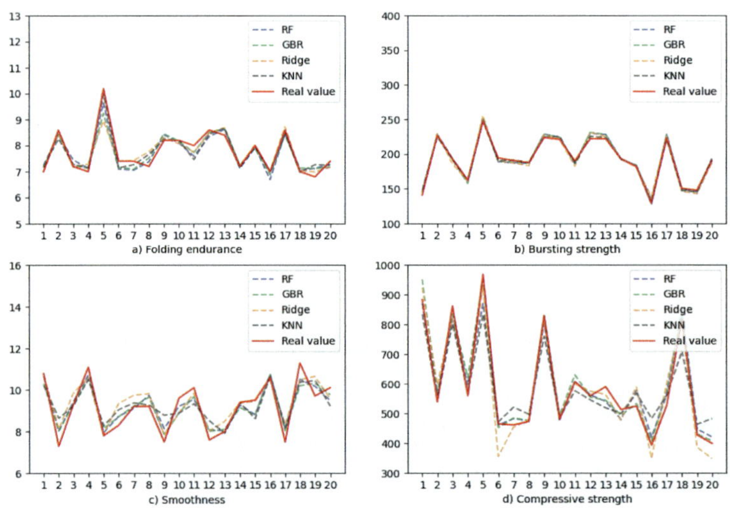

Figure 6: Comparison of sort sensors' output values with real sample variables

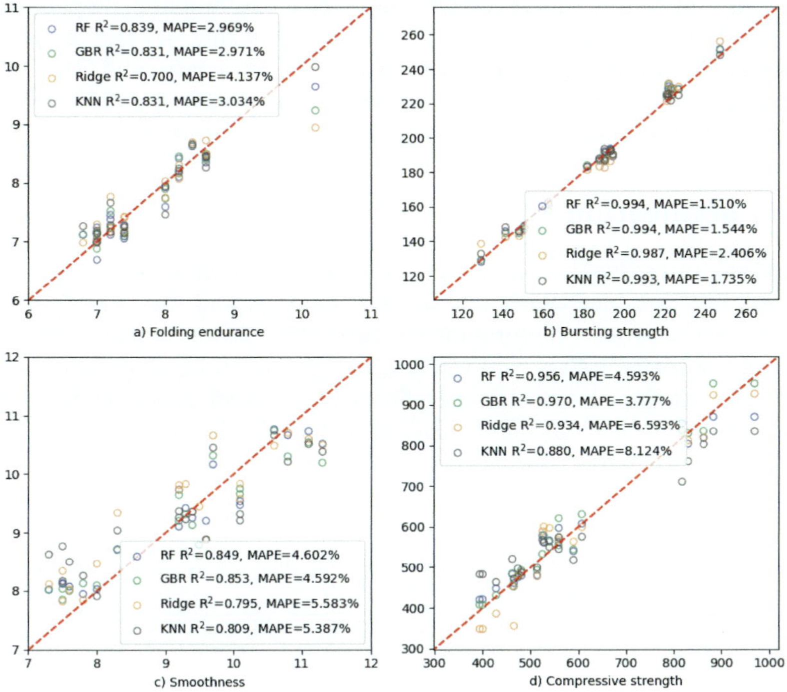

Figure 7: Linear fitting correlation coefficients of soft sensors predicted values versus real data

According to the linear fitting correlation coefficients of soft sensors predicted values versus real data (demonstrated in Fig.7) given in Fig.6, it is found that RF model (R^2=0.839, $MAPE$=2.967%) and GBR model (R^2=0.831, $MAPE$=2.971%) are apparently more stable and suitable than ridge regression (R^2=0.700, $MAPE$=4.137%) and KNN models (R^2=0.831, $MAPE$=3.034%) in folding endurance of papermaking process. Similar conclusion can also be obtained in other comparisons. This may not only attribute to the features of data (e.g., unbalance of sample data) collected and processed in this study, but also due to the disadvantage of ridge regression and KNN preventing over-fitting which could also cause under-fitting. The lower accuracy of them arises the risk of application in the complicated environment such as papermaking process. These results also indicate that both

RF can GBR can well deal with the interaction of variables and are comparatively more stable in nonlinear modeling in the papermaking process, and the chosen models, RF for folding endurance and bursting strength, and GBR for smoothness and compressive strength, are confirmed as the optimal soft sensors to be used in the following process optimization applications.

4.4 Model application

The formulated multi-objective optimization problem in papermaking process constraint to quality variables which soft sensors monitor, and the objective functions are related to process performance in three dimensions, namely production cost, energy consumption, and GHG emission. As 10 input variables are applied in the developed sensors, the effects of these variables on production cost, energy consumption and GHG emission should be minimized as lower as possible, therefore, the corresponding effects are analyzed and concluded.

Table 5: Factors of chemicals applied in papermaking process on cost, energy consumption and GHG emission

	Production cost	Energy consumption	GHG emission
Retention aids	1.4	11.79	1.03
Fillers	4.5	3.53	0.23
Dry strength agents / surface sizing agents	6.3	32.2	1.05

* $Cost(cny/t)$; $Energy(MJ/t)$; $GHG(kgCO_2eq/t)$.

4.4.1 Process performance

Due to the fact that a range of chemical agents are used in the papermaking process and are highly related to the quality, cost, energy consumption and GHG emission as a whole. Moreover, the line pressure indirectly determines the lifespan of papermaker's felt. Specifically, the impacts of materials are generally linear to the objectives depending on factors which can be found in literatures[11;56;57] as listed in Table5. And according to the experience of factory, the effects of line pressure on the lifespan of papermaker's felt (in consumption of grams felt by tons of paper production) basically follows the rules of Eq.5. The cost of felt is 274.65 cny/kg in average[1;58]. Its life cycle energy consumption and GHG emission are around 75 MJ/kg[59] and 2.5701 $kgCO_2eq/kg$[60;61]respectively.

$$C(g/t) = \begin{cases} 0.1 \cdot (P_{w1} + P_{w2}) + 96 \\ 68.54 \cdot e^{0.0034(P_{w1}+P_{w2})} \\ 0.012 \cdot (P_{w1} + P_{w2})^2 + 6.5 \cdot (P_{w1} + P_{w2}) + 1084 \end{cases} \tag{5}$$

Therefore, in the application of the developed soft sensors, the related production cost (cny), energy consumption (MJ) and GHG emission ($kgCO_2eq$) can be summarized as Eq.6 to Eq.8:

$$Cost = 1.4 \cdot a_{t2} + 4.5 \cdot a_{t3} + 6.3 \cdot (a_{ts} + a_{t1}) + 0.27465C \tag{6}$$

$$Energy = 11.79 \cdot a_{t2} + 3.53 \cdot a_{t3} + 32.2 \cdot (a_{ts} + a_{t1}) + 0.075C \tag{7}$$

$$GHG = 1.03 \cdot a_{t2} + 0.23 \cdot a_{t3} + 1.05 \cdot (a_{ts} + a_{t1}) + 2.5701C \tag{8}$$

4.4.2 Quality constraints

It is worth mentioning that the data of pulp fiber concentration and sizing quantity are not considered in present case, so that the basis weight is regarded constant in the optimization and therefore the total amount of chemical inputted plus pulp fibers is remain constant at the same time. Moreover, according to the quality standards provided by papermaking factory, the constraints in the present study could be drawn as Eq.9 to Eq.13:

$$A = A' \tag{9}$$

$$R_f \geq 7 \ times \tag{10}$$

$$\frac{F_b}{BW} \geq 1.6 \ kPa \cdot m^2/g \tag{11}$$

$$S \geq 7 \ s \tag{12}$$

$$\frac{F_t}{BW} \geq 16 \ N \cdot m/g \tag{13}$$

where A and A' are the total amount of original and optimized chemical agents' usage.

4.4.3 Case applications

The above three optimization algorithms are set to have a population of 500 individuals in practical applications, with an upper limit of 500 evolution times. The relevant parameters of the basic genetic operators (selection, crossover, mutation) are all set to default values.

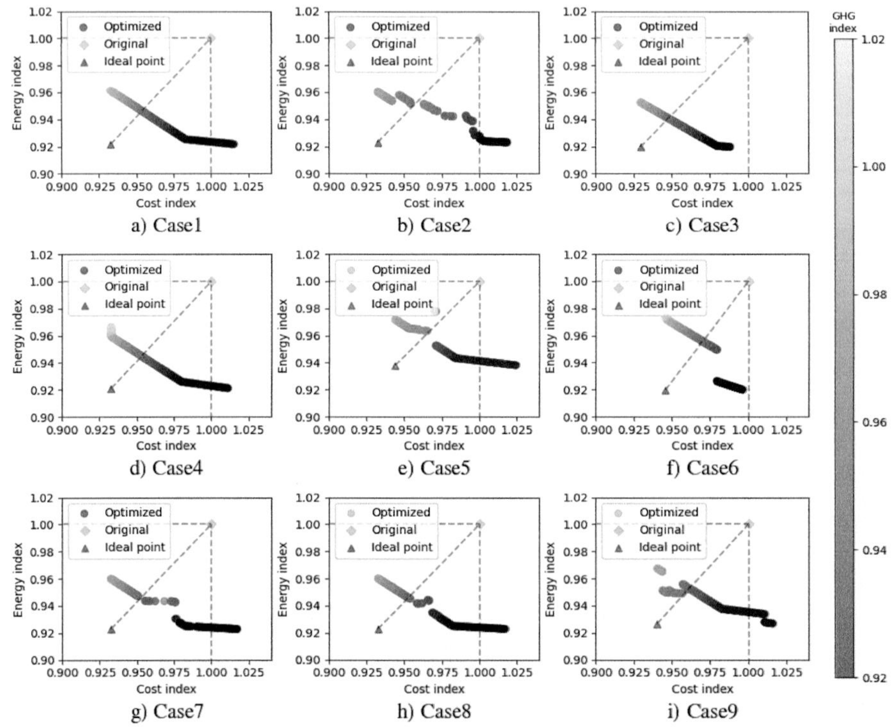

Figure 8: Pareto frontier for different cases calculated by NSGA-II

In this section, 9 different samples were randomly selected as optimization cases, and the Pareto frontiers of these different samples were obtained by solving them with the NSGA-II algorithm, as shown in Fig.8. The horizontal axis represents the material consumption cost per unit product, the vertical axis represents the energy consumption per unit product, and the third dimension is represented by color, which indicates the greenhouse gas emissions per unit product. To ensure data desensitization and enable unified comparison, further processing was carried out on the actual values. The results shown in the figure are all normalized. Specifically, the calculation results of the initial sample on each target are the original points, with coordinate values of 1, as indicated by the orange diamond in Fig.8. Furthermore, the triangle identification (ideal point) in the figure represents the optimal result on all three objectives, which cannot be achieved in practical situations and is only used for observation and analysis. Typically, the intersection formed by the connecting line between the original point and the ideal point and the Pareto front surface is used as the actual optimization solution.

By observing Fig.8, it can be noticed that the energy consumption per unit product generally shows a positive correlation with the greenhouse gas emissions per unit product. For instance, in Case 1, a decrease in Energy index results in a gradual decrease in GHG index. However, both are typically negatively correlated with the cost index per unit product. In Case 1, a decrease in Energy index and GHG index results in an increase in Cost index. From a trend perspective, the Energy index and Cost index exhibit two consecutive linear change stages. In the range of 0.925 to 0.98, increasing Cost index leads to a rapid decrease in Energy index. However, when the Cost index increases by more than 0.98, the downward trend of Energy index significantly slows down, indicating a weakening effect of increased cost input on the decrease in energy consumption, i.e. a decrease in cost-effectiveness.

The situations of Case 3, Case 4, and Case 1 are similar, with Case 4 being the only Pareto frontier with three complete and continuous linear stages. Observing the first linear stage of Case 4, it can be seen that the Energy index and GHG index exhibit the same trend of change, while the Cost index shows almost no change. However, according to the relevant theories of Pareto optimality frontier, it can be inferred that the third objective must be negatively correlated, indicating that at this stage, only a small cost index investment can significantly reduce the Energy index and GHG index, resulting in higher cost-effectiveness.

The shape of the Pareto front is influenced by the objective function, decision variables, constraints, interference variables, and solving algorithms. The remaining cases do not have continuous and linear front lines like Case 1, Case 3, and Case 4, but instead exhibit characteristics of discontinuity and jumping.

A feasible explanation is that in the cases of Case 1, Case 3, and Case 4, the constraints do not play a significant role in model solving due to the combined influence of interference variables and decision variable range settings. In these cases, the original problem weakens into a multi-objective optimization problem with few or no constraints, and the shape of the Pareto front is mainly influenced by the objective function. The objective function in this work is composed of both linear and nonlinear parts. When constraints are not considered, the blanket consumption per unit product is the main factor causing nonlinear partial optimization numerical changes, and the impact trend of consumption on the three objectives is the same. Therefore, the algorithm tends to adjust the blanket consumption per unit product to the lowest value during execution, and there should be a unique optimal value for the decision variable within a given range. Consequently, for the same case, the nonlinear calculation results of each Pareto optimal solution are the same. At this point, the shape of the Pareto front is mainly determined by the linear part of the objective function.

Furthermore, if the impact of the usage of various papermaking additives on different targets is analyzed separately, the same conclusion can be drawn as the consumption of blankets, i.e., the impact trend on different targets is the same. However, there is a strong constraint in this study that the quantity remains unchanged before and after optimization, so the minimum amount of various paper additives cannot be achieved simultaneously. At this point, the shape of the leading edge is jointly determined by the linear objective function and linear constraints, ultimately resulting in a smooth and continuous Pareto optimal leading edge.

Therefore, in the cases of Case 5, Case 6, Case 7, and Case 8, the impact is limited, and it is still possible to clearly distinguish whether they exhibit linear or continuous characteristics overall. However, the remaining cases of Case 2 and Case 9 have a greater impact, mainly manifested as discrete and discontinuous characteristics. By examining Fig.8, it was discovered that the initial points were not on the Pareto front, indicating that further optimization was possible in all optimization cases. This corresponds to the actual production of papermaking, where parameter adjustment based on expert experience remains the most widely used process optimization method. Due to human factors, overshooting often occurs in actual production to ensure the safe and stable operation of the production line, resulting in waste of resources and energy. This indirectly reflects the significant value that process optimization models can bring to manufacturing production.

For better observation and quantitative analysis, Fig.9 were drawn. These figures display three extreme solutions for specific cases, namely the Minimize

Cost index scheme, Minimize Energy index scheme, and Minimize GHG index scheme. In Fig.9, these solutions are represented by orange dotted lines, green dotted lines, and blue dotted lines, respectively, while the gray dotted lines represent the original situation. Similarly, the data in Fig.9 has also been normalized.

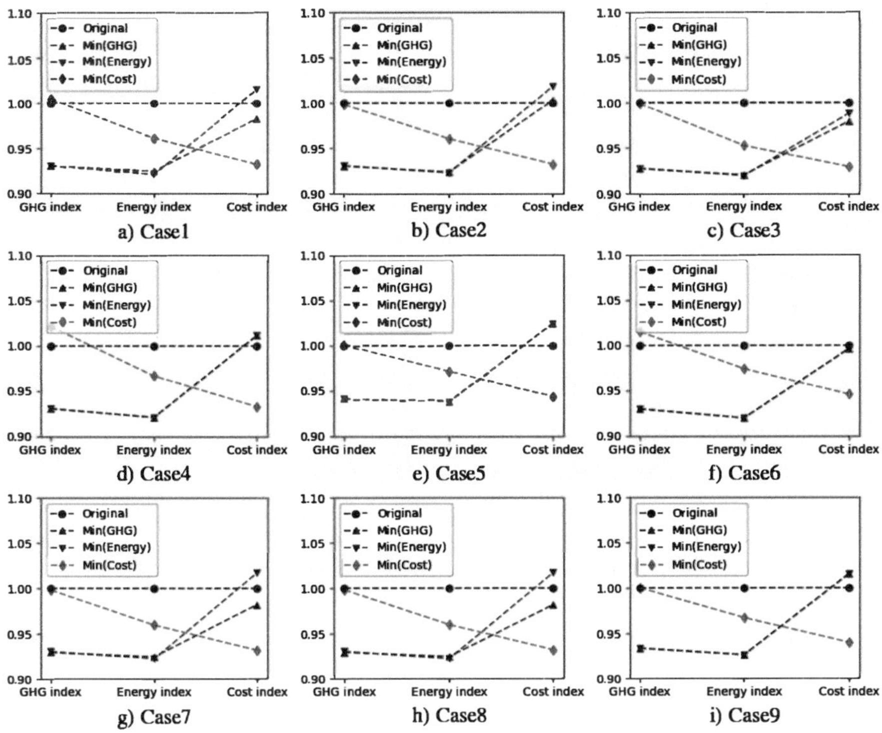

Figure 9: Extreme solutions obtained by NSGA-II solution for different cases

Taking Case 1 as an example for a detailed optimization case analysis, the orange dotted line represents the optimization results of material consumption cost, energy consumption, and greenhouse gas emissions under the minimized Cost index scheme. It can be observed that the lowest cost can be reduced to around 0.93 in the initial situation. However, the optimization effect of this plan on energy consumption is limited, and the greenhouse gas emissions at this point even exceed the initial situation. During the process of moving up the Cost index, where the optimization level of the Cost index decreases until it transitions to the blue dotted line, the Energy index decreases from 0.96 to about 0.92, while the GHG index decreases significantly from around 1 to 0.93, representing a decrease of about 0.07. At this stage, there is a significant positive correlation between the Energy index and the GHG index, and a significant negative correlation with the Cost index. Furthermore, as the Cost index continues to move upwards, transitioning from a blue dotted line to an green dotted line, the Energy index shows a negative correlation with the GHG index. In addition, the difference between the minimizing Energy index scheme and the minimizing GHG index scheme in terms of Energy index and GHG index is not significant, and the conversion between the two needs to come at the cost of a higher Cost index. A similar process to Case1 also occurs in the remaining cases, while the minimizing Energy index scheme and minimizing GHG index scheme of Case 4, Case 5, Case 6, and Case 9 exhibit significant overlap.

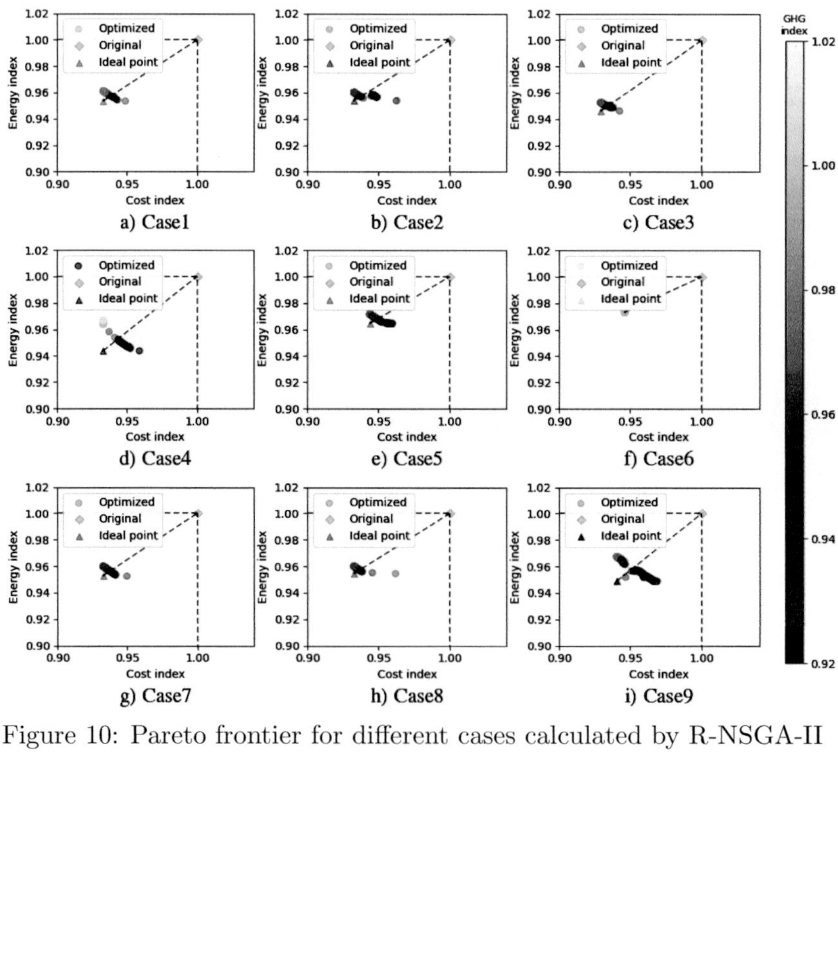

Figure 10: Pareto frontier for different cases calculated by R-NSGA-II

When using the R-NSGA-II algorithm in this study, we assumed that the initial values (i.e., the original calculation values on the three objectives) represented an optimization result approved by the decision-maker. Hence, we employed the original point as the sole reference point for the algorithm's optimization solutions. Figure10 depict the Pareto front obtained from R-NSGA-II on the nine samples.

It can be observed that the individuals generated by R-NSGA-II have a relatively concentrated distribution at the forefront. However, if the leading edge surface calculated by NSGA-II (as shown in Fig.8) is considered the true leading edge surface of the actual problem, R-NSGA-II would only obtain a partial leading edge situation of the problem. Generally, solution methods for modeling and optimizing complex industrial processes have limitations. Traditional planning methods require converting multiple objectives into a single objective for solution, leading to the optimization of a single result under the corresponding weight vector. On the other hand, solution methods based on intelligent algorithms pose potential risks in the implementation and application of optimization schemes due to issues such as stability and credibility of the solution.

R-NSGA-II provides a compromise solution by using an interactive optimization process with artificially given reference points to guide the evolution direction of the population (feasible solution), ultimately obtaining the Pareto optimal frontier that is closest to the reference point. From a practical application perspective, R-NSGA-II can provide decision-makers with higher credibility candidate solutions that are more easily accepted and adopted than non-reference point optimization algorithms such as NSGA-II. From the perspective of frontier integrity, R-NSGA-II only provides partial information on the real frontier due to the presence of reference points. However, this limitation is mainly related to the application scenarios considered during algorithm design and cannot be seen as a disadvantage of R-NSGA-II in computational solution.

Figure11 depict the Pareto frontier of NSGA-III. It's worth noting that both NSGA-III and R-NSGA-II are reference point optimization algorithms. However, their approach to utilizing reference points differs slightly. NSGA-III directs population evolution by setting a broad range of reference points, ensuring an even distribution of individuals on the frontier. Its main objective is to guarantee diversity in the population distribution. On the other hand, R-NSGA-II aims to compute a sequence of Pareto optimal frontiers within the decision-maker's preference region, using reference points.

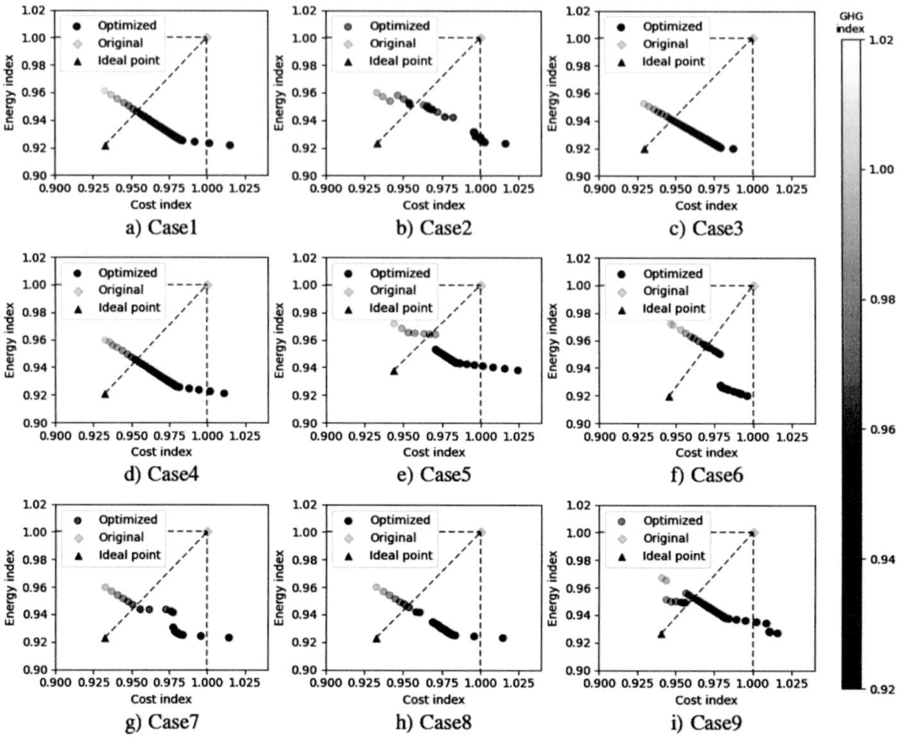

Figure 11: Pareto frontier for different cases calculated by NSGA-III

In addition, upon observing the leading edge surfaces of Fig.8, Fig.10, and Fig.11, it can be observed that the distribution of leading edge individuals in NSGA-III is uniform and sparse, whereas NSGA-II and R-NSGA-II are denser. This is caused by different selection mechanisms of the algorithms. NSGA-II and R-NSGA-II use individuals as a benchmark to compare the sorting situation (relative position) of different individuals and combine local quantity control strategies to ultimately affect the overall distribution. NSGA-III, on the other hand, is the only algorithm among the three that considers the absolute distribution of the population. It prioritizes controlling the overall distribution based on reference points that are not affected by the relative positions between individuals. For the multi-objective problem proposed in this chapter, although NSGA-III cannot provide more candidate solutions than NSGA-II under the same population size and evolution frequency, there is no significant difference in the frontier span between the two.

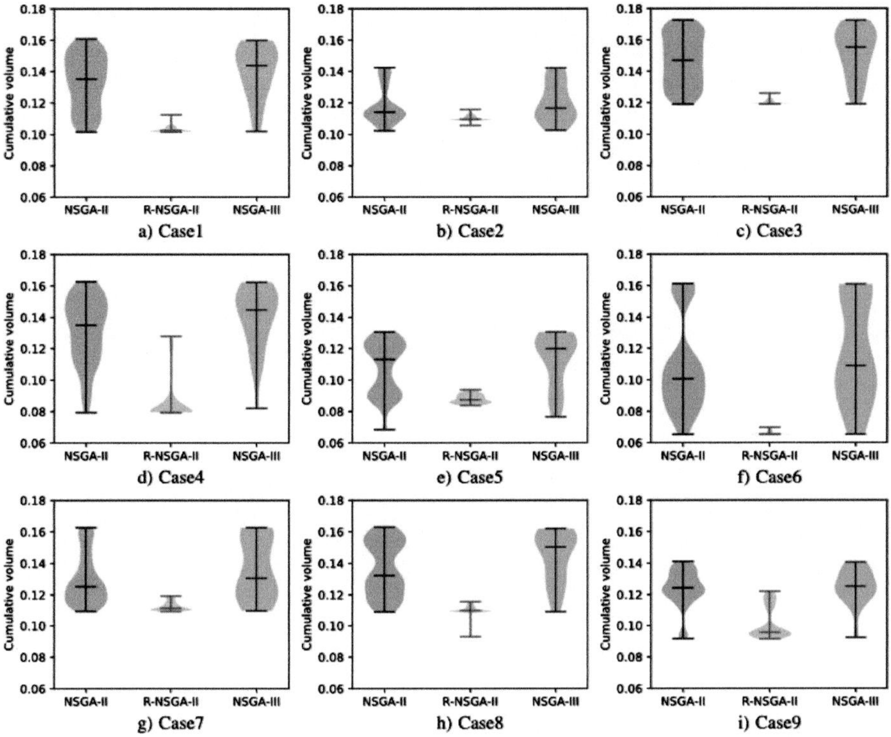

Figure 12: Cumulative volume distribution of solutions obtained by different algorithms

There is a non-dominated relationship between feasible solutions for multi-objective problems, meaning that there is no single solution that can optimize all objectives simultaneously. Additionally, to protect the privacy of enterprise data, this chapter uses a standardized processing method. Therefore, to better compare the advantages and disadvantages of each feasible solution and the differences in the results obtained by different algorithms, Fig.12 were created. These figures show the cumulative optimization distribution of feasible solutions obtained by NSGA-II, R-NSGA-II, and NSGA-III algorithms in different cases. As discussed earlier, since NSGA-III and NSGA-II have similar frontier spans, their cumulative optimization upper and lower limits are also similar. The difference in distribution density between the two is due to their different critical layer individual selection mechanisms. The analysis results for R-NSGA-II are consistent with the previous analysis, which only reflects a portion of the NSGA-II solution results.

Table 6: Comparison between original condition and optimization results

Case	Type	Decision variables						Optimization objectives		
		a_{ts}	a_{t3}	a_{t2}	a_{t1}	P_{w1}	P_{w2}	$Cost$	$Energy$	GHG
1	Original	2.9	108.7	115.0	71.4	97.0	130.0	1393.4	5183.2	246.0
	Optimized	2.6	117.7	103.5	64.3	97.0	129.9	1369.6	4795.8	229.0
2	Original	2.7	94.0	117.1	54.4	99.0	125.0	1214.5	4559.1	226.8
	Optimized	2.4	107.8	105.4	49.0	99.1	124.8	1218.3	4212.4	211.0
3	Original	4.8	98.1	121.6	113.3	63.0	120.0	1211.7	5047.6	218.6
	Optimized	4.3	113.6	109.4	101.9	63.0	120.0	1186.9	4645.9	202.7
4	Original	2.8	98.3	100.6	70.5	90.0	128.0	1508.7	5677.7	263.2
	Optimized	2.5	106.0	90.5	63.5	90.3	127.8	1478.8	5255.8	245.1
5	Original	2.2	96.2	102.5	69.5	85.0	125.0	1427.3	5351.5	252.3
	Optimized	2.3	100.9	92.3	62.6	85.0	125.1	1404.8	5048.0	237.6
6	Original	2.1	98.9	90.9	116.2	89.9	124.8	1980.9	7844.5	334.2
	Optimized	1.9	109.8	81.8	104.6	89.7	124.3	1941.5	7262.2	311.8
7	Original	2.9	94.1	116.0	54.4	99.0	125.0	1231.6	4639.6	228.8
	Optimized	2.6	101.6	104.4	48.9	98.8	124.7	1209.8	4291.8	212.8
8	Original	2.9	94.1	117.3	54.4	99.0	125.0	1231.1	4642.9	229.7
	Optimized	2.6	101.5	105.6	48.9	98.6	124.3	1209.1	4295.1	213.6
9	Original	2.3	95.8	102.5	70.3	85.0	125.0	1449.6	5460.4	256.3
	Optimized	2.3	101.5	92.2	63.3	85.1	125.0	1425.0	5119.0	240.6

* $Cost(cny/t)$; $Energy(MJ/t)$; $GHG(kgCO_2eq/t)$.

It is widely accepted that the NSGA-III algorithm can effectively address issues related to poor computational convergence and lack of population diversity that may arise while handling high-dimensional and complex problems with three or more optimization objectives, which are challenging to handle using the NSGA-II crowding concept. However, the results of the nine optimization case studies presented in this chapter demonstrate that NSGA-II still performs well when dealing with problems having three optimization objectives and multiple nonlinear constraints, which is in line with the findings of previous studies. Nevertheless, it is essential to note that as more factors, such as air pollution and water pollution, are incorporated into process optimization research, the mathematical representation of the cardboard manufacturing process is likely to become increasingly complex. Consequently, NSGA-III may become a more effective solution tool in such scenarios.

Table6 presents the final optimal values of cost, energy consumption, and GHG emissions, which are compared with the results of the original production scenario. The table shows that the optimal cost can be reduced by a range of -3.79 to 39.43 cny/t, which can be up to 2.0% lower than the original scenario. Similarly, the energy consumption can be decreased by 14.68 MJ/t to 22.43 MJ/t, which are 5.8% and 7.3% lower than the original values, respectively. Moreover, the optimal GHG emissions are reduced by 5.7% compared to the original condition, and reached 303.52 $kgCO_2eq/t$. Furthermore, it is noted that the sum of optimization in standardization is reduced by 14.1% to 17.3% at the optimal state, compared to the original mode.

In conclusion, the results of this study demonstrate the effectiveness of multi-objective optimization algorithms in improving the process optimization of cardboard manufacturing. NSGA-II, R-NSGA-II, and NSGA-III all provide solutions that outperform the original conditions, highlighting the potential for further optimization in this industry. NSGA-II excels at generating a large number of diverse solutions, achieving an optimization effect of up to 8% on each objective, and a cumulative maximum optimization amount of 17.3%. R-NSGA-II, although sometimes unable to obtain a complete Pareto optimal frontier, provides reliable solutions that are easily accepted by decision-makers due to the introduction of prior knowledge. NSGA-III, while providing a more uniform solution distribution, did not demonstrate significant advantages over NSGA-II in this particular problem. However, with the inclusion of more complex factors in the optimization process, NSGA-III may prove to be a more effective solution tool. Overall, the findings of this work provide valuable insights for future process optimization research in the cardboard manufacturing industry.

5 Conclusions and future works

This work aimed to develop data-driven soft sensors to monitor key paper properties and apply the models to optimize the papermaking process. Four different data-driven methods, including random forest, gradient boosting regression, ridgeregression, and K-nearest neighbor, were employed to establish soft sensor models for monitoring folding endurance, bursting strength, smoothness, and transverse ring compressive strength. Using industrial data, the performance of each model was tested, and the optimal models were selected based on their ability to accurately predict the target variables. The selected models showed R^2 values ranging from 0.853 to 0.994, $MAPE$ values from 4.59% to 1.51%, and $RMSE$ values from 29.534 to 0.333. The developed soft sensors were utilized to support process adjustments in the papermaking process, leading to a possible reduction in product costs, energy consumption, and greenhouse gas emissions. The calculation demonstrated that the optimal cost can be saved up to 39.43 cny/t, which is about 2.0% reduced from the original production scenario. And the corresponding results for energy consumption and GHG emission is 22.43 MJ/t in 7.3% and 303.52 $kgCO_2eq/t$ in 5.7% respectively. The optimized results overall show that, when process is qualified with soft sensors' support, the possible reduction of cost, energy consumption and GHG emission could be up to 17.3% in total. The developed soft sensors can make significant contribution to the papermaking industry.

References

[1] C. P. Association, The annual report of china papermaking industry in 2021(in chinese), China Paper Newsletters (2022) 6–17.

[2] C. P. Association, Papermaking industry "14th five-year plan" and the outline of medium and long-term high-quality development(in chinese), China Pulp and Paper Industry (2022) 12–17.

[3] P. Bajpai, Brief description of the pulp and papermaking process, Biotechnology for pulp and paper processing (2018) 9–26.

[4] Y. Man, M. Hong, J. Li, S. Yang, Y. Qian, H. Liu, Paper mills integrated gasification combined cycle process with high energy efficiency for cleaner production, Journal of Cleaner Production 156 (2017) 244–252.

[5] X. Chen, Y. Man, Q. Zheng, Y. Hu, J. Li, M. Hong, Industrial verification of energy saving for the single-tier cylinder based paper drying process, Energy 170 (2019) 261–272.

[6] Z. He, J. Xu, K. P. Tran, S. Thomassey, X. Zeng, C. Yi, Modeling of textile manufacturing processes using intelligent techniques: a review, The International Journal of Advanced Manufacturing Technology 116 (2021) 39–67.

[7] Q. Sun, Z. Ge, A survey on deep learning for data-driven soft sensors, IEEE Transactions on Industrial Informatics 17 (2021) 5853–5866.

[8] C. P. Association, The annual report of china papermaking industry in 2016(in chinese), China Paper Newsletters (2017) 8–18.

[9] J. Li, M. Mei, Y. Han, M. Hong, Y. Man, Life cycle cost assessment of recycled paper manufacture in china, Journal of Cleaner Production 252 (2020) 119868.
[10] Y. Zhang, Y. Han, Y. Man, Sustainability evaluation of tissue paper under different production paths, ACS Sustainable Chemistry & Engineering 9 (2021) 7341–7351.
[11] Y. Man, Y. Han, J. Li, M. Hong, W. Zheng, Life cycle energy consumption analysis and green manufacture evolution for the papermaking industry in china, Green Chemistry 21 (2019) 1011–1020.
[12] M. Sun, Y. Wang, L. Shi, J. J. Klemeš, Uncovering energy use, carbon emissions and environmental burdens of pulp and paper industry: A systematic review and meta-analysis, Renewable and Sustainable Energy Reviews 92 (2018) 823–833.
[13] Y. Hu, J. Li, M. Hong, J. Ren, R. Lin, Y. Liu, M. Liu, Y. Man, Short term electric load forecasting model and its verification for process industrial enterprises based on hybrid ga-pso-bpnn algorithm—a case study of papermaking process, Energy 170 (2019) 1215–1227.
[14] Y. Man, Y. Hu, J. Ren, Forecasting cod load in municipal sewage based on arma and var algorithms, Resources, Conservation and Recycling 144 (2019) 56–64.
[15] Y. Man, J. Li, M. Hong, Y. Han, Energy transition for the low-carbon pulp and paper industry in china, Renewable and Sustainable Energy Reviews 131 (2020) 109998.
[16] L. Kong, A. Hasanbeigi, L. Price, H. Liu, Energy conservation and co2 mitigation potentials in the chinese pulp and paper industry, Resources, Conservation and Recycling 117 (2017) 74–84.
[17] E. Małachowska, A. Lipkiewicz, M. Niemczyk, M. Dubowik, P. Boruszewski, P. Przybysz, Influences of fiber and pulp properties on papermaking ability of cellulosic pulps produced from alternative fibrous raw materials, Journal of Natural Fibers 18 (2021) 1751–1761.
[18] Y. Liu, M. Xie, Rebooting data-driven soft-sensors in process industries: A review of kernel methods, Journal of Process Control 89 (2020) 58–73.
[19] G. N. Vieira, M. Olazar, J. T. Freire, F. B. Freire, Real-time monitoring of milk powder moisture content during drying in a spouted bed dryer using a hybrid neural soft sensor, Drying Technology 37 (2019) 1184–1190.
[20] M. Fallmann, A. Poks, M. Kozek, Hybrid model-based online estimation of air temperature in mobile small-scale cooling chambers, Applied Thermal Engineering 208 (2022) 118147.

[21] B. H. da Silva Melo, R. F. Sales, L. da Silva Bastos Filho, J. S. P. da Silva, A. G. C. de Almeida Sousa, D. M. C. Peixoto, M. F. Pimentel, Handheld near infrared spectrometer and machine learning methods applied to the monitoring of multiple process stages in industrial sugar production, Food Chemistry 369 (2022) 130919.

[22] Y. Meng, T. Yao, S. Yu, J. Qin, J. Zhang, J. Wu, Data-driven modeling for crystal size distribution parameters in cane sugar crystallization process, Journal of Food Process Engineering 44 (2021) e13648.

[23] L. Strani, R. Vitale, D. Tanzilli, F. Bonacini, A. Perolo, E. Mantovani, A. Ferrando, M. Cocchi, A multiblock approach to fuse process and near-infrared sensors for on-line prediction of polymer properties, Sensors 22 (2022) 1436.

[24] D. C. de Souza, L. Cabrita, C. F. Galinha, M. S. Reis, Pat soft sensors for wide range prediction of key properties of diesel fuels and blending components for the oil industry, Computers & Chemical Engineering 153 (2021) 107449.

[25] Y.-L. He, Z.-Q. Geng, Q.-X. Zhu, Data driven soft sensor development for complex chemical processes using extreme learning machine, Chemical Engineering Research and Design 102 (2015) 1–11.

[26] K. Funatsu, Soft sensors: Chemoinformatic model for efficient control and operation in chemical plants, Molecular Informatics 35 (2016) 549–554.

[27] S. Taheri, G. Brodie, D. Gupta, Optimised ann and svr models for online prediction of moisture content and temperature of lentil seeds in a microwave fluidised bed dryer, Computers and Electronics in Agriculture 182 (2021) 106003.

[28] M. S. Zaghloul, G. Achari, A review of mechanistic and data-driven models of aerobic granular sludge, Journal of Environmental Chemical Engineering (2022) 107500.

[29] W. Yan, R. Xu, K. Wang, T. Di, Z. Jiang, Soft sensor modeling method based on semisupervised deep learning and its application to wastewater treatment plant, Industrial & Engineering Chemistry Research 59 (2020) 4589–4601.

[30] X. Zhang, J. Li, Y. Zhang, W. Cai, H. Liu, et al., Data-driven method for pulp properties estimation in stock preparation, BioResources 11 (2016) 4947–4963.

[31] X. Zhang, J. Li, H. Liu, P. Zhou, Soft sensors for pulp freeness and outlet consistency estimation in the alkaline peroxide mechanical pulping (apmp) high-consistency refining process., BioResources 11 (2016).

[32] Y. Liu, Z. Zhang, J. Chen, Ensemble local kernel learning for online prediction of distributed product outputs in chemical processes, Chemical Engineering Science 137 (2015) 140–151.
[33] D. Raffaele, T. Ondruch, Data-driven soft sensor for continuous production monitoring: an application to paper strength, in: 2020 25th IEEE International Conference on Emerging Technologies and Factory Automation (ETFA), volume 1, IEEE, pp. 1335–1338.
[34] M. Li, P. Zhou, L. Wang, Y. Yuan, Data-driven multiobjective predictive optimal control of refining process with non-gaussian stochastic distribution dynamics, IEEE Transactions on Industrial Informatics 17 (2021) 7269–7278.
[35] H. Liu, Y. Zhang, H. Zhang, Prediction of effluent quality in papermaking wastewater treatment processes using dynamic kernel-based extreme learning machine, Process Biochemistry 97 (2020) 72–79.
[36] H. Liu, C. Xin, H. Zhang, F. Zhang, M. Huang, Effluent quality prediction of papermaking wastewater treatment processes using stacking ensemble learning, IEEE Access 8 (2020) 180844–180854.
[37] Y. Jiang, S. Yin, J. Dong, O. Kaynak, A review on soft sensors for monitoring, control, and optimization of industrial processes, IEEE Sensors Journal 21 (2020) 12868–12881.
[38] V. Krishna, N. Pappa, S. Vasantharani, Realization of deep learning based embedded soft sensor for bioprocess application., Intelligent Automation & Soft Computing 32 (2022).
[39] J. Chen, S. Fan, C. Yang, C. Zhou, H. Zhu, Y. Li, Stacked maximal quality-driven autoencoder: Deep feature representation for soft analyzer and its application on industrial processes, Information Sciences 596 (2022) 280–303.
[40] P. Cao, X. Luo, X. Song, Modeling and identification for soft sensor systems based on the separation of multi-dynamic and static characteristics, Chinese Journal of Chemical Engineering 26 (2018) 137–143.
[41] Z. Li, H. Jin, S. Dong, B. Qian, B. Yang, X. Chen, Semi-supervised ensemble support vector regression based soft sensor for key quality variable estimation of nonlinear industrial processes with limited labeled data, Chemical Engineering Research and Design 179 (2022) 510–526.
[42] Y. Zhang, S. Ma, H. Yang, J. Lv, Y. Liu, A big data driven analytical framework for energy-intensive manufacturing industries, Journal of Cleaner Production 197 (2018) 57–72.

[43] M. Liukkonen, J. Silvennoinen, Y. Hiltunen, Toward better control of water-intensive processes: Case study in fluting manufacturing, IFAC-PapersOnLine 51 (2018) 683–688.

[44] H. Liu, C. Yoo, A robust localized soft sensor for particulate matter modeling in seoul metro systems, Journal of hazardous materials 305 (2016) 209–218.

[45] H. Liu, C. Yang, M. Huang, D. Wang, C. Yoo, Modeling of subway indoor air quality using gaussian process regression, Journal of hazardous materials 359 (2018) 266–273.

[46] K. Deb, J. Sundar, Reference point based multi-objective optimization using evolutionary algorithms, in: Proceedings of the 8th annual conference on Genetic and evolutionary computation, pp. 635–642.

[47] K. Deb, H. Jain, An evolutionary many-objective optimization algorithm using reference-point-based nondominated sorting approach, part i: solving problems with box constraints, IEEE transactions on evolutionary computation 18 (2013) 577–601.

[48] H. Jain, K. Deb, An evolutionary many-objective optimization algorithm using reference-point based nondominated sorting approach, part ii: Handling constraints and extending to an adaptive approach, IEEE Transactions on evolutionary computation 18 (2013) 602–622.

[49] J. Blank, K. Deb, P. C. Roy, Investigating the normalization procedure of nsga-iii, in: Evolutionary Multi-Criterion Optimization: 10th International Conference, EMO 2019, East Lansing, MI, USA, March 10-13, 2019, Proceedings 10, Springer, pp. 229–240.

[50] N. Srinivasan, K. Deb, Multi-objective function optimisation using nondominated sorting genetic algorithm, Evolutionary Comp 2 (1994) 221–248.

[51] H. Zhang, J. Li, M. Hong, Y. Man, Z. He, Cost optimal production-scheduling model based on vns-nsga-ii hybrid algorithm—study on tissue paper mill, Processes 10 (2022) 2072.

[52] H. Chen, Y. Cao, Y. Liu, Y. Qin, L. Xia, Enhancing the durability of concrete in severely cold regions: Mix proportion optimization based on machine learning, Construction and Building Materials 371 (2023) 130644.

[53] Z. Wang, X. Mo, P. Qin, Z. Zhao, T. Ouyang, Multi-dimensional assessment and multi-objective optimization of electricity-cooling cogeneration system driven by marine diesel engine waste heat, Journal of Cleaner Production 334 (2022) 130187.

[54] I. Das, J. E. Dennis, Normal-boundary intersection: A new method for generating the pareto surface in nonlinear multicriteria optimization problems, SIAM journal on optimization 8 (1998) 631–657.
[55] Z. Luo, M. Hasanipanah, H. Bakhshandeh Amnieh, K. Brindhadevi, M. Tahir, Ga-svr: a novel hybrid data-driven model to simulate vertical load capacity of driven piles, Engineering with Computers 37 (2021) 823–831.
[56] N. Development, R. C. of China, Guidelines on Accounting methods and Reporting of Greenhouse Gas Emissions from Paper and Paper Products Manufacturing Enterprises (Trial)(In Chinses), 2015.
[57] E. Centre, Swiss centre for life cycle inventories ecoinvent database version 3.1, https://www.ecoinvent.org, 2010.
[58] B. Club, Real-time quotes, https://www.100ppi.com, 2022.
[59] X. Ma, Life Cycle Assessment of Terylene Textile, Master's thesis, Donghua University, 2007.
[60] L. Shen, E. Worrell, M. K. Patel, Comparing life cycle energy and ghg emissions of bio-based pet, recycled pet, pla, and man-made cellulosics, Biofuels, bioproducts and biorefining 6 (2012) 625–639.
[61] N. Zhao, Carbon footprint assessment and low carbon measures of the polyester textile, Master's thesis, Donghua University, 2012.

3. Kriging Model Based Greenhouse Gas Emissions Model of Papermaking Wastewater Treatment Process

Hongze Zheng,[1] Zhenglei He,[1] Mengna Hong[1] and Yi Man[1,2,*]

1 Introduction

China's paper and paperboard production was 121.1 million tons in 2021, ranking first in the world[1]. At the same time, the wastewater treatment capacity of China's paper industry reached 1.697 billion tons, accounting for 6.6% of the industrial wastewater treatment capacity[2]. Papermaking wastew-ater contains as many as 90 kinds of pollutants such as organic carbon and nitrogen[3]. Papermaking enterprises need to collect and treat the wastewater centrally so that the wastewater meets the requirements of standards before discharge. However, microorganisms are used in the papermaking wastewater treatment process to decompose pollutants such as organic carbon and nitro-gen into CO_2, N_2O and CH4[4]. The global warming potential of CH_4 is 25 kg CO_2 eq/kg, while that of N_2O is up to 298kg CO_2 eq/kg[5]. which is often directly discharged without any treatment.

With the goal of carbon peaking and carbon neutrality[6], the papermaking industry has been listed as one of the eight key carbon emission control

[1] State Key Laboratory of Pulp and Paper Engineering, South China University of Technology, Guangzhou, China,

[2] Pazhou Lab, Guangzhou, China.

* Corresponding author: manyi@scut.edu.cn

industries in China[7]. At present, some GHG emissions reduction strategies have been put forward for the wastewater treatment process. For example, intermittent aeration and hydraulic load regulation can improve the effluent quality without affecting the CH4 discharge[8]. Additionally, combining natural and engineered systems can save energy and reduce GHG emissions during the natural treatment process[9]. It is also one of the strategies to recycle the ammonia nitrogen generated in the wastewater treatment process to generate electricity[10]. Although the above methods have performed effectively, there are also problems such as extra costs and high electricity consumption, which decelerate the development of GHG emission reduction. Therefore, based on regulating the control parameters, it will be quite possible to achieve GHG emission reduction in the existing treatment process[11].

In order to achieve this objective, it is necessary to realize an accurate quantification of GHG in the wastewater treatment process and setting it as the target parameter. However, the GHG emissions of wastewater treatment in the papermaking industry are mostly calculated accounting form[12], lacking systematic calculation and analysis methods, which results in a large deviation from the actual quantity of emissions[13]. Thus, the corresponding emission reduction control is impossible. Therefore, how to accurately quantify the GHG emissions in the papermaking wastewater treatment and achieve GHG emission reduction control is an important issue currently facing the papermaking industry.

From the perspective of mechanism, the quantification of GHG emissions in the wastewater treatment process requires a complete mechanism model. The BSM1 developed by the International Water Quality Association (IWA) provides the basis for the treatment treatment mechanism description of wastewater, which has wide applicability in engineering design[14]. Moreover, according to the characteristics of papermaking wastewater, BSM1 could be successfully applied to the activated sludge wastewater treatment of paper mills after improvement[15]. BSM2, BSM2d and BSM3 are improved models based on BSM1 and have relatively better accuracy. Among these, BSM2 added equipment such as primary sedimentation tank, anaerobic digester and dewatering device, which could successfully evaluate the plant-wide control performance. In addition, BSM2 can simulate the biological phosphorus removal process in more detail. However, the model involves a large amount of calculations and multi variable measurement. Also, there is a lack of measuring instruments for some variables on the actual papermaking wastewater treatment site[16]. Compared with BSM2, BSM2d describes the biological process of phosphorus accumulation and denitrification under the influence of temperature[17], which has a lesser impact on the papermaking wastewater treatment. BSM3 introduces a more realistic description of the decay process and focuses on the storage process of organic substrates. However, when BSM3 is applied to industrial processes, it requires a large amount of experimental data to determine model parameters[18]. Therefore, based on the characteristics of papermaking wastewater, BSM1 will be used to simulate the wastewater treatment process in this study.

Nevertheless, BSM1 doesn't calculate GHG emissions. The mechanism of GHG emissions in activated sludge wastewater treatment still needs further research. At present, the GHG emissions of wastewater treatment plants can be divided into direct and indirect emissions[19]. The direct emissions are mainly related to the reaction of microorganisms that can be calculated through material conservation[20]. Indirect emissions are mainly generated by external activities such as electricity consumption, which can be calculated through corresponding emission factors[21]. Therefore, the current research mainly estimates the GHG emissions of wastewater treatment plants through the calculation model. Then GHG emissions are reduced through automatic control, such as control of dissolved oxygen[22] and nitrate nitrogen[23]. Dissolved oxygen concentration is a key parameter for GHG emissions from wastewater treatment, which has a great influence on the degradation rate of organic pollutants, the sedimentation performance and of the nitrification activity[24]. The control strategy is to stabilize the dissolved oxygen concentration to a lower set value while satisfying the effluent quality[25]. However, the influent compositions and flow are complex and changeable, which often lead to excessive aeration while ensuring the effluent quality[26]. Therefore, the traditional control strategy cannot achieve better GHG emissions reduction effects due to the change of the optimal set point under different conditions[27]. Furthermore, other manipulative variables such as circulating flow in wastewater treatment affect GHG emissions[28]. Also, there is a strong coupling between variables, which makes it difficult for the calculation model of GHG to analyze the relationship between variables in the wastewater treatment and their specific impact on GHG emissions, so it is also difficult to achieve multivariable GHG emissions reduction control.

To achieve multi-parameter optimization, it is necessary to combine the mechanism model of wastewater treatment with the calculation method of

GHG emissions. However, the complex mechanism model makes the calculation cost of simulation too expensive, which also greatly increases the difficulty of optimization design[29]. As an effective tool for optimization design, a surrogate model can approximate the original model[30]. Moreover, its function structure is relatively simple, which can reduce the optimization design of calculation cost[31]. Currently, the most widely used surrogate model is the polynomial response surface model, whose structure is relatively simple. Nevertheless, it requires a general understanding of the functional relationship between variables. Additionally, the model is not applicable to nonlinear and high-dimensional engineering problems[32]. A neural network has better prediction accuracy in complex engineering problems[33]. However, its modeling process requires a large number of sample points, which makes its training consume a lot of computing time[34]. The Kriging model not only has enough accuracy in nonlinear function fitting, but also takes less computational resources[35], which is suitable for the modeling of the papermaking wastewater treatment process.

In summary, this study established a model of GHG emissions according to the functional relationship between wastewater treatment process parameters and GHG emissions, aiming to analyze the impact of various parameters on GHG emissions under different conditions. In this way, the optimal control parameters can be found, which can provide a basic theoretical reference for the multivariable emission reduction control scheme of GHG. Therefore, this study first applied BSM1 to simulate the wastewater treatment process. Then the GHG emissions were quantified in combination with the calculation model. As a result wastewater treatment acquires the characteristics of high dimensionality, nonlinearity and instability, which make its GHG emissions model complex and computationally costly. The Kriging surrogate model technology was applied to establish a CHG emissions model in the papermaking wastewater treatment, which aimed to reduce the complexity and calculation cost while ensuring the accuracy of the model. Finally, according to the actual papermaking wastewater influent data, the impact of different manipulation parameters on GHG emissions is analyzed. The GA algorithm is used to optimize the manipulation parameters to achieve GHG emission reduction.

2 Papermaking wastewater treatment process

Activated sludge is the most widely used wastewater treatment process in papermaking enterprises. Papermaking wastewater first passes through the anaerobic and aerobic tanks, where the growth characteristics of microorganisms are used to decompose and adsorb the organic pollutants. The aerobic tank continually passes enough oxygen through the aerator, which promotes the aerobic reaction[36]. Then, the biochemically treated wastewater passes through the secondary sedimentation tank to settle the refractory pollutants to the bottom and discharge them in the form of sludge[37]. Moreover, in order to prevent the accumulation of wastewater for a long time, papermaking wastewater treatment is a continuous process. The wastewater treatment process is an unstable, high-dimensional, nonlinear and complex system due to a great variety of pollutants, the large variation range of wastewater flow and pollutant concentration, and the impact of many factors on the microbial reaction process[38]. The wastewater treatment process described in BSM1 is more consistent with the papermaking wastewater treatment process, which can realize the simulation of the original process with high precision.

According to the monitoring system of the activated sludge process in the paper mill, the influent data includes COD, TSS, TN, ammonia nitrogen and flow rate. While BSM1 has relatively clear and detailed requirements for the pollutant types, which cannot directly use the actual data as its input. Therefore, it is necessary to divide the COD and TN components into variables in BSM1 according to the actual characteristics of the incoming water of papermaking wastewater. Among them, COD can be divided into four components: S_s, S_i, X_s and X_i, accounting for 18.11%, 4.11%, 19.97% and 57.81% of COD respectively[15]. TN can be divided into three components: S_{no}, X_{nd} and S_{nd}, accounting for 2.23%, 91.63% and 6.14% respectively[39]. The exact meaning of these components could be found in Table 1. In addition, the pH is basically stable at about neutral 7. Also, there is no microorganism in the influent of papermaking wastewater.

3 Benchmark simulation model 1

BSM1 consisted of two anaerobic tanks, three aerobic tanks and one sedimentation tank, accompanied with internal and external reflux. The model layout was shown in Figure 2. The Activated Sludge Model No.1 (ASM1) and

Takács double sedimentation index function were adopted for the reactor and sedimentation tank respectively[40].

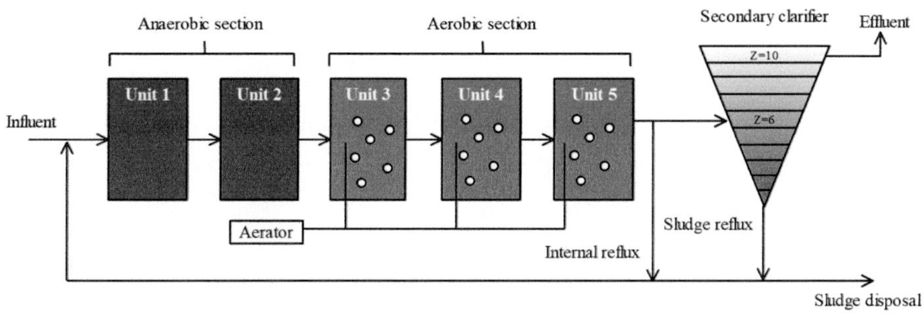

Figure 1: General overview of the BSM1 plant

ASM1 contained 13 components variables, as shown in Table 1. The reaction mainly included organic oxidation, denitrification and nitrification, which could be divided into eight processes, including hydrolysis, microbial growth and death.

What's more, influent variables of BSM1 could be cut down according to the characteristics of papermaking wastewater inflow. To begin with, the papermaking wastewater did not contain microorganisms. Its pH was controlled to 7. Therefore, the influent variables of BSM1 could be reduced to 8. In addition, the influent flow, the dissolved oxygen transfer coefficient, the internal and external reflux flow were the remaining variables of BSM1. The value range of variables is referred to the actual data collected in the activated sludge process of the paper mill.

4 GHG emissions model

The GHG emission sources in the papermaking wastewater treatment process include direct and indirect emissions, whose framework was shown in Figure 3. Due to different mechanisms of GHG emissions under anaerobic and aerobic environments, this study calculates the GHG emissions of anaerobic anaerobic

Table 1: Variables of papermaking wastewater influent

Number	Notation	Definition
1	S_i	Soluble inert organic matter
2	S_s	Readily biodegradable substrate
3	X_i	Particulate inert organic matter
4	X_s	Slowly biodegradable substrate
5	X_{bh}	Active heterotrophic biomass
6	X_{ba}	Active autotrophic biomass
7	X_p	Particulate products arising from biomass decay
8	S_o	Oxygen
9	S_{no}	Nitrate and nitrite nitrogen
10	S_{nh}	NH_4^+ and NH_3 nitrogen
11	S_{nd}	Soluble biodegradable organic nitrogen
12	X_{nd}	Particulate biodegradable organic nitrogen
13	S_{alk}	Alkalinity

and aerobic reactors respectively. In general, the sources of GHG in wastewater treatment are mainly divided into Biochemical Oxygen Demand (BOD) removal, endogenous respiration, nitrification and denitrification[41]. Among them, CH_4 is mainly produced in the anaerobic reactor. CO_2 is consumed in the synthesis of organic nitrogen and other substances in the nitrification process. N_2O is the intermediate product of the denitrification process, which is usually calculated by the empirical emissions factor because of the unclear generation mechanism[42]. Indirect GHG emissions are mainly generated by electrical consumption and sludge disposal wherein, the GHG generated by the electrical consumption of aerators and pumps are calculated by the corresponding emission factors. For sludge disposal, CO_2 emissions are calculated by material balances.

4.1 Anaerobic reaction

In the anaerobic reactor, part of the BOD in wastewater participates in the synthesis of microbial cells, and the other part decomposes into CO_2 ($CO_{2,BOD,AN}$) and CH_4 ($CH_{4,BOD,AN}$) under anaerobic conditions. In addition, some microorganisms further converts to CO_2 ($CO_{2,EDG,AN}$) and

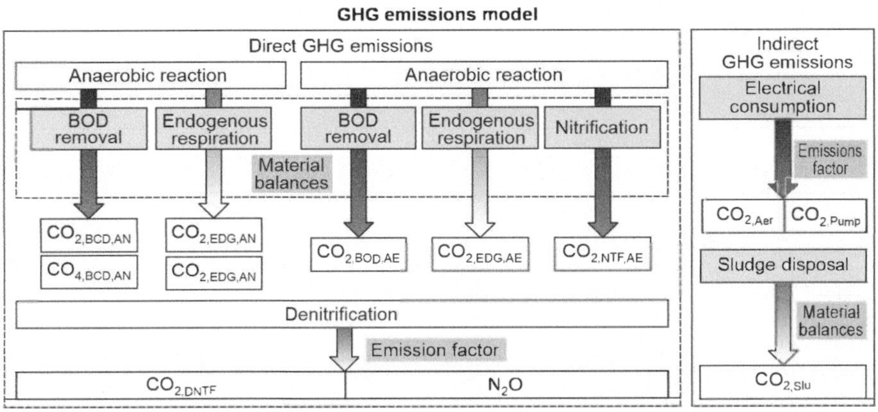

Figure 2: Framework for the GHG emissions model

CH_4 ($CH_{4,EDG,AN}$) through endogenous respiration. Chemical equations of BOD removal and endogenous respiration in an anaerobic environment are presented by Equations (1) and (2). The soluble substrate and biomass in wastewater are expressed by $C_{10}H_{19}O_3N$ and $C_5H_7O_2N$.

$$0.02C_{10}H_{19}O_3N + 0.094H_2O \rightarrow 0.004C_5H_7O_2N + 0.049CO_2 + 0.115CH_4 + 0.016HCO_3^- + 0.016NH_4^+ \quad (1)$$

$$0.05C_5H_7O_2N + 0.2H_2O \rightarrow 0.075CO_2 + 0.125CH_4 + 0.05NH_4^+ + 0.05HCO_3^- \quad (2)$$

$CO_{2,BOD,AN}$ and $CH_{4,BOD,AN}$ emissions are shown in Eqs. (3) and (4).

$$M^{An}_{CO_2,BOD} = 0.27 \times (\frac{BOD^{An}_{in} - BOD^{An}_{out}}{f} - 1.42 \times P^{An}_{X,bio}) \quad (3)$$

$$M^{An}_{CH_4,BOD} = 0.25 \times (\frac{BOD^{An}_{in} - BOD^{An}_{out}}{f} - 1.42 \times P^{An}_{X,bio}) \quad (4)$$

where $P_{X,bio}$ is the microbial yield during BOD removal.

$CO_{2,EDG,AN}$ and $CH_{4,EDG,AN}$ emissions are presented by the following Eqs. (5) and (6).

$$M^{An}_{CO_2,decay} = 0.58 \times M^{An}_{x,degraded} \tag{5}$$

$$M^{An}_{CH_4,decay} = 0.35 \times M^{An}_{x,degraded} \tag{6}$$

where $M^{An}_{x,degraded}$ is the biomass decayed.

4.2 Aerobic reaction

In the aerobic reaction, CO_2 is also generated from BOD removal ($CO_{2,BOD,AE}$) and endogenous respiration ($CO_{2,EDG,AE}$), but CH_4 is not produced in this process. The reaction equations are presented by the following Equations (7) and (8) that follow.

$$\begin{aligned} 0.02C_{10}H_{19}O_3N + 0.01NH_4^+ + 0.01HCO_3^- + 0.1O_2 \\ \rightarrow 0.03C_5H_7O_2N + 0.11H_2O + 0.06CO_2 \end{aligned} \tag{7}$$

$$\begin{aligned} 0.05C_5H_7O_2N + 0.25O_2 \rightarrow 0.2CO_2 + 0.05NH_4^+ + 0.05HCO_3^- \\ + 0.05H_2O \end{aligned} \tag{8}$$

The calculation formulae for $CO_{2,BOD,AE}$ and $CO_{2,EDG,AE}$ emissions are shown in Eqs. (9) and (10).

$$M^{Ae}_{CO_2,BOD} = 0.33 \times (\frac{BOD^{Ae}_{in} - BOD^{Ae}_{out}}{f} - 1.42 \times P^{An}_{X,bio}) \tag{9}$$

$$M^{Ae}_{CO_2,decay} = 1.56 \times M^{Ae}_{x,degraded} \tag{10}$$

Under the condition of sufficient oxygen, autotrophic bacteria grow well and undergo nitrification. Additionally, the ammonia nitrogen in the reactor eventually gets transformed to nitrate, and CO_2 is consumed at the same time ($CO_{2,NTF,AE}$). The nitrification process in the aerobic reactor is presented by Eq. (11) that follows.

$$\begin{aligned} &0.127NH_4^+ + 0.24O_2 + 0.01CO_2 + 0.0025HCO_3^- \\ &\quad \rightarrow 0.0025C_5H_7O_2N + 0.125NO_3^- + 0.25H^+ + 0.1175H_2O \end{aligned} \tag{11}$$

In this process, part of ammonia nitrogen will be oxidized to nitrate nitrogen, while the other part will be transformed into microbial components. Therefore, the oxidized ammonia nitrogen needs to be calculated by subtracting the total consumption from the amount used by microorganisms. The calculation formula is shown in Eq. (12).

$$NH_{OX} = (TKN_{in}^{Ae} - TKN_{out}^{Ae}) - 0.12 \times P_{X,bio}^{Ae} \tag{12}$$

where TKN is the Total Kjeldahl Nitrogen.

The $CO_{2,NTF,AE}$ can be calculated according to the amount of oxidized ammonia nitrogen obtained in Eq. (12), as shown in Eq. (13).

$$M_{CO_2,nitri}^{An} = 0.25 \times NH_{OX} \tag{13}$$

4.3 Denitrification reaction

Under anaerobic conditions, heterotrophic bacteria will grow anaerobically and carry out denitrification. Among them, nitrate will be reduced by a multi-step reaction during the process to generate gases such as NO, N_2O and N_2. The CO_2 produced in this process ($CO_{2,DNTF}$) can be calculated by Eq. (14), and the calculation formula is shown in Eq. (15).

$$\begin{aligned} &0.02C_{10}H_{19}O_3N + 0.193NO_3^- + 0.19H^+ \rightarrow \\ &\quad 0.001C_5H_7O_2N + 0.2NH_4^+ + 0.096N_2 + 0.232H_2O \\ &\quad + 0.173CO_2 + 0.02HCO_3^- \end{aligned} \tag{14}$$

$$M_{CO_2,denit}^{An} = 2.82 \times (NO_{in}^{An} - NO_{out}^{An}) \tag{15}$$

N_2O is the intermediate product of denitrification, which is also a greenhouse gas that cannot be ignored. In the process of nitrogen removal, the emissions of N_2O can be obtained by multiplying the empirical emissions factor by the nitrogen-containing substances removed. The calculation formula is presented by Eq. (16) that follows.

$$M_{N_2O} = R_{N_2O} \times (TN_{in} - TN_{out}) \tag{16}$$

where R_{N_2O} is the empirical emissions factor, which can be set according to the specific field conditions. TN is the total nitrogen.

4.4 Indirect GHG emissions

The main sources of indirect GHG emissions in the papermaking wastewater treatment are four processes: electrical consumption of aerators and pumps, sludge disposal.

(1) Electrical consumption

Electrical consumption of aerators is obtained as:

$$AE = \frac{S_O^{sat}}{1800 \times T} \int_{t0}^{t} \sum_{t=1}^{n} V_i \times K_L a_i dt \tag{17}$$

Electrical consumption of pumps can be calculated from Eq. (18).

$$PE = \frac{1}{T} \int_{t0}^{t} (0.004 \times Q_{int} + 0.008 \times Q_r + 0.05 \times Q_w) dt \tag{18}$$

The CO_2 emissions generated by electrical consumption ($CO_{2,Aer}$ and $CO_{2,Pump}$) can be obtained by multiplying the electrical energy by the annual average emission factor of the national grid. The calculation of $CO_{2,Aer}$ and $CO_{2,Pump}$ emissions are shown in Eq. (19).

$$M_{CO_2,electric} = EF_e \times (AE + PE) \tag{19}$$

(2) Sludge disposal

The discharged sludge needs to be anaerobically treated, which produces CO_2 and CH_4 ($CO_{2,Slu}$). The chemical reaction equation is represented by Eq. (20).

$$C_5H_7O_2N + 4H_2O \rightarrow 1.5CO_2 + 2.5CH_4 + NH_4^+ + HCO_3^- \tag{20}$$

Additionally, $CO_{2,Slu}$ emissions can be calculated from the conservation of substances, as shown in Eq. (21).

$$M_{CO_2,sludge} = (0.584 + 0.354 \times GWP_{CH_4}) \times X_{sludge} \tag{21}$$

5 Kriging model

The surrogate model technology aims to construct the functional relationship between the independent and dependent variables by regression fitting according to the functional values of the sample points. The surrogate model can not only ensure the accuracy of the original model, but also has a simple structure, which can solve the problem of high computational cost and reduce the difficulty of optimization design[43]. Among them, the Kriging model is an unbiased estimation method with the minimum estimation variance, which is also an interpolation estimation method. Based on mathematical regression and the covariance function, it realizes the optimal unbiased estimation of the entire optimization space. After the selection of different covariance functions, it realizes approximate fitting and completes the spatial modeling and prediction of the random process. The Kriging model was first applied to geostatistics, showing good application prospects in nonlinear problems, and later became an optimization design method widely used in various disciplines. The basic expression of the Kriging model is as follows[44]:

$$y(x) = f(x)\beta + z(x) \tag{22}$$

where $f(x)$ is the regression function; β is the coefficient; $z(x)$ is a Gaussian process with mean value,0, variance σ^2, and its covariance is:

$$Cov(z(x_i), (z(x_j)) = \sigma^2 \mathrm{R}[R(x_i, x_j)] \tag{23}$$

where R is the symmetric correlation matrix; $R(x_i, x_j)$ is the Gaussian correlation function, which can be expressed as:

$$R(x_i, x_j) = exp(-\sum_{k=1}^{n} \theta_k |x_i^k - x_j^k|^2) \tag{24}$$

where x_i^k and x_j^k corresponds to the k_{th} element of x_i and x_j respectively; n is the number of input variables; θ_k is the correlation parameter, whose value determines the correlation between sample points.

From the above formulas, it can be seen that the construction of the Kriging model can be completed as long as the basis function $f(x)$, the regression coefficient β, the variance σ^2 and θ_k are determined. Among them, the basis function determines the type of Kriging model. For example, when the basis function is 1, it is ordinary Kriging. The estimated values of the regression coefficient and variance can be obtained from Eq. (25) and Eq. (26) respectively.

$$\tilde{\beta} = (f^T R^{-1} f) f^T \mathrm{R}^{-1} y \tag{25}$$

$$\tilde{\sigma} = \frac{(y - f\tilde{\beta})\mathrm{R}^{-1}(y - f\tilde{\beta})}{n} \tag{26}$$

where y is the vector formed by the response function; $\tilde{\beta}$ and $\tilde{\sigma}$ are the estimated values of β and σ respectively.

θ_k can be obtained by the maximum likelihood estimation method:

$$L(\theta) = (2\pi\tilde{\sigma}^2)^{-\frac{n}{2}}|R|^{-\frac{1}{2}}exp(-\frac{n}{2}) \tag{27}$$

Thus, the predicted value of Kriging model at any point in the design space can be obtained:

$$\tilde{y}(x) = f(x)\tilde{\beta} + \tilde{r}(x)^T R^{-1}(y - f\tilde{\beta}) \tag{28}$$

where $\tilde{r}(x)$ is the correlation vector between the prediction points and the sample points.

The construction of the surrogate model is generally divided into four steps: experimental design, model selection, model fitting, and accuracy verification[45]. In this study, Latin Hypercube Sampling (LHS) was adopted in the experimental design[46]. Then, according to the output value of BSM1 under steady state, the GHG emissions generated in the process were calculated. Additionally, Kriging surrogate model technology was used to establish the functional relationship between input variables and GHG emissions. Finally, R^2, relative average absolute error (RAAE) and relative maximum absolute error (RMAE) were used to verify the accuracy of the model[47], as shown in Eq. (29), Eq. (30) and Eq. (31).

$$R^2 = 1 - \frac{\sum_{i=1}^{n}(y_i - \hat{y}_i)^2}{\sum_{i=1}^{n}(y_i - \bar{y}_i)^2} \tag{29}$$

$$RAAE = \frac{\sum_{i=1}^{n}|y_i - \hat{y}_i|}{n \times STD} \tag{30}$$

$$RMAE = \max_{1 \le i \le n}(\frac{|y_i - \hat{y}_i|}{STD}) \tag{31}$$

where y_i is the truth value of the test sample point; $\hat{y}_i$ is the fitted value of the surrogate model; $\bar{y}_i$ represents the mean values of all test sample points; n represents the number of sample points; STD is the standard deviation of the sample points.

Table 2: Inflow variables of papermaking wastewater

Notation	Definition	Limit
1	COD	$< 90\ g/m^3$
2	BOD	$< 20\ g/m^3$
3	S_{nh}	$< 8\ g/m^3$
4	TN	$< 12\ g/m^3$
5	TSS	$< 30\ g/m^3$

6 Optimization of GHG emissions

There are many important process parameters in the process of papermaking wastewater treatment. These variables not only play an important role in the pollutant removal efficiency, but also affect the GHG emissions. Among them, the inflow flow and composition are uncontrollable variables, which are constantly changing with time in the actual papermaking production process. In order to make the effluent quality of wastewater treatment meet the national standards, it is necessary to reasonably control the controllable variables such as dissolved oxygen transfer coefficient and circulating flow. However, the current control scheme mainly focuses on improving the effluent quality and reducing the operation cost, which does not consider the GHG generated in the process, resulting in a large amount of GHG emissions from the paper mill. Since GHG emissions reduction is increasingly being targeted in the wastewater treatment process, higher requirements have been put forward for the control of manipulation variables.

Based on the Kriging surrogate model, this study optimizes the controllable variables to achieve the goal of GHG emissions reduction on the premise that the effluent quality reaches the standard. The discharging standards of water pollutants for the papermaking industry are shown in Table 2. Taking them as constraints, the GA algorithm was applied to optimize the dissolved oxygen transfer coefficients, and the internal and external circulation flow rates to obtain the optimal control values under different water inflow conditions.

Table 3: The accuracy of Kriging model

Number	R^2	$RAAE$	$RMAE$
1	0.9655	0.1299	0.7656
2	0.9696	0.1300	0.7665
3	0.9696	0.1350	0.6792
4	0.9651	0.1384	0.6488
5	0.9633	0.1493	0.6455
6	0.9662	0.1365	0.7111

7 Results and discussion

In this study, there are 9 variables in the simulation model of papermaking wastewater treatment. The Latin hypercube sampling has been used to collect 1000 groups of data samples in the multi-dimensional sample space, which were input into the BSM1 for simulation. After the GHG emissions in the wastewater treatment process were calculated, Kriging surrogate model was used for approximate fitting to establish the model. Finally, 300 groups of data were collected to test the accuracy of the surrogate model. Table 3 showed the precision results of the model after 6 tests. It could be seen that the fitting effect of Kriging surrogate model was good, whose R^2 was basically stable at 96-97%. Also the global error and local error were small, which meant that the model could accurately predict GHG emissions under different conditions. In addition, the regression fitting results of the Kriging model in the six groups of tests were shown in Figure 4. From the figure, it could be found that GHG emissions fluctuate between 200-1000 kg CO_2-eq/h. Moreover, the predicted values of the Kriging model were globally close to the original values, which indicated that it had a high fitting accuracy.

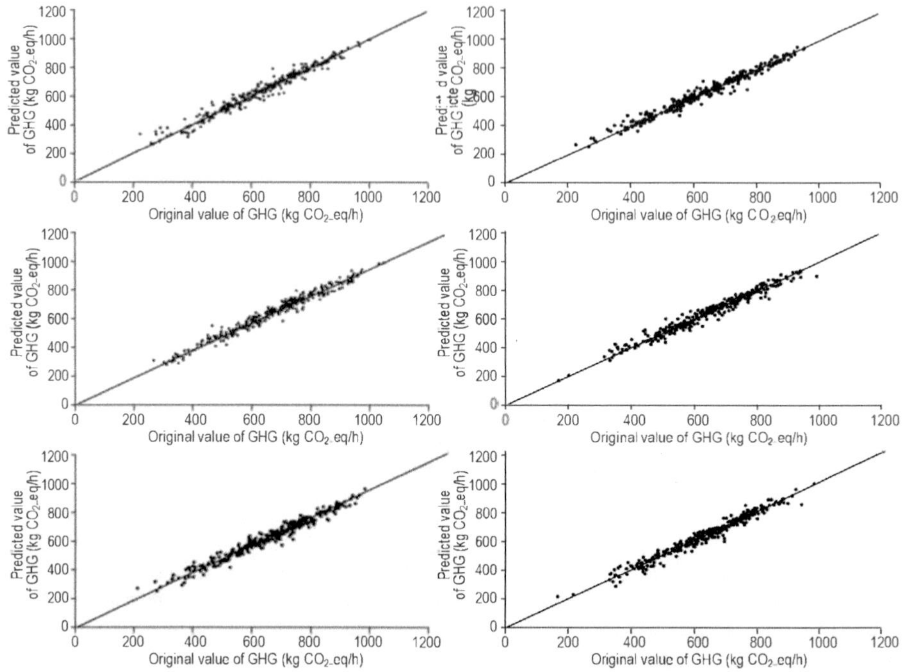

Figure 3: Regression fitting diagram of Kriging model

Figure 5 showed the influent data of activated sludge wastewater treatment in a paper mill, with a sampling period of 1 h and a total of 140 sets of data. Among them, the average influent flow was 7868.45 m^3/d, and the average concentration of influent COD, Snh and TN were 608.39 g/m^3, 31.06 g/m^3, 17.35 g/m^3 respectively. In the actual papermaking wastewater treatment, the influent flow Qin, COD, Snh and TN were uncontrollable and changing in real time. Therefore, this study mainly focuses on the impact of controllable variables such as dissolved oxygen transfer coefficient and internal and external circulation flows on GHG emissions.

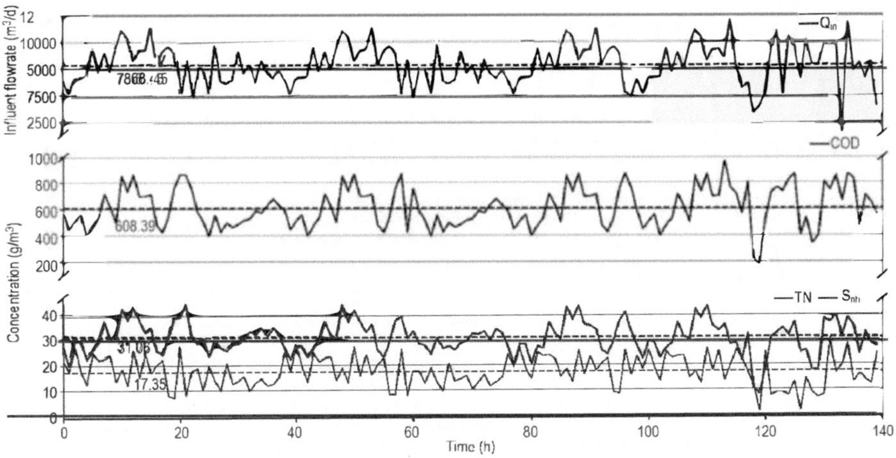

Figure 4: Influent data of papermaking wastewater

In the aerobic reactor, it was necessary to transfer oxygen through the aerator to adjust the dissolved oxygen transfer coefficient in the reactor and promote the aerobic reaction. If the dissolved oxygen transfer coefficient was too high, the aerator would consume a lot of electrical energy and lead to indirect GHG emissions. On the other hand, if the dissolved oxygen transfer coefficient was too low, the biochemical reaction would be incomplete and the effluent quality would be reduced. In this study, the dissolved oxygen transfer coefficients were adjusted to 60 d^{-1}, 120 d^{-1}, 180 d^{-1}, 240 d^{-1} and 300 d^{-1} respectively. The GHG emissions were shown in Figure 6. An increasing dissolved oxygen transfer coefficient generally raised GHG emissions. Wherein, when K_La heightened from 60 to 120 d^{-1}, the GHG emissions increased quickly. The main reason was that when K_La reached 120 d^{-1}, the most organic pollutants could be completely degraded. Further increasing the dissolved oxygen transfer coefficient cannot significantly improve the degradation rate of organic pollutants, but the power consumption will increase. Moreover, K_La of the three aerobic tanks have different effects on GHG emissions, because they are progressive in the process of biochemical reactions.

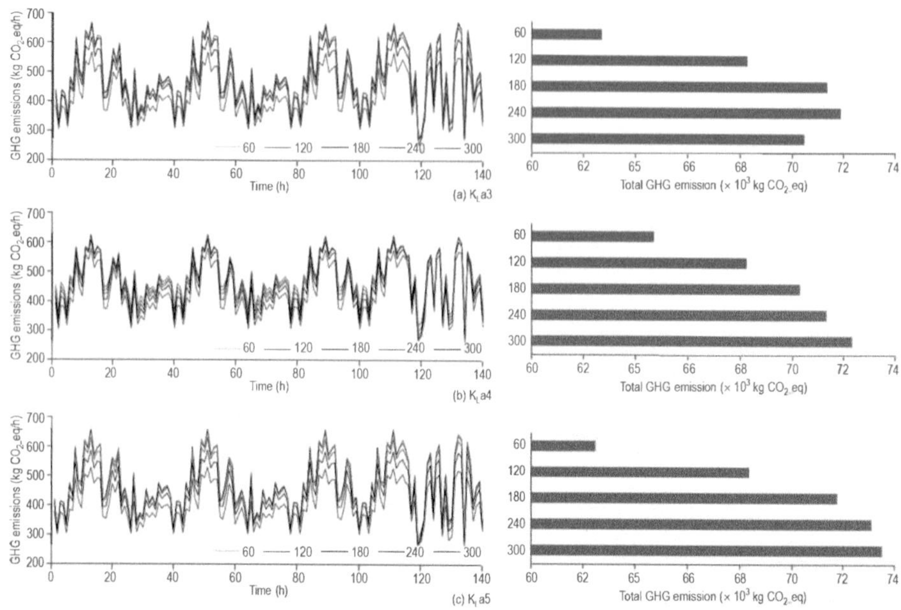

Figure 5: GHG emissions under different dissolved oxygen transport coefficients of (a) K_La3, (b) K_La4 and (c) K_La5

There were two circulation backflows in the papermaking wastewater treatment, namely internal and external circulation. The internal circulation was that part of the effluent stream from the last reactor which directly returns to the first reactor, which was mainly used to deeply reduce the nitrate nitrogen through multiple biochemical reactions. The external circulation serves to return the sludge at the bottom of the secondary sedimentation tank to the reactor inlet, which could ensure the microbial content in the reactor and promote the biochemical reaction. In this study, the internal circulation flow was adjusted to 10000 m^3/d, 20000 m^3/d, 30000 m^3/d, 40000 m^3/d and 50000 m^3/d, while the external circulation flow was adjusted to 5000 m^3/d, 10000 m^3/d, 15000 m^3/d, 20000 m^3/d and 25000 m^3/d. The GHG emissions under different circulation flows were shown in Figure 7. With the increment in circulating flows, GHG emissions will also increase accordingly. Among them, the external circulation flow had a greater impact on GHG emissions, because there were a large number of microorganisms in the external circulation, which directly affected the activity of biochemical reactions. The main purpose of internal circulation was to control the nitrate nitrogen content in the system, which mainly affected indirect greenhouse gas emissions. Similarly, when the basic reaction of organic pollutants was complete, increasing the external circulation flow would only enhance the power consumption of the pump, which increases the indirect GHG emissions.

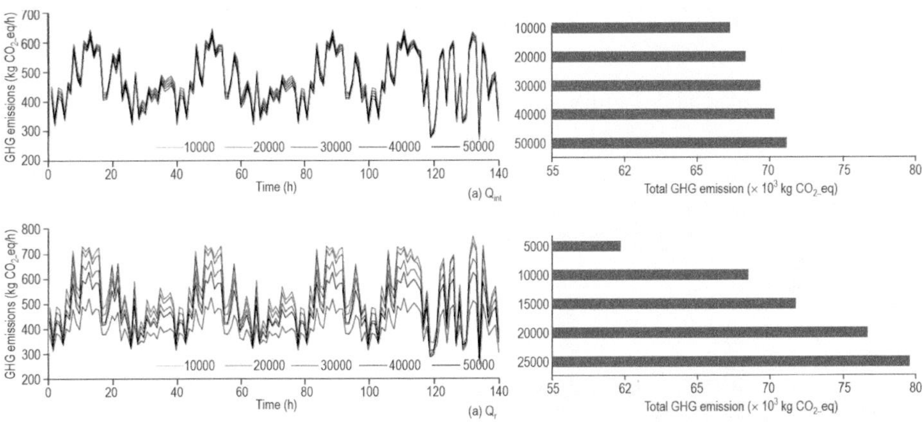

Figure 6: GHG emissions under different circulating flows of (a) Q_{int} and (b) Q_r

Therefore, under the scenario set in this study, increasing the dissolved oxygen transfer coefficient and circulating flow would increase GHG emissions, as shown in Figures 8. The main reason was that when these operating parameters were too small, the pollutants in the wastewater could not react completely, which led to less direct GHG emissions from biochemical reactions. However, this would cause the effluent quality to be too low to meet the discharge standard. When the parameters were too high, it would also lead to excessive unnecessary indirect carbon emissions, such as electrical consumption of aerators and pumps. And from Figure 9, we could see that the average change rate of GHG emissions was decreasing with the rise in variables. Because when the variables reached a certain threshold, the biochemical reaction efficiency did not rise, which led to the direct emission of GHG will not enhance.

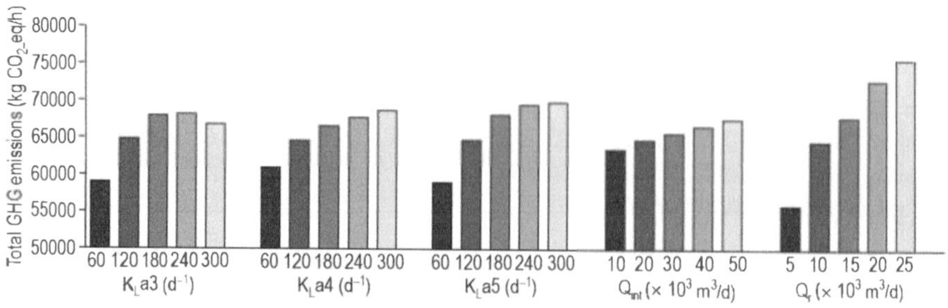

Figure 7: Total GHG emissions under different conditions

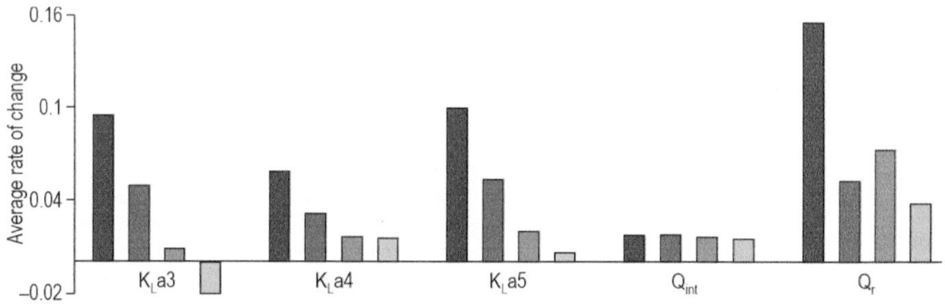

Figure 8: Average change rate of GHG emissions under different conditions

Finally, according to the above results, the default operating parameters were set as appropriate values to make the effluent quality meet the national standards wherein the dissolved oxygen transfer coefficient was set as 180 d^{-1}, the internal circulation flow was set as 20000 m^3/d, and the external circulation flow was set as 10000 m^3/d. The GA algorithm was used to optimize the multi parameters of GHG emissions. The effluent quality before and after optimization was shown in Figure 10, where the black line is the emission limit of each component (Table 2). Most of the time, the effluent quality could meet the discharge standard. However, due to the high content of influent pollutants, the quality of effluent at some times would exceed the standard, which did not affect the compliance discharge of the accumulated wastewater. Besides that, the effluent quality had improved after optimization.

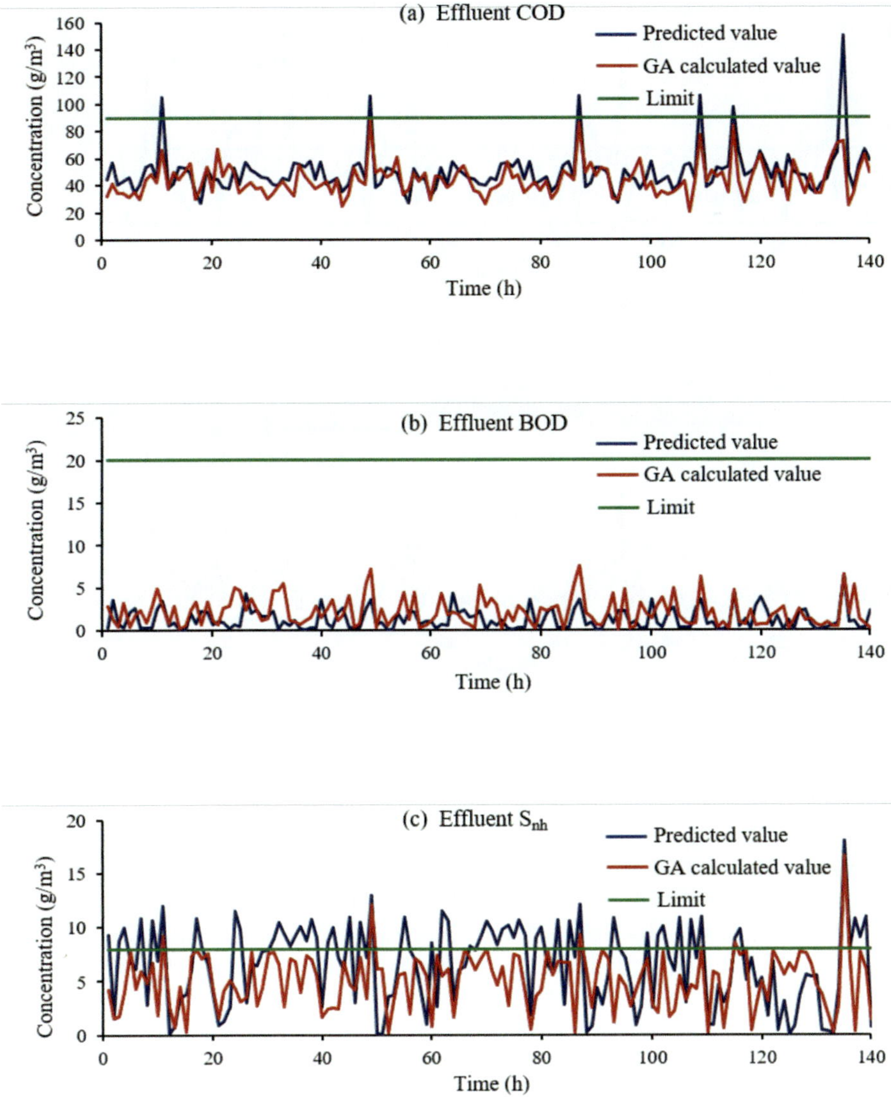
(a) Effluent COD
Predicted value
GA calculated value
Limit
Concentration (g/m3)
Time (h)
(b) Effluent BOD
Predicted value
GA calculated value
Limit
Concentration (g/m3)
Time (h)
(c) Effluent Snh
Predicted value
GA calculated value
Limit
Concentration (g/m3)
Time (h)

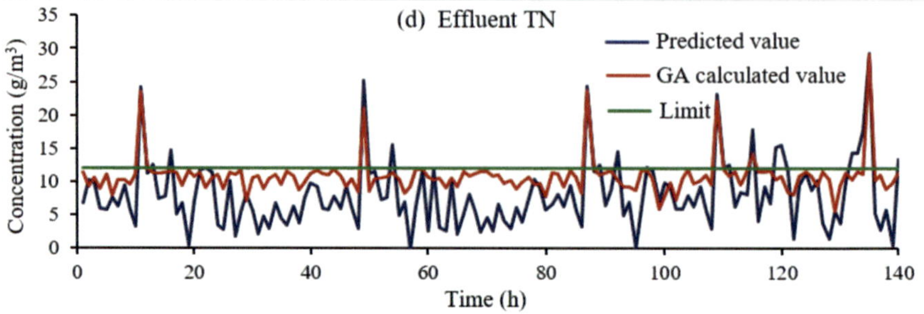

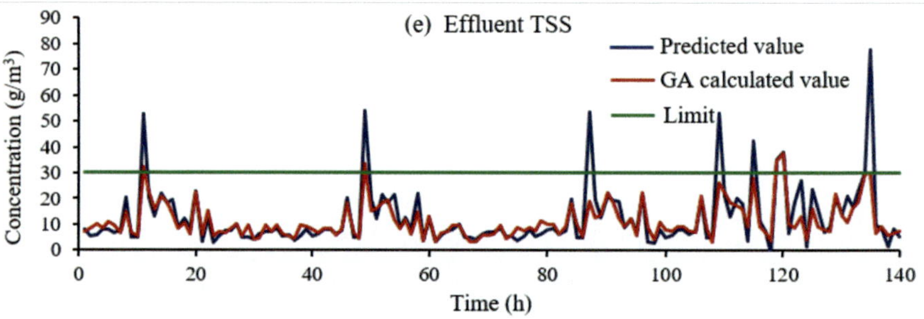

Figure 9: Effluent water quality before and after optimization of (a) Effluent COD, (b) Effluent BOD, (c) Effluent S_{nh}, (d) Effluent TN and (e) Effluent TSS

The GHG emissions were shown in Figure 11, where the blue line was the original value, the red line was the optimization, and the reduction was at the bottom of the figure. The original values of GHG emissions fluctuated between 320-730 kg CO_2-eq/h, while the optimizations were between 300-600 kg CO_2-eq/h. On the premise of meeting the effluent quality, the optimization of control parameters could achieve GHG emission reduction most of the time, with a maximum reduction of about 340 kg CO_2-eq/h. Furthermore, it could reduce GHG emissions by about 35.3% as a whole, which perfectly met the goal of GHG emission reduction.

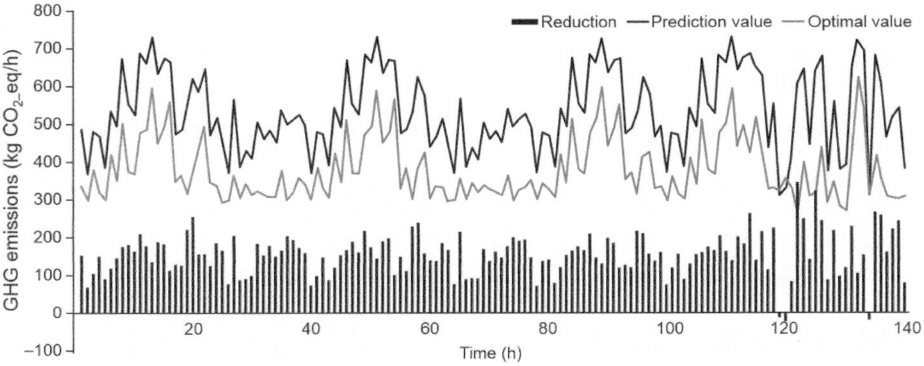

Figure 10: GHG emissions before and after optimization

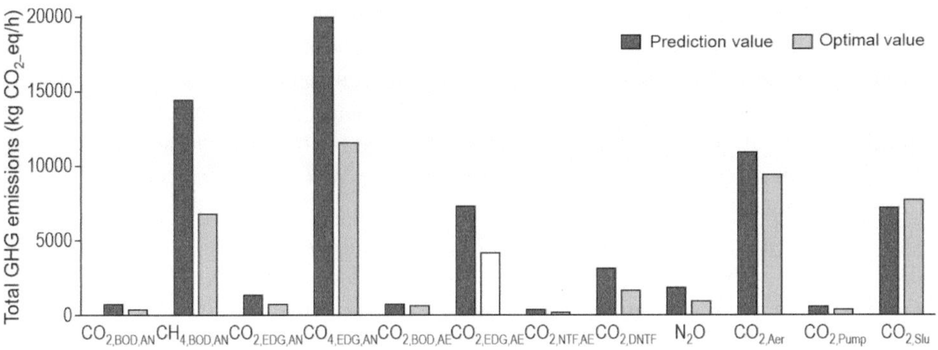

Figure 11: Total GHG emissions of part before and after optimization

This study also calculated the GHG emissions of each part before and after optimization. The results were shown in Figure 12, which showed the comparison of GHG emissions before and after optimization. It was observed that all parts of GHG emissions have been reduced to varying degrees. However, the GHG emissions from sludge disposal had increased, which was mainly due to the rise of discharged sludge after adjusting the variables. The total $CH_{4,BOD,AN}$ and $CH_{4,EDG,AN}$ decreased by about 7600 and 9000 kg CO_2-eq respectively. The total direct CO_2 emissions of all parts were reduced by about 5800 kg CO_2-eq. As for indirect GHG emissions, $CO_{2,Aer}$ and $CO_{2,Pump}$ were reduced by about 1500 and 200 kg CO_2-eq respectively. The results proved that reasonable control of operating parameters can significantly reduce direct GHG emissions, especially CH_4 with a high GWP.

8 Conclusions

The mechanism of GHG emissions generated in papermaking wastewater treatment is complex and unclear, which also leads to the difficulty of controlling the multi-variables to achieve GHG reduction. In this study, the papermaking wastewater treatment process was simulated based on BSM1. The GHG emissions under steady-state conditions were calculated by the conservation of substances. According to the functional relationship between the input parameters of the papermaking wastewater treatment and GHG emissions, the Kriging surrogate model with high precision was established.

Based on the Kriging model of GHG emission, this study also analyzed the impact of different operating parameters on GHG emissions. Under the 140 h papermaking influent data and the scenario set in this study, the increase of the dissolved oxygen transfer coefficient and circulating flow would raise GHG emissions. With the increment of dissolved oxygen transfer coefficient and circulating flow, its change rate decreased, mainly because the organic pollutants were completely reacted.

Finally, this study optimized the operating parameters in the wastewater treatment process, which could reduce GHG emissions by about 35.3%. Among them, CH_4 with a high GWP produced by anaerobic reactions is expected to significantly reduce. This was enough to prove that reasonable optimization and control of parameters could significantly reduce direct GHG emissions in the biochemical reaction process.

In conclusion, the Kriging surrogate model established in this study could quantify GHG emissions under different conditions. Moreover, the manipulated parameters were optimized to achieve GHG emission reduction, which would provide a basic reference for the optimization and control of GHG emission reduction in the papermaking wastewater treatment.

References

[1] The annual report of china papermaking industry in 2021, China Paper Newsletters 5 (2022) 6–17.
[2] H. Hu, S. Sun, China Ecological Environment Statistics Annual Report in 2021, China Statistics Press, 2022.
[3] L. Ma, Z. Chen, C. Xu, F. Li, H. Jin, L. Shi, H. Hu, Water meta-cycle model and indicators for industrial processes- the pulp & paper case in china, Resources, Conservation and Recycling 139 (2018) 228–236.
[4] W. Hu, J. Tian, L. Chen, Greenhouse gas emission by centralized wastewater treatment plants in chinese industrial parks: Inventory and mitigation measures, Journal of Cleaner Production 225 (2019) 883–897.
[5] X. Wu, S. Zang, D. Ma, J. Ren, Q. Chen, X. Dong, Emissions of co2, ch4, and n2o fluxes from forest soil in permafrost region of daxing'an mountains, northeast china, International journal of environmental research and public health 16 (2019) 2999.
[6] L. Sun, H. Cui, Q. Ge, Will china achieve its 2060 carbon neutral commitment from the provincial perspective?, Advances in Climate Change Research 13 (2022) 169–178.
[7] Y. Man, J. Li, M. Hong, Y. Han, Energy transition for the low-carbon pulp and paper industry in china, Renewable and Sustainable Energy Reviews 131 (2020) 109998.
[8] J. Pang, J. Pan, D. Tong, X. Fu, B. Sun, M. Yang, H. Li, How do hydraulic load and intermittent aeration affect pollutants removal and greenhouse gases emission in wastewater ecological soil infiltration systems?, Ecological Engineering 146 (2020) 105747.
[9] P. Stathatou, P. Dedousis, G. Arampatzis, H. Grigoropoulou, D. Assimacopoulos, et al., Energy savings and reduced emissions in combined natural and engineered systems for wastewater treatment and reuse: the wwtp of antiparos island, greece, Desalination and Water Treatment 159 (2019) 13–23.
[10] O. Grasham, V. Dupont, M. A. Camargo Valero, P. García Gutiérrez, T. Cockerill, Combined ammonia recovery and solid oxide fuel cell use at wastewater treatment plants for energy and greenhouse gas emission improvements, Applied Energy 240 (2019) 698–708.
[11] A. Kumar, A. Thanki, H. Padhiyar, N. K. Singh, S. Pandey, M. Yadav, Z. Yu, Greenhouse gases emission control in wwts via potential operational strategies: A critical review, Chemosphere 273 (2021) 129694.

[12] Y. Li, J. Liu, L. Kong, Accounting and application of greenhouse gas emissions in paper enterprises, China Pulp and Paper 36 (2019) 24–29.

[13] D. Caniani, G. Esposito, R. Gori, C. Caretti, G. Bellandi, I. M. Mancini, M. Caivano, R. Pascale, A. Cosenza, H. Abouissa, et al., Toward a new plant-wide experimental and modeling approach for reduction of greenhouse gas emission from wastewater treatment plants, Journal of Environmental Engineering 145 (2019) 04019043.

[14] D. Andraka, Reliability analysis of activated sludge process by means of biokinetic modelling and simulation results, Water 12 (2020) 291.

[15] Y. Man, W. Shen, X. Chen, Z. Long, M. N. Pons, Modeling and simulation of the industrial sequencing batch reactor wastewater treatment process for cleaner production in pulp and paper mills, Journal of Cleaner Production 167 (2017) 643–652.

[16] Q. Liu, R. K. Dereli, D. Flynn, E. Casey, Demand response through reject water scheduling in water resource recovery facilities: A demonstration with bsm2, Water Research 188 (2021) 116516.

[17] X. Zhang, J. Nan, T. Liu, Q. Xiao, B. Liu, X. He, H. H. Ngo, A. Ding, Modeling and simulation of an extended asm2d model for the treatment of wastewater under different cod: N ratio, Journal of Water Process Engineering 40 (2021) 101831.

[18] M. Ayoub, Performance evaluation of an oily industrial wastewater treatment system using the application of activated sludge model no. 3, Water and Environment Journal (2013) 1–9.

[19] H. Nayeb, M. Mirabi, H. Motiee, A. Alighardashi, A. Khoshgard, Estimating greenhouse gas emissions from iran's domestic wastewater sector and modeling the emission scenarios by 2030, Journal of Cleaner Production 236 (2019) 117673.

[20] F. Huang, W. Li, W. Shen, P. Seferlis, Y. Man, J. P. Corriou, A hierarchical intelligent control strategy for greenhouse gas reduction in wastewater treatment process of paper mill, Journal of Cleaner Production 379 (2022) 134818.

[21] W. Yang, C. Yuan, W. Chen, Y. Yang, C. Hung, et al., Diurnal variation of greenhouse gas emission from petrochemical wastewater treatment processes using in-situ continuous monitoring system and the associated effect on emission factor estimation, Aerosol and Air Quality Research 17 (2017) 2608–2623.

[22] I. Santín, M. Barbu, C. Pedret, R. Vilanova, Dissolved oxygen control in biological wastewater treatments with non-ideal sensors and actuators, Industrial & Engineering Chemistry Research 58 (2019) 20639–20654.
[23] J. Qiao, Y. Hou, H. Han, Optimal control for wastewater treatment process based on an adaptive multi-objective differential evolution algorithm, Neural Computing and Applications 31 (2019) 2537–2550.
[24] J. Wen, M. W. LeChevallier, W. Tao, Nitrification kinetics and microbial communities of activated sludge as a full-scale membrane bioreactor plant transitioned to low dissolved oxygen operation, Journal of Cleaner Production 252 (2020) 119872.
[25] F. Huang, W. Shen, X. Zhang, P. Seferlis, Impacts of dissolved oxygen control on different greenhouse gas emission sources in wastewater treatment process, Journal of Cleaner Production 274 (2020) 123233.
[26] X. Liao, Y. Tian, Y. Gan, J. Ji, Quantifying urban wastewater treatment sector's greenhouse gas emissions using a hybrid life cycle analysis method-an application on shenzhen city in china, Science of the Total Environment 745 (2020) 141176.
[27] H. Lu, H. Wang, Q. Wu, H. Luo, Q. Zhao, B. Liu, Q. Si, S. Zheng, W. Guo, N. Ren, Automatic control and optimal operation for greenhouse gas mitigation in sustainable wastewater treatment plants: A review, Science of The Total Environment 855 (2023) 158849.
[28] T. Nguyen, H. Ngo, W. Guo, S. Chang, D. Nguyen, L. Nghiem, T. Nguyen, A critical review on life cycle assessment and plant-wide models towards emission control strategies for greenhouse gas from wastewater treatment plants, Journal of environmental management 264 (2020) 110440.
[29] K. B. Newhart, R. W. Holloway, A. S. Hering, T. Y. Cath, Data-driven performance analyses of wastewater treatment plants: A review, Water research 157 (2019) 498–513.
[30] Y. Jin, H. Wang, T. Chugh, D. Guo, K. Miettinen, Data-driven evolutionary optimization: An overview and case studies, IEEE Transactions on Evolutionary Computation 23 (2018) 442–458.
[31] A. Bhosekar, M. Ierapetritou, Advances in surrogate based modeling, feasibility analysis, and optimization: A review, Computers & Chemical Engineering 108 (2018) 250–267.
[32] J. Kudela, R. Matousek, Recent advances and applications of surrogate models for finite element method computations: a review, Soft Computing 26 (2022) 13709–13733.

[33] E. Haghighat, M. Raissi, A. Moure, H. Gomez, R. Juanes, A physics-informed deep learning framework for inversion and surrogate modeling in solid mechanics, Computer Methods in Applied Mechanics and Engineering 379 (2021) 113741.
[34] L. Sun, H. Gao, S. Pan, J. Wang, Surrogate modeling for fluid flows based on physics-constrained deep learning without simulation data, Computer Methods in Applied Mechanics and Engineering 361 (2020) 112732.
[35] T. Chugh, Y. Jin, K. Miettinen, J. Hakanen, K. Sindhya, A surrogate-assisted reference vector guided evolutionary algorithm for computationally expensive many-objective optimization, IEEE Transactions on Evolutionary Computation 22 (2016) 129–142.
[36] J. Ouyang, C. Li, G. Zhang, D. Wei, L. Wei, C. Chang, Activated sludge and other aerobic suspended culture processes, Water Environment Research 91 (2019) 992–1000.
[37] E. Hong, A. M. Yeneneh, T. K. Sen, H. M. Ang, A. Kayaalp, A comprehensive review on rheological studies of sludge from various sections of municipal wastewater treatment plants for enhancement of process performance, Advances in Colloid and Interface Science 257 (2018) 19–30.
[38] M. Sadeghassadi, C. J. Macnab, B. Gopaluni, D. Westwick, Application of neural networks for optimal-setpoint design and mpc control in biological wastewater treatment, Computers & Chemical Engineering 115 (2018) 150–160.
[39] J. Pengkang, S. Li, R. Wuang, Transformation characteristics of different forms of nitrogen nutrients in process of wastewater treatment, Chinese Journal of Environmental Engineering 9 (2015) 4193–4198.
[40] E. Tejaswini, S. Panjwani, E. Torfs, I. Nopens, S. R. Ambati, Design and evaluation of hierarchical control strategies for wastewater treatment plants using bürger-diehl settler model in bsm1 framework, Journal of Environmental Engineering 148 (2022) 04022042.
[41] M. B. Shahabadi, L. Yerushalmi, F. Haghighat, Estimation of greenhouse gas generation in wastewater treatment plants-model development and application, Chemosphere 78 (2010) 1085–1092.
[42] Y. Ren, H. H. Ngo, W. Guo, B. Ni, Y. Liu, Linking the nitrous oxide production and mitigation with the microbial community in wastewater treatment: a review, Bioresource Technology Reports 7 (2019) 100191.
[43] P. Ye, A review on surrogate-based global optimization methods for computationally expensive functions, Software Engineering 7 (2019) 68–84.

[44] M. Xiao, J. Zhang, L. Gao, A system active learning kriging method for system reliability-based design optimization with a multiple response model, Reliability Engineering & System Safety 199 (2020) 106935.
[45] K. Ma, N. V. Sahinidis, S. Amaran, R. Bindlish, S. J. Bury, D. Griffith, S. Rajagopalan, Data-driven strategies for optimization of integrated chemical plants, Computers & Chemical Engineering 166 (2022) 107961.
[46] R. Ferrari, D. Froio, E. Rizzi, C. Gentile, E. N. Chatzi, Model updating of a historic concrete bridge by sensitivity-and global optimization-based latin hypercube sampling, Engineering Structures 179 (2019) 139–160.
[47] X. Li, Q. Yang, Y. Wang, X. Han, Y. Cao, L. Fan, J. Ma, Development of surrogate models in reliability-based design optimization: A review, Mathematical Biosciences and Engineering 18 (2021) 6386–6409.

4. Peculiarities of BPG-Based Automatic Lossy Compression of Noisy Images

Vladimir Lukin,[1,*] Bogdan Kovalenko[1] and Sergii Kryvenko,[1] Victoriya Naumenko[1] and Benoit Vozel[2]

1 Introduction

Images of different origins have become an essential part of our life. Photos and other images in social networks[1], medical[2] and remote sensing (RS)[3] images are the most known examples where their amount grows rapidly. Simultaneously, the image average size grows as well. In the first order, this is due to technological achievements – the resolution of imaging devices improves steadily for practically all types of images. In addition, images are acquired more frequently, and their exchange has become a habit[4] or necessity (like in telemedicine or between hospitals for better diagnostics[5]. There are many other applications[6;7] where an image is acquired in one place (e.g., on-board a satellite or at some emergency object), and then it has to be transferred to another place (e.g., a center of RS image processing, land monitoring or emergency control) for processing and/or storage. In all these cases, a problem of image size might arise dealing with a limited bandwidth of communication

[1] National Aerospace University, Kharkov, Ukraine.
[2] University of Rennes, CNRS, Institut d'Electronique et des Technologies du numéRique, UMR 6164, Lannion, France.
* Corresponding author: v.lukin@khai.edu

lines and/or restricted memory for data storage. Then, one has to apply compression[8;9;10].

As known, there exist techniques of lossless and lossy image compression[11;12]. An advantage of the former ones is that they preserve all information contained in images. A drawback is that the compression ratio (CR) for these methods can be inappropriately large (too small). This explains practical restrictions for using lossless compression tools. Then, lossy compression methods are employed[7;13;14;15;16;17]. For them, original (subject to compression) and decompressed images differ from each other, i.e. certain distortions are inevitably introduced, and a typical task is to provide a trade-off between compressed image quality and CR reached[14;15;16;17;18].

Depending on the application, there could be different priorities of requirements to lossy compression. First, it can be needed to ensure a CR which is not less than some threshold. Then, JPEG2000[16;19] or its analogs[18] can be useful. Sometimes, it is stressed that some standard (e.g., JPEG) has to be used with providing the desired quality[20]. Additional requirements could be to reach the trade-off as quickly as possible or save computational resources[18;20]. Full automation of image lossy compression is almost always needed[10;21;22]. It is often desired to use a compression technique that provides a higher quality than other counterparts and a wide range of CR values, for most images. In these senses, the better portable graphics (BPG) coder[23] has several advantages that have stimulated its use in different areas[24;25;26;27]. In particular, BPG encoder produces a higher quality (in terms of PSNR) for a wide range of CR compared to JPEG and many modern coders. It is supported by most Web browsers. Besides, BPG supports the same formats as JPEG (such as grayscale, YCrCb 4:2:0, 4:2:2, 4:4:4) and the most popular color spaces such as RGB, YCgCo, and CMYK. The available versions[24] are able to process images having from 8 to 14 bits per channel. It will be also shown later that the coder can produce the desired quality quite easily. These are the main reasons why we pay attention just to this coder below.

In lossy image compression, it is usually assumed that images are noise-free (of ideal quality) or possible noise presence is ignored (not taken into consideration)[9;10;25]. However, there are numerous practical situations when it is not true. The most typical case is, probably, speckle in radar and ultrasound images[28;29;30;31]. Noise is often quite intensive in astronomic[32] and medical images[33]. It is also visible in images acquired in bad illumination conditions[34;35]. Finally, some component images in multispectral and hyperspectral remote sensing data are corrupted by intensive noise and have a low input peak signal-to-noise ratio (PSNR)[36].

It has been found more than 20 years ago that lossy compression of noisy images has certain peculiarities[37;38;39;40;41]. First, the noise filtering effect due to lossy compression was discovered for astronomic and synthetic aperture radar images[37;38;39]. Later[40;41], the term optimal operation point was introduced where OOP is such a PCC value for a given coder (bits per pixel (BPP), quantization step (QS), quality factor (QF) for JPEG so that the quality of the compressed image is the closest to the corresponding noise-free image (according to a certain metric) and also closer than the quality of the uncompressed (noisy) image. At the first stage of studies[37;38;39;40;41], such conventional metrics as the mean square error (MSE) or peak signal-to-noise ratio ($PSNR$) were used. Then, OOP relates to MSE global minimum and for $PSNR$ – to the global maximum where, recall, the metrics are calculated between the compressed and noise-free images (only available in simulations). Later it has been also established[42] that OOP can be observed for visual quality metrics[43;44] such as $PSNR-HVS-M$[45] and $MS-SSIM$[46]. Note that visual quality metrics are nowadays widely used because of two reasons. First, most images are subject to visual perception or analysis[43;44]. Second, visual perception highly correlates with classification accuracy for the types of classes that are quite heterogeneous (correspond to textures like Urban or Forest)[47]. The use of visual quality metrics is the second specific aspect of this chapter.

If OOP for a given image exists, it is reasonable to carry out image compression in OOP because of two reasons. First, the quality of the compressed image happens to be better than for any other PCC value. Second, CR usually happens to be rather large. In addition, the classification accuracy of images compressed in OOP is usually better than for the corresponding original images[48].

Although earlier papers on lossy compression dealt with multiplicative and signal-dependent noise[37;38;39], it was later shown that OOP might exist for images corrupted by additive noise, both white[49;50] and spatially correlated[51]. For the case of signal-dependent noise, one can rely on two possible approaches. According to the first one, the corresponding variance stabilizing transform (VST) is applied, and, then, just compression is used (the decompression is inverse)[52;53]. For the second approach, compression is applied directly without VST[31;52;53]. Both approaches have their advantages and drawbacks. Logarithmic, Anscombe, and generalized Anscombe transforms are used depending on the type of signal-dependent noise[52;53]. Note that it has been shown that lossy compression of noisy images in OOP (if it exists) can be performed by coders based on different orthogonal transforms including discrete cosine transform (DCT) and wavelets[31;54]. For the BPG coder, the possible existence of OOP has been recently shown in our papers[55;56;57].

One problem with OOP is that it can be accurately determined only if the true image exists. But, in practice, the true image is absent. Fortunately, an approach to OOP prediction was put forward[58] for the DCT-based coders[59;60]. Recently, it has been modified to carry out the prediction of OOP existence and compression parameters in OOP for the BPG encoder[57;61]. This chapter reconsiders the results of the papers[55;56;57;61]. It concentrates more on two aspects – consideration of the approach's applicability to images of different origins and fields and possible improvement of prediction accuracy.

The chapter structure is as follows. First, we show how to determine Q that might correspond to OOP using a set of test images with artificially simulated additive white Gaussian noise (AWGN) having a set of variance values. Second, a general approach to the prediction of OOP existence is discussed. Within this approach, there is room for variation and selection of the approach's parameters. Some of them can be easily calculated in the DCT domain although other statistical parameters can be helpful for improving the prediction accuracy. Practical recommendations on Q setting in the absence of OOP are presented. Aspects of compression automation are discussed as well. Besides, we present initial data on OOP existence for the case of direct (without VST) lossy compression of grayscale images corrupted by Poisson noise. Finally, the conclusions are given.

2 Image/Noise Model and Quantitative Criteria of Compression Efficiency

In studies dealing with image processing (lossless and lossy compression, denoising), it is important to take into account the variability of properties of images and noise. Properties of images should reflect the common types, shapes, and contrasts of large-area objects, features of textures and small-sized objects. Images used in the analysis should be of different complexity since

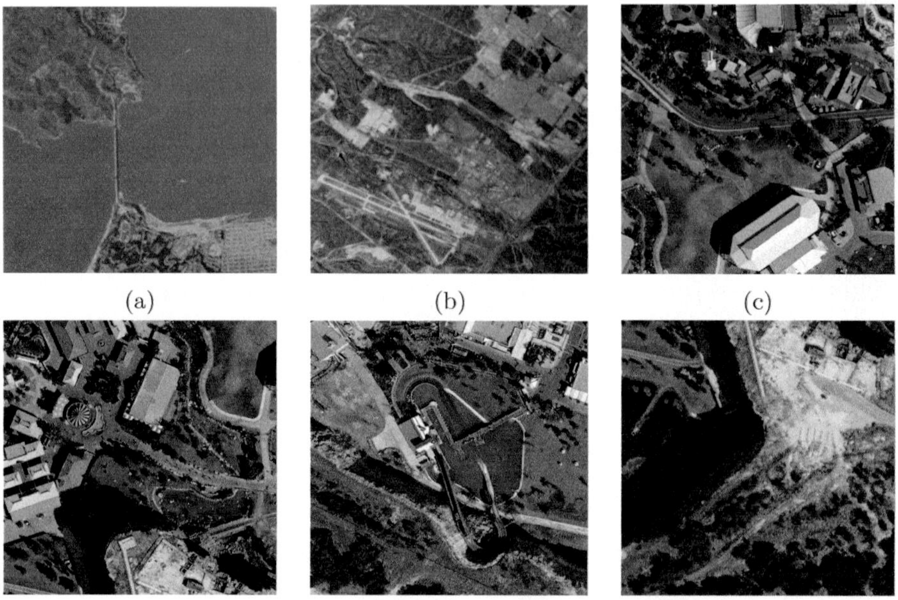

Figure 1: 512x512 pixel images having simple (a), complex (b), and middle complexity (c-f) structures (namely, Frisco (a), Diego (b), Fr01 (c), Fr02 (d), Fr03 (e), and Fr04 (f))

compression efficiency depends on image complexity sufficiently[62]. Keeping this in mind, below we consider three groups of grayscale images. The first one contains six RS images shown in Figure 1. As one can see, one of them (Frisco) is quite simple and has a large-area background (quasi-homogeneous area). The other five images are of middle and high complexity.

Another group consists of well-known standard optical grayscale test images such as Lena, Baboon, Peppers and Goldhill. Finally, the third group includes six medical images. Two of them are given as examples in Figure 3.

As mentioned above, we focus on AWGN which is the simplest noise model. There are, at least, two reasons behind this. First, AWGN is often used as a starting point in studies dealing with image denoising[63;64] and lossy compression[49;50]. Then, after getting the initial knowledge and conclusions,

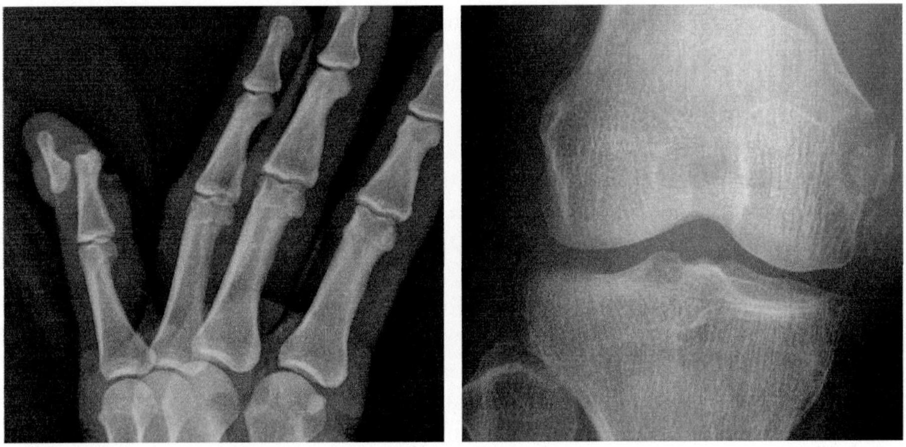

(a) (b)

Figure 2: Examples of true (noise-free) test medical images used in our studies: a – hand_512.png, b – knee_joint_512.png

it becomes possible to pass to more complex and/or adequate noise models. Second, noise becomes almost AWGN after VST under conditions that, in original images, it was spatially uncorrelated and had a probability density function close to Gaussian (this takes place for images corrupted by Poisson noise and multiplicative noise in multi-look radar images[31;53;62]). Then, an observed grayscale image can be presented as follows in equation 1

$$I_{ij}^{n} = I_{ij}^{true} + n_{ij}, \tag{1}$$

where I_{ij}^{true}, $i = 1, .., I_{\Im}, j = 1, .., J_{\Im}$ denotes the true or noise-free image, n_{ij} denotes AWGN in the ij-th pixel, $I_{\Im}$ and $J_{\Im}$ define the image size. It is usually assumed that the noise mean is equal to zero and its variance equals to σ^2. Furthermore, let us assume that σ^2 is a priori known or is pre-estimated with appropriately high accuracy (note that there are many blind (automatic) methods for estimation of noise statistical characteristics[65;66;67;68].

A standard way to characterize the quality of an original (noisy) image is to calculate $PSNR^n$ as follows in equation 2

$$PSNR^n = 10 \log_{10} \frac{255^2}{\sigma^2}, \tag{2}$$

where 8−bit representation is assumed for images we deal with. Note that noise becomes visible for such images if $PSNR^n$ is about 36 dB or less[44], *i.e.* noise variance is about 20 or more. Another way to characterize the quality of original images is to apply visual quality metrics. Note that there are numerous visual quality metrics designed so far[43;44;69] including those based on trained neural networks[69;70]. The existing visual quality metrics take into account different aspects of a human vision system (HVS) and image perception by humans. They have different ranges of variation. Their testing has been performed for different databases. Because of this, it is difficult to find (choose, recommend) the best metric for characterizing the quality of compressed images. It is shown in[71] that there are many visual quality metrics (VQMs), for which Spearman rank order correlation coefficient (SROCC) between metrics and mean opinion score (MOS) for images with distortions typical for medical and remote sensing imaging exceeds 0.92. These "good" metrics belong to different "families" including SSIM-based, PSNR-based, and others. Although SROCC values are high, none of the HVS-based metrics can be considered perfect. Because of this, it is reasonable to use several of them in analysis to come to reliable and motivated conclusions. Hence, we basically rely on the metrics $PSNR{-}HVS{-}M$[45] which belongs to the "PSNR family", $MS-SSIM$[46] which is one of the best representatives of the "SSIM family", and MDSI[72].

The metric $PSNR - HVS - M$ is calculated as

$$PSNR - HVS - M^n = 10\log_{10}\left(\frac{255^2}{MSE - HVS - M^n}\right), \tag{3}$$

where $MSE - HVS - M^n$ is determined in a set of 8x8 pixel blocks in the DCT domain. It takes into consideration the higher sensitivity of a human vision to distortions in low spatial frequencies than in high spatial frequencies as well as the masking effect. Similarly to $PSNR$, $PSNR - HVS - M$ is expressed in dB where larger values of both metrics correspond to a better quality. For such degradations as AWGN and similar ones, $PSNR - HVS - M$ is by a few dB larger than PSNR because of the aforementioned masking effect. It is useful that distortion visibility thresholds have been established for these two metrics[44]. These thresholds are approximately equal to 36 dB for PSNR and 41 dB for $PSNR - HVS - M$. From (2), it follows that this takes the place for the noise variance of about 20. Thus, we focus on examining the cases of $PSNR^n < 36dB$ or, equivalently, $\sigma^2 > 20$.

The metric MS-SSIM is based on extracting and using structural information from an image. Similarly to other SSIM-based metrics, it varies between zero and unity where its larger values relate to a better visual quality. Distortions are usually invisible if MS-SSIM exceeds 0.99. The MDSI metric employs a gradient magnitude to measure structural distortions and chrominance features to assess color distortions. The obtained maps are then combined and the final quality score is computed. MDSI varies from very small values to about 0.6 for severely distorted images where the invisibility threshold is about 0.17[73]. Opposite to the previously mentioned metrics, smaller values for MDSI correspond to a better visual quality.

Suppose that a lossy compression technique is applied to an original (noisy) image obtaining the corresponding compressed image I^c_{ij}, $i = 1, .., I_{Im}$, $j = 1, .., J_{Im}$ that, certainly, differs from I^n_{ij}, I^c_{ij}, $i = 1, .., I_{Im}$, $j = 1, .., J_{Im}$. This difference depends on a used coder and compression controlling parameter (CCP) employed in it. As CCP, the quality factor, scaling factor, quantization step and bits per pixel can be used[20;22;25] the BPG coder employs its own strategy of compression control. The quality parameter Q (which is an integer in the range of 1 to 51) allows for varying the compressed image quality and compression ratio[74] where a smaller Q corresponds to a better quality and smaller CR. Then, rate-distortion curves $Metr^{nc}(Q)$ can be obtained for test images where $Metr^{nc}$ is a metric of interest determined for the noisy (original) and compressed images.

Conventionally, rate-distortion curves behave in such a manner that the metric values monotonously become worse (smaller for $PSNR$, $PSNR-HVS-M$ and $MS-SSIM$, and larger for $MDSI$) if CR increases (in our case, Q increases). This property is supported by all six plots given in Figure 4. Figures 4a and 4b show the dependences of $PSNR^{nc}(Q)$ for the six test RS images in Figure (3) for two values of the noise variance equal to 64 and 100, respectively.

If $Q < 7$, the noisy and compressed images practically are almost identical, i.e. compression can be treated as near-lossless. Then, for $Q \geq 7$, the dependencies have a practically linear part for all test images. This is a very useful property of the BPG coder that allows for an easy setting Q to provide a desired $PSNR^{nc}$.

The part of dependencies starts at $Q \simeq 7$ and ends at such Q where one has $PSNR^{nc}(Q) \simeq PSNR^n$ (recall that we assume that the noise variance is known in advance or accurately pre-estimated, $PSNR^n$ is known in advance, too).

For larger values of Q, the dependences for different test images start to diverge. For this (third) interval of Q values, larger $PSNR^{nc}$ is observed for simpler structure images (like the image Frisco for the plots in Figures 4a and 4b).

For the middle interval of Q, the following approximation is valid

$$PSNR^{nc}(Q) \approx 63 - Q, dB. \quad (4)$$

For $\sigma^2 = 64$, the middle interval ends at $Q \approx 33$ ($PSNR^{nc} \approx 30dB$), whilst for $\sigma^2 = 100$, it ends at $Q \approx 35$ ($PSNR^{nc} \approx 28dB$).

The dependency $PSNR - HVS - M^{nc}(Q)$ (Figure 4c) has a similar structure as the rate-distortion curves $PSNR^{nc}(Q)$: almost linear dependency starts at $Q \approx 8$ and ends for Q for which $PSNR^{nc}(Q)$ is slightly smaller than $PSNR^n$.

Then, the curves $PSNR - HVS - M^{nc}(Q)$ diverge where $PSNR - HVS - M^{nc}(Q)$ values are slightly larger for the simple structure images.

One difference in the behavior of $PSNR - HVS - M^{nc}(Q)$ compared to $PSNR^{nc}(Q)$ is that for most Q, $PSNR - HVS - M^{nc}(Q)$ is larger than $PSNR^{nc}(Q)$ for the same Q due to the masking effect.

Another difference is that, for the middle interval, there are differences in $PSNR - HVS - M^{nc}(Q)$ for different images depending on their complexity.

Analysis of the plots in Figure 4d shows that differences in visual image quality are negligible till Q of about 25. Then, for larger Q, differences can be noticed, especially for simple structure images, for which the masking effect is smaller. For complex structure images, the distortions introduced by the lossy compression start to be noticeable even for larger Q $(27 - 30)$. These conclusions are confirmed by analysis of data for other visual quality metrics (Figure 4c, 4d and 4f).

Therefore, it is possible to compress noisy images having complex structures using Q of about 28 without losing their visual quality if there is no optimal operation point for them.

One more interesting observation can be drawn from the analysis of the plots in Figure 4e.

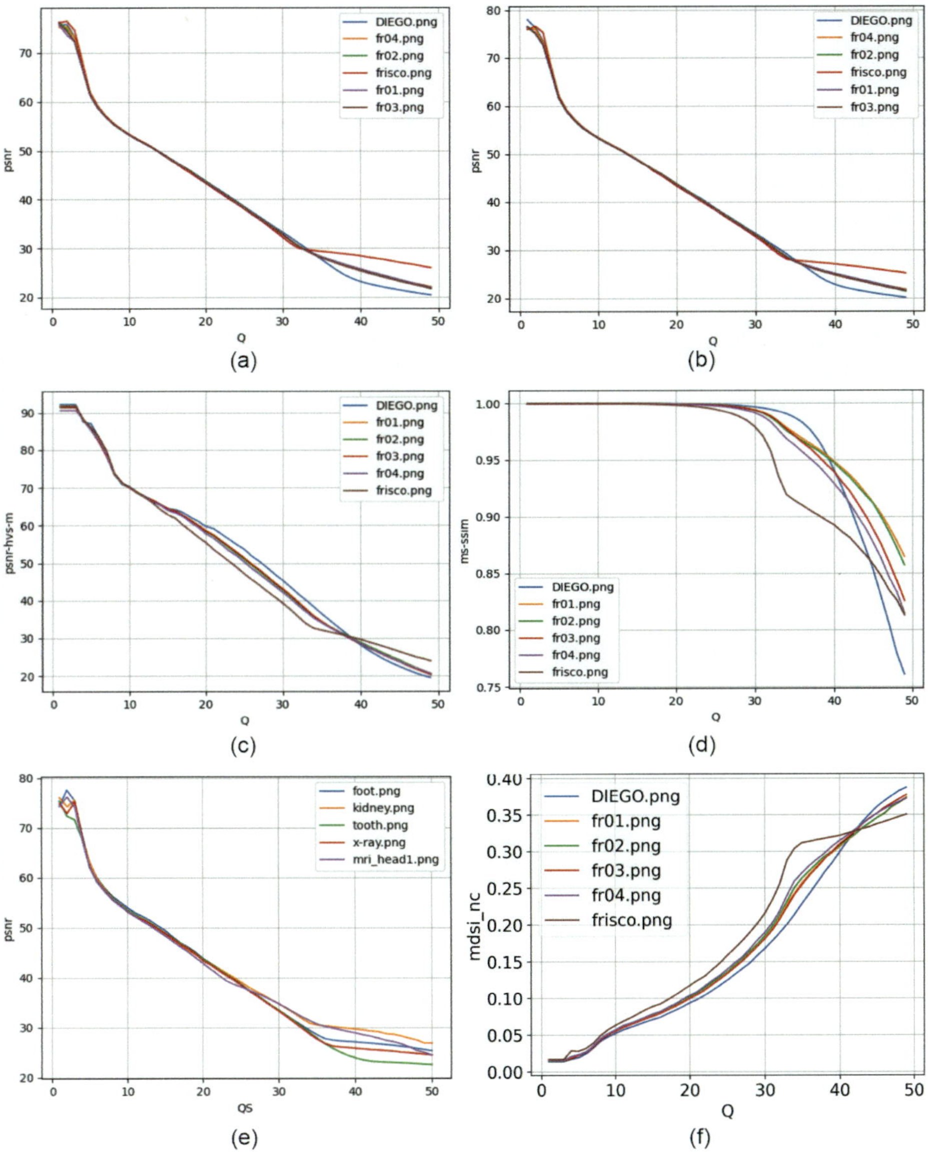

Figure 3: Dependences of the considered metrics (calculated between the original and compressed images on Q: $PSNR^{nc}(Q)$ for noise variance equal to 64 (a), $PSNR^{nc}(Q)$ for noise variance equal to 100 (b), $PSNR - HVS - M^{nc}(Q)$ (c) and $MS - SSIM^{nc}(Q)$ (d) for noise variance equal to 64, $PSNR^{nc}(Q)$ for medical images corrupted by Poisson noise (e), $MDSI^{nc}(Q)$ for RS images corrupted by the noise with variance equal to 64 (f).

Although noise is not AWGN here (images are corrupted by Poisson noise) and the images are medical ones, the plots $PSNR^{nc}(Q)$ behave in a very similar manner to the ones in Figures 4a and 4b.

Again, we have three intervals with different behavior where linear approximation (4) is valid for the middle interval. This indicates the generality of the main properties of the rate-distortion curves for the considered task of noisy image lossy compression by the BPG coder.

For the middle interval, the behavior of the curves can be easily predicted and it is easy to provide a desired value of a metric $Metr^{nc}$[73].

However, the metrics $Metr^{nc}$ produce information only on how different are the original and the corresponding compressed images. Meanwhile, it could be also interesting to see how different are the compressed and noise-free images.

Certainly, this can be analyzed only for simulated data. Then, one has a true image and adds noise to it. After this, the obtained image is compressed using different Q values, and it becomes possible to calculate any metric $Metr^{tc}$ between the compressed and true images with obtaining the dependency $Metr^{tc}(Q)$.

Four examples of such dependencies are presented in Figure 5. As one can see, for $Q < 29$, the image quality is practically "constant", and it is the same as for the original image. Then, if Q increases, three variants of behavior are possible: 1) quality starts to become better and, after reaching OOP (associated with $PSNR$, $PSNR-HVS-M$, and $MS-SSIM$ maxima as well as $MDSI$ minimum), the quality starts to become worse; 2) the quality continues to be practically the same (it is even possible that local maxima or minima occur) and then a fast decrease is observed; 3) the quality starts to reduce. An example of the first situation is the dependence of the metric $MS-SSIM^{tc}$ for five RS images (except for the test image Diego) in Figure 5c. The same relates to five dependencies $MDSI^{tc}(Q)$ in Figure 5d. Examples of the second situation are presented by the metric $PSNR - HVS - M^{tc}$ for several middle complexity RS test images (Figure 5b). In turn, a good example of the third situation is the dependence $PSNR - HVS - M^{tc}(Q)$ for the complex structure image Diego in Figures 5a and 5b.

For the first situation, it is reasonable to carry out compression in OOP. For the second situation, lossy compression in the neighborhood of local maxima (minimum of MDSI, Figure 5d) seems reasonable. It is slightly unclear how to act in the third situation. However, it is at least possible to perform compression with Q equal to 28 or 29, for which the introduced distortions are practically invisible.

More examples of dependencies $Metr^{tc}(Q)$ can be found in our papers[55;56;57;61] as well as in the Ukrainian language paper[75] where standard optical test images are considered. Certainly, the OOP can exist according to PSNR (see an example in Figure 6a for RS test images) and this happens more frequently than according to visual quality metrics. Moreover, the aforementioned properties of $Metr^{tc}(Q)$ take place for medical images corrupted by AWGN or Poisson noise (see an example in Figure 6b).

A positive feature of the coder BPG is that, if OOP exists, it is provided for the same Q for all considered quality metrics and all images. In particular, Q_{OOP} is equal to 35 for $\sigma^2 = 100$ (see, e.g., data in Figure 5a, to 33 for $\sigma^2 = 64$ (see the plots in Figure 6a), and to 38 for $\sigma^2 = 196$ (Figure 5c).

A detailed analysis has shown that, for a known noise standard deviation, it is possible to calculate the optimal Q as

$$Q_{OOP} \approx 14.9 + 20 \log_{10}(\sigma). \tag{5}$$

The situation for Poisson noise is not so clear (Figure 6b) since the maxima for different images are observed not for the same Q. However, the main observation is that Q_{OOP} moves towards larger values if $PSNR^n$ (which is approximately equal to $PSNR^{tc}$ $(Q = 1)$ decreases. Thus, we hope to develop a way to set Q_{OOP} for images corrupted by Poisson noise in the future.

Summarizing the presented data and preliminary conclusions, we can state the following. It is possible to calculate (predict) Q_{OOP} having accurate information on AWGN standard deviation at our disposal. Meanwhile, we do not know in advance wheter OOP exists for a given image or not. If one has a priori information about OOP existence, then it becomes clear what Q to set. Since the true image is absent, it is impossible to check for a given noisy image subject to further compression if OOP exists. Fortunately, we have already shown that OOP existence can be predicted, at least, for the metrics $PSNR$ and $PSNR-HVS-M$[57;61].

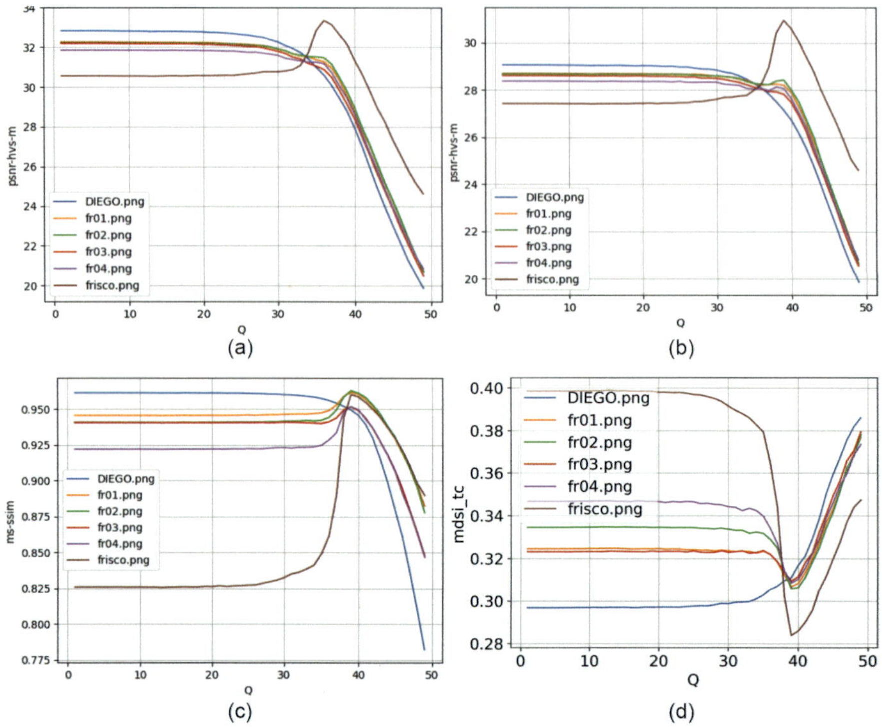

Figure 4: Dependencies of the considered metrics (calculated between the true and compressed images on Q: $PSNR-HVS-M^{tc}(Q)$ for noise variance equal to 100 (a), $PSNR-HVS-M^{tc}(Q)$ for noise variance equal to 196 (b), $MS-SSIM^{\ tc}(Q)$ for noise variance equal to 196 (c), $MDSI^{\ tc}(Q)$ for AWGN variance equal to 196 (d).

3 Basic Idea of Prediction and Preliminary Results

The idea of prediction of OOP existence and compressed image parameters in it is not new. It was first put forward in the paper[58] for DCT-based coders where the quantization step plays the role of PCC. In turn, the approach[58] was based on the earlier papers[64;76;77] dealing with denoising efficiency prediction for the DCT-based filters. It has been established that denoising efficiency depends on noise intensity and image complexity where they can be characterized jointly by one parameter determined in the DCT domain (recall that in our previous analysis, it has also been shown that OOP is more probable for higher noise intensity and more simple structure images where image complexity can be, in general, characterized in different ways depending on a task to be solved[62]).

The idea is that, having such a parameter or several parameters, it is possible to predict the performance of image processing including improvement of metric values due to filtering, CR for a given PCC, and increase or reduction of a metric value for a compressed noisy image with respect to the corresponding metric values for the original image. Within this framework, a lot of questions appear:

1. What should be an input parameter (or input parameters) for such a prediction?

2. How accurate can be such a prediction? Is this accuracy appropriate? How is it possible to improve accuracy if needed?

3. How fast is such a prediction? How to accelerate it if needed?

4. What factors influence prediction accuracy? What are the main ones and which factors can be neglected?

5. What are good output parameters? How to get the dependence between output and input parameters? What are the best tools for this (regression, neural networks, support vector machines)?

Below we partly answer these questions and give the corresponding recommendations.

So, we assume that it is possible to predict (estimate) not $Metr^{tc}(Q_{OOP})$ but the difference $\Delta Metr = Metr^{tc}(Q_{OOP}) - Metr^{n}$.

In fact, we need just the difference to know has a metric improved in OOP assuming that the OOP exists if this difference is positive (for the metrics $PSNR$, $PSNR-HVS-M$, and $MS-SSIM$) or negative (for $MDSI$). If the difference is negative (positive for $MDSI$) but close to zero – then one has situation 2 described in the previous Section. Finally, if the difference is negative (or positive for $MDSI$) and its magnitude is quite large enough, then we have situation 3, i.e. the compression with Q_{OOP} (5) for such an image is not recommended.

To follow such rules, our prediction should be accurate enough for avoiding wrong decisions and we should know what is "quite a large magnitude" (it is different depending on a metric considered since the metrics have different ranges of variation).

From the practical viewpoint, the prediction has to be fast enough, e.g., sufficiently faster than just the compression.

In our previous papers[58;76;77], two statistical parameters, the probabilities $P_{2\sigma}$ and $P_{2.7\sigma}$, have been used. Their advantage is that they can be easily calculated in a limited number (500, 1000) of 8x8 pixel blocks in the DCT domain. In[53], several expressions that allow computing $\Delta PSNR$ and $\Delta PSNR-HVS-M$ for a given $P_{2\sigma}$ or $P_{2.7\sigma}$ used as an argument have been obtained. In this chapter, we exploit and develop those ideas.

The probability $P_{2\sigma}$ is estimated as

$$P_{2\sigma} = \sum_{m=1}^{M} \frac{P_{2\sigma}(m)}{M}, P_{2\sigma}(m) = \left(\sum_{k=0}^{7} \sum_{l=0}^{7} \delta(k,l,m) \right\}, \tag{6}$$

$$\delta(k,l,m) = \begin{cases} 1, & if\ |D(k,l,m)| < 2\sigma, \\ 0, & otherwise \end{cases} \tag{7}$$

Here M denotes the number of blocks used in the analysis, $D(k,l,m)$ denotes the $kl-th$ DCT coefficient for the $m-th$ block, $m = 1, \ldots, M$. $P_{2\sigma}$ can be treated as an estimate of the probability that magnitudes of DCT coefficients in blocks are smaller than 2σ. Note that similar to $P_{2\sigma}$ it is possible to calculate any $P_{\alpha\sigma}$ using the threshold $\alpha\sigma$ in (7).

The probability $P_{2.7\sigma}$ (further denoted as $P_{2.7\sigma_n}$) is estimated as

$$P_{2.7\sigma_n} = \sum_{m=1}^{M} \frac{P_{2.7\sigma}(m)}{M}, P_{2.7\sigma}(m) = \left(\sum_{k=0}^{7} \sum_{l=0}^{7} \delta(k,l,m) \right\} - 1, \tag{8}$$

$$\delta(k,l,m) = \begin{cases} 1, & \text{if } |D(k,l,m)| > 2.7\sigma, \\ 0, & \text{otherwise} \end{cases} \tag{9}$$

Similar to $P_{2.7\sigma_n}$, it is possible to calculate any $P_{\alpha\sigma_n}$ using the threshold $\alpha\sigma$ instead of 2.7σ in (9).

The statistical parameter $P_{2\sigma}$ has come from the famous Sigma filter proposed by J.S. Lee whilst $P_{2.7\sigma_n}$ has arisen from the theory of DCT-based filtering[76]. It was demonstrated in[64] that such parameters have the positive feature of jointly characterizing image complexity and noise intensity. For instance, $P_{2\sigma}$ occurs to be relatively large for simple structure images and/or high-intensity noise.

Then, we have to solve the task of obtaining dependencies connecting predicted parameters such as $\Delta PSNR$ and $\Delta PSNR-HVS-M$ or $\Delta MDSI$ with input parameters such as $P_{2\sigma}$ or $P_{2.7\sigma_n}$. Our first attempts to solve this task dealt with regression (although it can be treated as specific training by specialists from other areas). Note that this task has to be solved offline and in advance. Similar to machine learning, we have to take into account a wide set of situations possible in practice, i.e. images of different complexity and noises of different intensity.

To establish the connection between output and input parameters, we have employed scatter-plots of a given $\Delta Metr$ on a given input parameter. The scatter-plots have been obtained in the following manner: noise having a given noise variance has been artificially added to a considered noise-free test image, which has been compressed using Q_{OOP} (5). The input parameter has been calculated before compression and the output parameter has been computed after compression. A given image for a given noise variance has been used to get a unique point on the scatter-plot. We used eleven test images (baboon, barbara, diego, fr01, fr02, fr03, fr04, frisco, goldhill, lenna, peppers) having different complexities, the noise variance varied in the very limits from 0.25 to 400. Examples of the obtained scatter-plots are presented in Figures 7-11.

As one can see, points for all five scatter-plots are placed in a compact manner indicating quite a strict connection between output and input parameters. This allows an easy selection of good candidates of function types for curve fitting. A visual analysis of scatter-plots in Figures 7-10 demonstrates that there are monotonic dependences of the output parameter on the input parameter. Image quality becomes better according to both $\Delta PSNR$ and $\Delta PSNR - HVS - M$ if $P_{2\sigma}$ increases and $P_{2.7\sigma_n}$ reduces. For the metric $\Delta MDSI$ (Figure 10), the fitted curve is formally not monotonous, but it is smooth and the results have a clear physical meaning – with high probability, there is no OOP if $P_{2\sigma} < 0.88$. This is in agreement with the results for $\Delta PSNR - HVS - M$ (Figure 8).

There are many scatter-plot points for which both $\Delta PSNR$ and $\Delta PSNR{-}HVS{-}M$ are negative. There are some points (e.g., for $P_{2\sigma}{\approx}0.8$ in Figures 7 and 8) where $\Delta PSNR$ are mainly positive and $\Delta PSNR - HVS - M$ are mainly negative, i.e. it is unclear wheter the image quality improved or not.

Finally, there exist also some points (for $P_{2\sigma} > 0.9$) where OOPs exist for both quantitative criteria and, thus, quality improvement due to lossy compression in OOP is obvious.

Prediction is carried out as follows. For a given image with known noise standard deviation, the input parameter is calculated. Then, using the presented fitted curve, the output value is calculated. Suppose, that the estimated $P_{2\sigma} = 0.6$. Then, according to data in Figure 7, $\Delta PSNR{\approx} - 1dB$ and, according to data in Figure 8, $\Delta PSNR - HVS - M{\approx} - 6.5dB$. Then, obviously, there are no OOPs for both metrics and image quality reduces due to lossy compression in OOP. Thus, a more "careful" compression (with $Q < Q_{OOP}$has to be performed.

As one can see, such a prediction happens to be quite accurate for $\Delta PSNR$ (points of the scatter-plot in Figure 7 for $P_{2\sigma}{\approx}0.6$ are characterized by $\Delta PSNR$ values close to – 1 dB). Meanwhile, it is less accurate for $\Delta PSNR - HVS - M$ (points of the scatter-plot in Figure 8 for $P_{2\sigma}{\approx}0.6$ differ more from – 6.5 dB). For smaller $P_{2\sigma}$ (e.g., $P_{2\sigma} = 0.2$, Figure 7), the diversity of scatter-plot points is even larger. Thus, we need some quantitative estimates of prediction error statistics for the proposed approach (this also relates to data in Figure 11).

What can be the preliminary assumptions concerning error statistics and the factors that influence it. Let us list these factors first. They are the following (at least, the main ones):

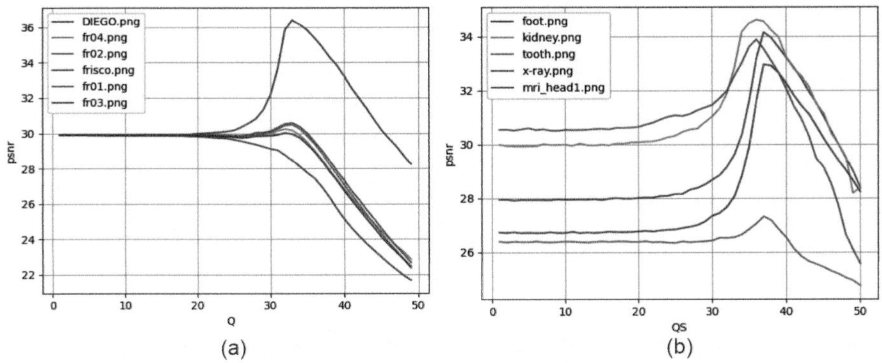

Figure 5: Dependencies $PSNR^{tc}(Q)$ for AWGN variance equal to 64 (a) and Poisson noise (b).

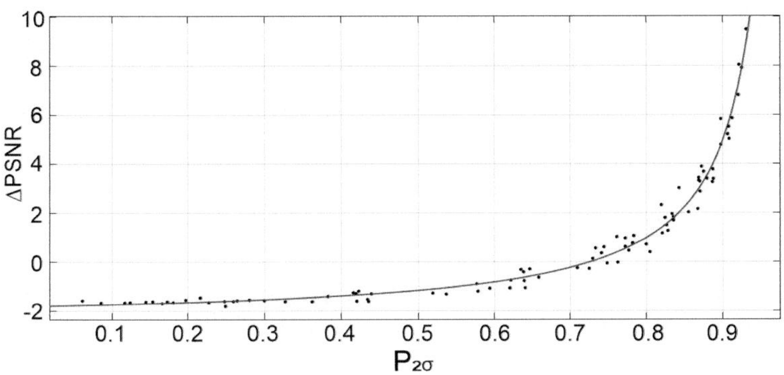

Figure 6: The scatter-plot $\Delta PSNR$ vs $P_{2\sigma}$ and the fitted curve

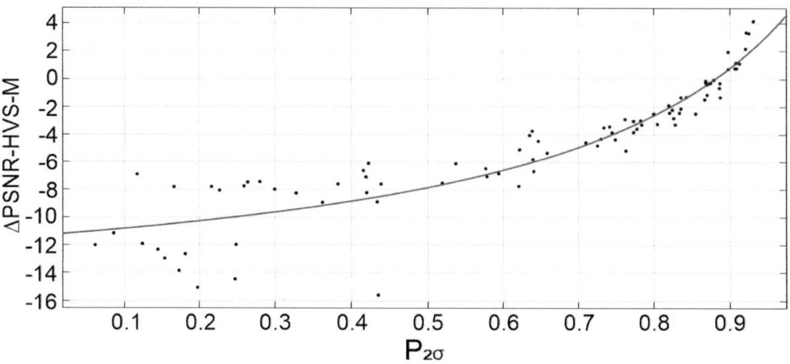

Figure 7: The scatter-plot $\Delta PSNR - HVS - M$ vs $P_{2\sigma}$ and the fitted curve

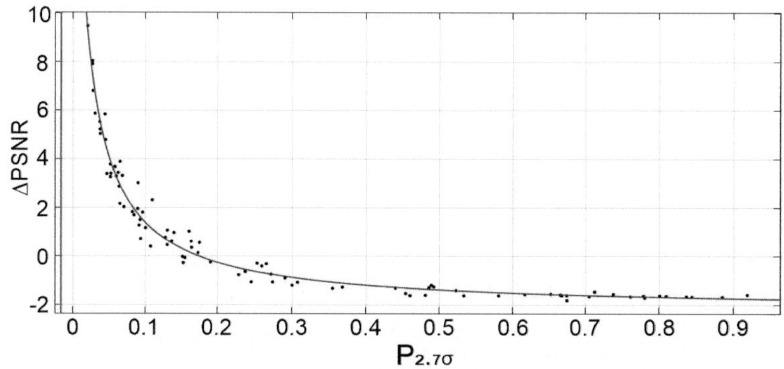

Figure 8: The scatter-plot $\Delta PSNR$ vs $P_{2.7\sigma_n}$ and the fitted curve

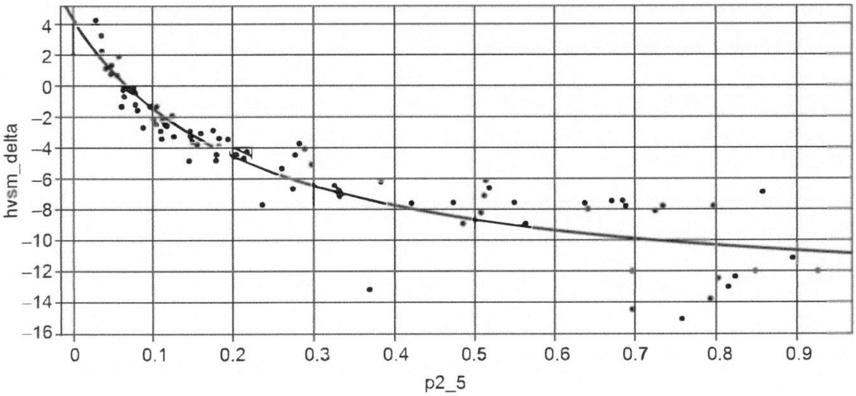

Figure 9: The scatter-plot $\Delta PSNR - HVS - M$ vs $P_{2.7\sigma_n}$ and the fitted curve

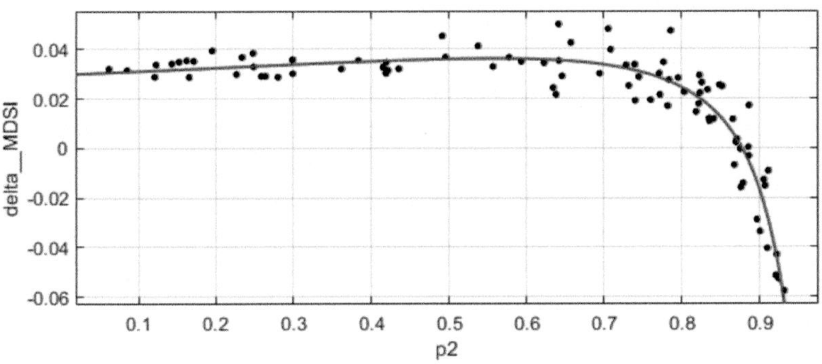

Figure 10: The scatter-plot $\Delta MDSI$ vs $P_{2\sigma}$ and the fitted curve

1. How well is the scatter-plot formed?
2. What is the accuracy of curve fitting?
3. What input parameter(s) to use?
4. How accurately is the input parameter estimated? How does the method of block selection influence this accuracy? How many blocks are needed?
5. Do noise realizations influence the input parameter?

Investigations carried out in[57] have already given answers to some of these questions. In particular, it has been shown that noise realization has a negligible impact on accuracy. It has been also established that the use of 500-1000 blocks randomly placed on an analyzed image allows estimating such input parameters as $P_{2\sigma}$ or $P_{2.7\sigma_n}$ with appropriately high accuracy. Thus, we have to answer the first three questions.

In general, these three questions have to be considered and answered jointly. First, we need to study possible fitting variants, to characterize their accuracy quantitatively and to find the best or, at least, appropriate fitting. Note that there exist good theories and practical solutions for least mean square error (LMSE) and robust fitting[78;79]. Visual inspection of the scatter-plots presented in Figures 6-10 demonstrates that there is no necessity to use robust fitting as no outliers are observed in the data. LMSE regression methods and algorithms are realized in numerous software packages including MATLAB. We have applied a MATLAB Curve Fitting Tool.

A typical way to characterize the fitting accuracy is to determine and analyze root mean square RMSE (it should as small as possible), goodness-of-the-fit R^2, and adjusted goodness-of-the-fit AdjR^2 (they should be as large as possible, the maximal possible value for both is unity).

Polynomials of a different order, exponentials or weighted sum of exponents, Fourier series, and rational functions are typical representatives of functions that can be tried. We have tested all of them. The fitted polynomials of the fourth and fifth orders as well as Fourier series produce quite good results according to quantitative criteria but they have undesired fluctuations (are non-monotonous or smooth enough). Because of this, we have come to the conclusion that it is better to apply either rational functions or the weighted sum of two exponentials that are characterized by approximately the same

Table 1: Quantitative characteristics of curve fitting for scatter-plots in Figures 7- 11.

Dependence	RMSE	R^2	Adjusted R^2
$\Delta PSNR$ vs $P_{2\sigma}$	0.37	0.98	0.98
$\Delta PSNR-HVS-M$ vs $P_{2\sigma}$	1.72	0.85	0.85
$\Delta PSNR$ vs $P_{2.7\sigma_n}$ vs $P_{2\sigma}$	0.43	0.975	0.975
$\Delta PSNR-HVS-M$ vs $P_{2.7\sigma_n}$	1.66	0.86	0.86
$\Delta MDSI$ vs $P_{2\sigma}$ vs $P_{2\sigma}$	0.0081	0.8905	0.8866

RMSE, R^2, and adjusted R^2, which are slightly better than for polynomials and Fourier series. Just the fitted rational functions are shown in Figures 7-11, the parameters of these functions for the scatter-plots in Figures 7-10 are given in Table 1 in[57].

Here we are more interested in just the fit accuracy. The corresponding data are given in Table 1 below. As one can see, fitting is almost perfect for dependences $\Delta PSNR$ versus $P_{2\sigma}$ and $\Delta PSNR$ versus $P_{2.7\sigma_n}$. The results for the metric $\Delta MDSI$ are worse (R^2 and Adjusted R^2 values are smaller, RMSE values cannot be compared since $\Delta MDSI$ varies in other limits). Meanwhile, the characteristics of fitting for $\Delta PSNR-HVS-M$ are even worse than for $\Delta MDSI$. R^2 and Adjusted R^2 values are smaller than 0.9; they are slightly better, if $P_{2.7\sigma_n}$ is exploited as the input parameter. Visual analysis of the scatter-plots in Figures 8, 10 and 11 shows that such relatively low values of R^2 and adjusted R^2 are explained not by bad fitting, but by the diversity of scatter-plot points.

In fact, it is more difficult to provide good predictions for visual quality metrics for image denoising too[80]. Thus, it is reasonable to try some other approaches to predict visual quality metrics, at least, $\Delta PSNR-HVS-M$.

We have not yet answered the question of how well was the scatter-plot formed. It can be answered in different ways. First, it is possible to take a "new" set of images, get a scatter-plot for them, obtain for them the fitted curve, and compare it to the fitted curve obtained earlier. Second, one can check for some images how accurately $\Delta PSNR$, $\Delta PSNR-HVS-M$, $\Delta MDSI$ are predicted for them for different values of noise variance. We have decided to follow the first way. For this purpose, we have chosen another set of test images (North Island NAS, Foster City, Couple, Aerial, Stream and bridge,

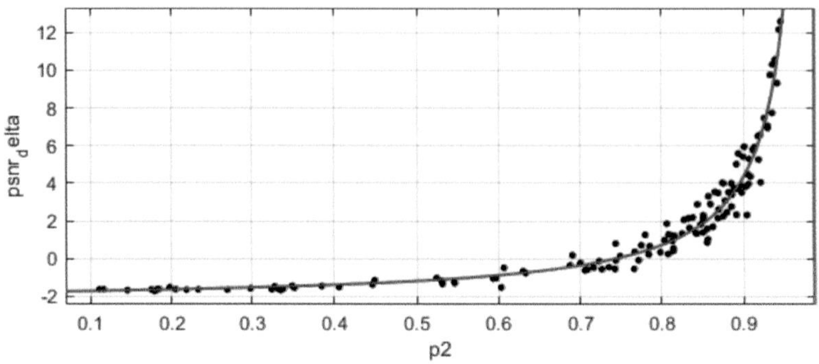

Figure 11: The scatter-plot $\Delta PSNR$ vs $P_{2\sigma}$ and the fitted curve for another set of test images

Truck, Airplane, Car and APCs, Truck and APCs, Fishing Boat), obtained scatter-plots for them, and fitted the curves.

Figure 12 shows the scatter-plot $\Delta PSNR$ vs $P_{2\sigma}$ that can be compared to data in Figure 7. As one can see, the scatter-plot and fitted curve properties are very similar. For example, for $P_{2\sigma}$≈0.6, the predicted values of $\Delta PSNR$ are in both cases about −0.8 dB. $\Delta PSNR(Q)$ crosses the zero level for $P_{2\sigma}$≈0.74 and so on. Figure13 presents the data for $\Delta PSNR - HVS - M$ vs $P_{2\sigma}$. Comparing them to data in Figure 8, we can state that the main properties are very similar. In particular, $\Delta PSNR - HVS - M$ crosses the zero level for $P_{2\sigma}$≈0.88 in both cases. For the newly obtained fitted curves, the fitting parameters are the following: for Δ $PSNR$, R ^{2}= 0.96, Adjusted R^2 = 0.96, RMSE = 0.67. For $\Delta PSNR - HVS - M$ one has: $R^2 = 0.85$, Adjusted $R^2 = 0.85$ and $RMSE = 1.9$. Comparing these data to the corresponding data in Table 1, we see that all fitting parameters are slightly worse. Meanwhile, again, the fitting (and, respectively, the prediction accuracy) for $\Delta PSNR$ is considerably better than for $\Delta PSNR - HVS - M$. This confirms our preliminary hypothesis that it is more important to look for ways to improve the prediction accuracy for $\Delta PSNR - HVS - M$ than for $\Delta PSNR$.

Having confirmed that the scatter-plots have been obtained correctly, we can discuss input parameters and their properties. As said, the parameters $P_{2\sigma}$ and $P_{2.7\sigma_n}$ have been taken empirically, based on previous experience.

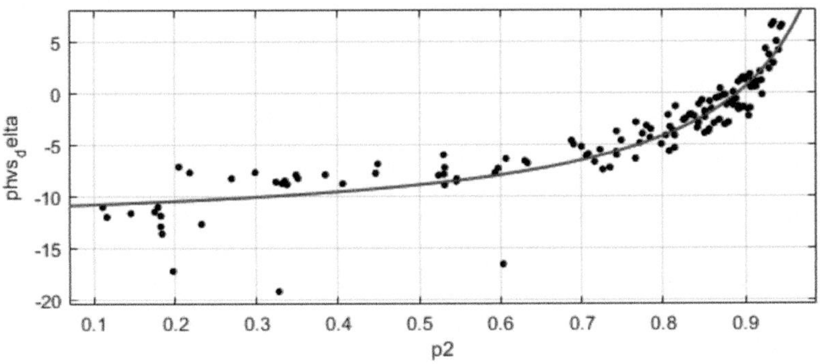

Figure 12: The scatter-plot $\Delta PSNR - HVS - M$ vs $P_{2\sigma}$ and the fitted curve for different set of images

Analysis of data in Table 1 shows that $P_{2\sigma}$ is slightly better (fitting is characterized by larger R^2 and smaller RMSE values respectively) for predicting $\Delta PSNR$ whilst $P_{2.7\sigma_n}$ is slightly better for predicting $\Delta PSNR - HVS - M$. This has run us into the idea that both parameters can be not optimal in the sense of producing maximal R^2 or minimal RMSE. Because of this, we have decided to check whether some other $P_{\alpha\sigma}$ and/or $P_{\alpha\sigma_n}$ are able to provide better results.

The first o bservation t hat w e h ave d iscovered i s t hat t he fi tting quality characteristics (R^2 and RMSE) are the same (for the case of using rational functions for fitting) if one uses either $P_{\alpha\sigma}$ or $P_{\alpha\sigma_n}$ as input parameters for the same α. This is proven by the scatter-plots in Figures 14 and 15 for $P_{2.5\sigma}$ and $P_{2.5\sigma_n}$, respectively. In both cases (for the first set of test images), one has the following fitting p arameters: $R^2 = 0.86$, Adjusted $R^2 = 0.86$, $RMSE = 1.68$, i.e. they are very close to the data for curve fitting i nto t he scatter-plots $\Delta PSNR - HVS - M$ vs $P_{2\sigma}$ and $\Delta PSNR - HVS - M$ vs $P_{2.7\sigma_n}$ in Table 1. This means that in our attempt to find the optimal input parameter, we can try either $P_{\alpha\sigma}$ or $P_{\alpha\sigma_n}$ and vary α for the chosen input parameter.

Let us present the results obtained for predicting $\Delta PSNR - HVS - M$ (note that it is more important to improve the prediction accuracy $\Delta PSNR - HVS - M$ than $\Delta PSNR$ since the prediction accuracy for $\Delta PSNR$ is already high enough).

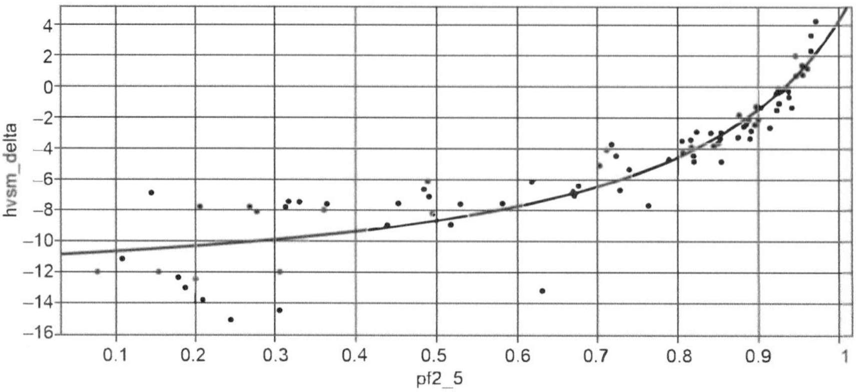

Figure 13: The scatter-plot $\Delta PSNR - HVS - M$ vs $P_{\alpha\sigma}$ and the fitted curve for α equal to 2.5

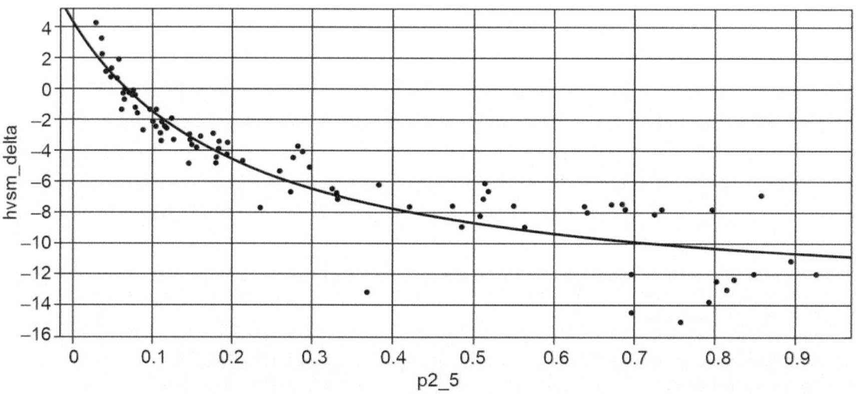

Figure 14: The scatter-plot $\Delta PSNR - HVS - M$ vs $P_{\alpha\sigma_n}$ and the fitted curve for α equal to 2.5

Table 2: Curve fitting parameters for the scatter-plots $\Delta PSNR - HVS - M$ vs $P_{\alpha\sigma}$ for different α.

Parameter	$P_{0.2\sigma}$	$P_{0.5\sigma}$	$P_{1.5\sigma}$	$P_{2\sigma}$	$P_{2.4\sigma}$
R^2	**0.8642**	0.864	0.8634	0.8622	0.8602
Adjusted R^2	**0.8594**	0.8591	0.8585	0.8573	0.8552
RMSE	**1.647**	1.648	1.652	1.659	1.671

Table 3: Curve fitting parameters for the scatter-plots $\Delta PSNR - HVS - M$ vs $P_{\alpha\sigma_n}$ for different α.

Parameter	$P_{2.5\sigma_n}$	$P_{2.7\sigma_n}$	$P_{3.0\sigma_n}$	$P_{3.5\sigma_n}$	$P_{4.0\sigma_n}$
R^2	**0.8595**	0.8581	0.8554	0.8504	0.8447
Adjusted R^2	**0.8545**	0.853	0.8503	0.845	0.8391
RMSE	**1.675**	1.683	1.699	1.729	1.761

Table 2 gives the fitting parameters for $P_{\alpha\sigma}$ with α equal to 0.2, 0.5, 1.5, 2.0, and 2.4 whilst Table 3 presents data for $P_{\alpha\sigma_n}$ with α equal to 2.5, 2.7, 3.0, 3.5 and 4.0.

Analysis of data in Table 2 shows that, among the considered values of α, $P_{0.2\sigma}$ is formally optimal, but the difference in the corresponding fitting accuracy parameters for different α is negligible. Thus, any of the considered $P_{\alpha\sigma}$ can be used as inputs. The results presented in Table 3 demonstrate two things.

First, the values of R^2, adjusted R^2, and RMSE for $P_{2.4\sigma}$ and $P_{2.5\sigma_n}$ are practically the same, *i.e.* there is no difference in what input parameters to apply - $P_{\alpha\sigma}$ or $P_{\alpha\sigma_n}$ for the same chosen α. Second, if α increases, the fitting parameters for $P_{\alpha\sigma_n}$ slowly become worse. One more conclusion is that it is impossible to get a sufficient improvement in predicting $\Delta PSNR - HVS - M$ if one uses $P_{\alpha\sigma}$ or $P_{\alpha\sigma_n}$ as only one input parameter. Note also that we have tried other ways of joint processing the local estimates $P_{2\sigma}(m), m = 1, \ldots, M$ or $P_{2.7\sigma}(m), m = 1, \ldots, M$.

In particular, we have calculated the median ($P_{2\sigma med}$) and mode ($P_{2\sigma mod}$) of these estimates as well as their variance ($Var P_{2\sigma loc}$). A part of the obtained results is given in Table 4. The results obtained for $P_{2\sigma med}$ are slightly worse than for $P_{2\sigma}$; prediction based on$P_{2\sigma mod}$ is, in general, possible but it is even less accurate. Surprisingly, there is a certain (non-zero) correlation between

Table 4: Curve fitting parameters for some other input parameters.

Parameter	$P_{2\sigma}$	$P_{2\sigma\mathrm{med}}$	$P_{2\sigma\mathrm{mod}}$	Var $P_{2\sigma}$
R^2	**0.9774**	0.9616	0.827	0.479
Adjusted R^2	**0.9766**	0.9603	0.8208	0.4604
RMSE	**0.4148**	0.5409	1.149	1.993

predicted metrics and $VarP_{2\sigma loc}$. This shows that, most probably, the use of two or more input parameters can be reasonable to improve prediction.

Such an idea was earlier successfully exploited in the prediction of performance for the DCT-based filters[77;80]. As the first step, we have tried to further improve the accuracy of prediction for the metric $\Delta PSNR$. Several combinations of two parameters have been considered. However, none of them produces better results than for one input parameter. Probably, there are two reasons. First, such a parameter as $P_{2\sigma med}$ is strongly correlated with $P_{2\sigma}$ and, thus, it does not give any "additional" information to improve prediction which is already very accurate. Second, the use of $VarP_{2\sigma loc}$, which is "quite noisy" (has low correlation with $\Delta PSNR$, does not provide additional benefits as well.

Hence, we have also taken into account another approach proposed in[80]. Recall that we assume noise variance is known in advance. This means that we also know $PSNR^n$. In[80], it was proposed to apply $PSNR^n$ as one additional joint input parameter. The results for the prediction of $\Delta PSNR - HVS - M$ for different possible pairs of input parameters are collected in Table 5. Here, $\mathrm{SymP}_{2\sigma_s}$ and $\mathrm{SymP}_{2.7\sigma_n}$ denote symmetry coefficients for the sets of local estimates $P_{2\sigma}(m)$ and $P_{2.7\sigma_n}(m), m = 1, \ldots, M$, respectively. Similarly, $\mathrm{KurtP}_{2\sigma_k}$ and $\mathrm{KurtP}_{2.7\sigma\mathrm{n}}$ are kurtosis estimates for the sets of local estimates $P_{2\sigma}(m)$ and $P_{2.7\sigma_n}(m), m = 1, \ldots, M$, respectively.

As it follows from the analysis of data in Table 5, all six combinations improve R^2 and reduce the RMSE compared to the best case of using one input parameter (see data in Table 2 in the prediction of $\Delta PSNR - HVS - M$. The information on higher-order statistics of the distributions of local estimates $P_{2\sigma}(m)$ and $P_{2.7\sigma_n}(m), m = 1, \ldots, M$ is beneficial. Similarly, the information about the distribution mode or median is useful, too. But the most important information is contained in $\mathrm{PSNR^n}$, the use of which as the second parameter leads to the largest increase in R^2 that approaches the R^2 value for the prediction of $\Delta PSNR$.

Table 5: Fitting parameters for regression based on two input parameters.

Statistical parameters	R^2	RMSE	Statistical parameters	R^2	RMSE
$P_{2\sigma}$ + $PSNR^n$	**0.9717**	0.784	$P_{2.7\sigma_n}$ + $PSNR^n$	**0.9633**	0.893
$P_{2\sigma}$ + Var $P_{2\sigma}$	0.8949	1.51	$P_{2.7\sigma_n}$ + Var $P_{2.7\sigma}$	0.8877	1.561
$P_{2\sigma}$ + $P_{2\sigma \mathrm{med}}$	0.9061	1.428	$P_{2.7\sigma_n}$ + $P_{2.7\sigma_n \mathrm{med}}$	0.8908	1.54
$P_{2\sigma}$ + $P_{2\sigma \mathrm{mod}}$	0.8884	1.556	$P_{2.7\sigma_n}$ + $P_{2.7\sigma_n \mathrm{mod}}$	0.8821	1.6
$P_{2\sigma}$ + Sym$P_{2\sigma_s}$	0.9187	1.329	$P_{2.7\sigma_n}$ + Sym$P_{2.7\sigma_n}$	0.9137	1.369
$P_{2\sigma}$ + Kurt$P_{2\sigma_k}$	0.8751	1.647	$P_{2.7\sigma_n}$ + Kurt$P_{2.7\sigma_n}$	0.8715	1.67

The studies on this approach have been started recently and have not been completed yet. We have tested only fitting based on 2D polynomials of the third order as

$$f(x,y) = p_{00} + p_{10} * x + p_{01} * y + p_{20} * x^2 + p_{11} * x * y + p_{02} * y^2 \\ + p_{30} * x^3 + p_{21} * x^2 * y + p_{12} * x * y^2 + p_{03} * y^3$$

and the sum of exponentials as

$$f(x,y) = a * \exp(-b * x) + c * \exp(d * y).$$

The data in Table 5 are given for the polynomial approximation. Quantitative characteristics of the fitting for the sum of two exponentials are practically the same as for the polynomial fitting (Table 5). The examples of approximating 2D functions (surfaces) by polynomial and exponential functions are given in Figure 16. Analysis shows the following. The 2D surface for exponential fitting seems smoother and more natural. The points are mostly placed "along a diagonal". Points with larger $P_{2\sigma}$ and smaller $P_{2.7\sigma_n}$ are mostly associated with the images with small $PSNR^n$, i.e. noisy ones. In turn, negative $\Delta PSNR - HVS - M$ with large magnitudes are observed for large $PSNR^n$, where image quality reduction due to lossy compression does not lead to appearing visually noticeable distortions. RMSE is about $0.5 - 0.6$ dB, which is acceptable for practice.

Below we give practical expressions for predicting $\Delta PSNR$ and $\Delta PSNR - HVS - M$

$$\Delta PSNR = (1.533e + 04 * x - 1.112e + 04)/(x^3 + 75.71 * x^2 - 6291 * x + 6139)$$

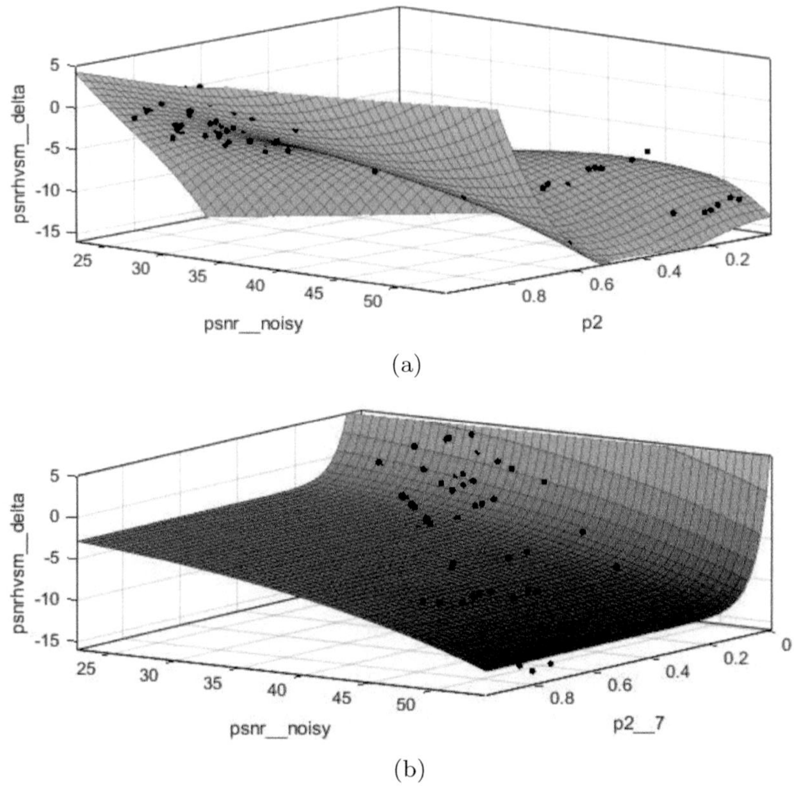

Figure 15: 2D scatter-plots and the fitted surfaces for the third order polynomial (a) and weighted sum of exponentials (b).

$$\Delta PSNR - HVS - M = 4.959e - 05 * exp(13.03 * x) - 0.3702 * exp(0.02888 * y)$$

$$\Delta MDSI = (-0.0594 * x + 0.05212)/(x^2 - 2.755 * x + 1.75)$$

Certainly, there are many ways to improve the prediction accuracy further. In particular, it is possible to jointly apply three or more input parameters using more sophisticated regression tools such as support vector machines or pre-trained neural networks. Meanwhile, more input parameters require more computations and make the prediction more complicated. However, all the parameters that have been employed in our analysis above are rather simple and can be calculated easily and quickly since they are based either on DCT (that has fast software and hardware implementations) or simple arithmetic operations. Recall also that the amount of blocks is limited.

4 Decision Rules and Compression Examples

Let us assume that, for a given image, the prediction has been performed. Then, one can get one parameter (e.g., $\Delta PSNR$) or two parameters (e.g., $\Delta PSNR$ and $\Delta PSNR - HVS - M$) or three (e.g, ΔMDSI in addition to the aforementioned two). What can be done then? Obviously, there are many options – to use one, two, or three parameters for the decision undertaking, apply some thresholds and rules, employ the majority voting, among others. Probably, a special study to attract customers to assess the quality of compressed images can be helpful.

Leaving the final answer concerning the best decision undertaking method, let us recall the practical algorithm (Algorithm 1) proposed by us in[57]:

- Case 1: If $\Delta PSNR + \Delta PSNR - HVS - M > 1dB$, we consider that OOP exists and, thus, an image is compressed using Q_{OOP};

- Case 2: If $-1dB \leq \Delta PSNR + \Delta PSNR - HVS - M \leq 1dB$, we consider that OOP might exist and recommend using $Q_{OOP} - 1$ (but not less than 28), the recommended Q is smaller than Q_{OOP} to avoid compressed image oversmoothing and introducing annoying artifacts;

- Case 3: If $\Delta PSNR + \Delta PSNR - HVS - M < -1dB$, we consider that OOP is absent and recommend using $Q = 28$ or 29 to ensure the invisibility of introduced distortions and the possibility of noise removal in it by applying post-filtering to decompressed images.

An example of Case 1 is shown in Figure 17. AWGN variance is equal to 64 and the noise is seen well, especially in homogeneous image regions (Figure 17a). The image compressed by the BPG coder with optimal Q is represented in Figure 17b. It is seen that noise is sufficiently suppressed whilst the useful information is preserved well.

An example for Case 2 is depicted in Figure 18. The noise is intensive and it can be easily noticed in different parts of the image in Figure 18a. The image compressed with the recommended Q is presented in Figure 18b. Visual inspection shows that noise is considerably suppressed although some smearing is observed, especially for low-contrast details.

Finally, an example for Case 3 is illustrated in Figure 19. AWGN is not intensive and, thus, it is almost not seen in Figure 19a. Then, the positive effects of noise suppression can be hardly noticed. Meanwhile, negative effects of some edge/detail smearing take place. To minimize their influence, compression with the recommended $Q = 29$ is applied, and the compressed image quality is high.

5 Conclusions and Future Work

In this chapter, the task of lossy compression of noisy images by the BPG coder is considered. The analysis performed has demonstrated that OOP is possible. OOP can take place according to different metrics including visual quality ones ($PSNR$, $MS - SSIM$, $PSNR - HVS - M$, $MDSI$) although the probability of OOP existence for visual quality metrics is smaller than for $PSNR$. The way to automatically determine Q_{OOP} is described for AWGN with known variance. The procedure for the prediction of parameters for images compressed in potential OOP is discussed. Based on such a prediction, an algorithm for setting the best Q is given and the examples confirming these recommendations are presented. Variants of input parameters are considered. It is shown that for predicting $\Delta PSNR - HVS - M$, it is reasonable to use two input parameters. The case of Poisson noise in the grayscale images is briefly studied. It is demonstrated that OOP is possible for this noise type as

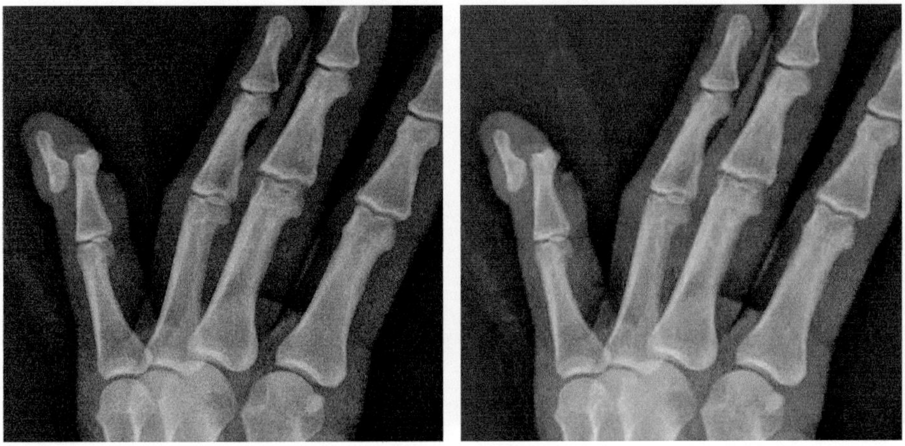

(a) (b)

Figure 16: Medical image "Hand" corrupted with AWGN with variance 64 (a) BPG coder in the optimal way (with Q_{OOP})

(a) (b)

Figure 17: Remote sensing image "Fr02" corrupted with AWGN with variance 196 (a) and the image compressed by the BPG coder with $Q = Q_{OOP} - 1$ (b)

(a) (b)

Figure 18: Remote sensing image "Fr03" corrupted with AWGN with variance 25 (a) and the image compressed by the BPG coder with $Q = 29$ (b)

well. Finally, it is shown that the proposed method of prediction is valid for images of different fields.

However, there are quite a few questions still looking for answers. In particular, the case of signal-dependent noise is of prime interest since AWGN is a common but not too realistic noise model. Spatially correlated noise has to be analyzed as well. Three more channel images have to be considered[81], especially with application to RS image classification. I mage p re- and post-filtering[82;83] m ight i mprove c ompressed i mage q uality a dditionally, and this approach should be addressed, too.

References

[1] Michele Zappavigna. Social media photography: construing subjectivity in Instagram images, aug 2016. ISSN 1470-3572. URL `http://journals.sagepub.com/doi/10.1177/1470357216643220`.

[2] Anthony B. Wolbarst, Patrizio Capasso, and Andrew R. Wyant. Medical Imaging, apr 2013. URL http://doi.wiley.com/10.1002/9781118480267.

[3] Siamak Khorram, Cynthia F. van der Wiele, Frank H. Koch, Stacy A. C. Nelson, and Matthew D. Potts. Future Trends in Remote Sensing. *Principles of Applied Remote Sensing*, 277–285:277–285, 2016. doi: 10.1007/978-3-319-22560-9_9. URL http://link.springer.com/10.1007/978-3-319-22560-9_9.

[4] Carolyn Pedwell. Mediated habits: images, networked affect and social change. *Subjectivity*, 10(2):147–169, jul 2017. ISSN 1755-6341. doi: 10.1057/s41286-017-0025-y. URL http://link.springer.com/10.1057/s41286-017-0025-y.

[5] S.G. Shini, Tony Thomas, and K. Chithraranjan. Cloud Based Medical Image Exchange-Security Challenges. *Procedia Engineering*, 38:3454–3461, 2012. ISSN 18777058. doi: 10.1016/j.proeng.2012.06.399. URL https://linkinghub.elsevier.com/retrieve/pii/S1877705812023120.

[6] Nataliia Kussul, Mykola Lavreniuk, Andrii Kolotii, Sergii Skakun, Olena Rakoid, and Leonid Shumilo. A workflow for Sustainable Development Goals indicators assessment based on high-resolution satellite data. *International Journal of Digital Earth*, 13(2):309–321, feb 2020. ISSN 1753-8947. doi: 10.1080/17538947.2019.1610807. URL https://www.tandfonline.com/doi/full/10.1080/17538947.2019.1610807.

[7] Neha Joshi, Matthias Baumann, Andrea Ehammer, Rasmus Fensholt, Kenneth Grogan, Patrick Hostert, Martin Jepsen, Tobias Kuemmerle, Patrick Meyfroidt, Edward Mitchard, Johannes Reiche, Casey Ryan, and Björn Waske. A Review of the Application of Optical and Radar Remote Sensing Data Fusion to Land Use Mapping and Monitoring. *Remote Sensing*, 8(1):70, jan 2016. ISSN 2072-4292. doi: 10.3390/rs8010070. URL http://www.mdpi.com/2072-4292/8/1/70.

[8] Miew Keen Choong, Rajasvaran Logeswaran, and Michel Bister. Improving Diagnostic Quality of MR Images Through Controlled Lossy Compression Using SPIHT. *Journal of Medical Systems*, 30(3):139–143, jun 2006. ISSN 0148-5598. doi: 10.1007/s10916-005-8374-4. URL https://link.springer.com/10.1007/s10916-005-8374-4.

[9] Ian Blanes, Enrico Magli, and Joan Serra-Sagrista. A Tutorial on Image Compression for Optical Space Imaging Systems. *IEEE Geoscience and Remote Sensing Magazine*, 2(3):8–26, sep 2014. ISSN 2168-6831.

doi: 10.1109/MGRS.2014.2352465. URL http://ieeexplore.ieee.org/document/6928587/.

[10] Guoxia Yu, Tanya Vladimirova, and Martin N. Sweeting. Image compression systems on board satellites. *Acta Astronautica*, 64(9-10): 988–1005, may 2009. ISSN 00945765. doi: 10.1016/j.actaastro.2008.12.006. URL https://linkinghub.elsevier.com/retrieve/pii/S0094576508004062.

[11] A.J. Hussain, Ali Al-Fayadh, and Naeem Radi. Image compression techniques: A survey in lossless and lossy algorithms. *Neurocomputing*, 300:44–69, jul 2018. ISSN 09252312. doi: 10.1016/j.neucom.2018.02.094. URL https://linkinghub.elsevier.com/retrieve/pii/S0925231218302935.

[12] Jarno Mielikainen. Linear prediction in lossless compression of hyperspectral images. *Optical Engineering*, 42(4):1013, apr 2003. ISSN 0091-3286. doi: 10.1117/1.1557174. URL http://opticalengineering.spiedigitallibrary.org/article.aspx?doi=10.1117/1.1557174.

[13] Miloš Radosavljević, Branko Brkljač, Predrag Lugonja, Vladimir Crnojević, Željen Trpovski, Zixiang Xiong, and Dejan Vukobratović. Lossy Compression of Multispectral Satellite Images with Application to Crop Thematic Mapping: A HEVC Comparative Study. *Remote Sensing*, 12(10):1590, may 2020. ISSN 2072-4292. doi: 10.3390/rs12101590. URL https://www.mdpi.com/2072-4292/12/10/1590.

[14] Giovanni Motta, Francesco Rizzo, and James A. Store. Hyperspectral Data Compression. *Hyperspectral Data Compression*, 3(1):1–417, 2006. doi: 10.1007/0-387-28600-4. URL http://link.springer.com/10.1007/0-387-28600-4.

[15] Bruno Aiazzi, Luciano Alparone, and Stefano Baronti. Near-lossless compression of 3-D optical data. *IEEE Transactions on Geoscience and Remote Sensing*, 39(11):2547–2557, 2001. ISSN 01962892. doi: 10.1109/36.964993. URL http://ieeexplore.ieee.org/document/964993/.

[16] Wenjun Zeng, Scott Daly, and Shawmin Lei. An overview of the visual optimization tools in JPEG 2000. *Signal Processing: Image Communication*, 17(1):85–104, jan 2002. ISSN 09235965. doi: 10.1016/S0923-5965(01)00029-7. URL https://linkinghub.elsevier.com/retrieve/pii/S0923596501000297.

[17] G.K. Wallace. The JPEG still picture compression standard. *IEEE Transactions on Consumer Electronics*, 38(1):xviii–xxxiv, 1992. ISSN

00983063. doi: 10.1109/30.125072. URL http://ieeexplore.ieee.org/document/125072/.

[18] Fangfang Li, Sergey Krivenko, and Vladimir Lukin. A Fast Method for Visual Quality Prediction and Providing in Image Lossy Compression by SPIHT. *Lecture Notes in Networks and Systems*, 188:17–29, 2021. ISSN 23673389. doi: 10.1007/978-3-030-66717-7_2. URL http://link.springer.com/10.1007/978-3-030-66717-7_2.

[19] David S. Taubman. *JPEG2000: Image Compression Fundamentals, Standards and Practice*, volume 11. Kluwer Academic Publishers, Boston, apr 2002. doi: 10.1117/1.1469618. URL http://electronicimaging.spiedigitallibrary.org/article.aspx?doi=10.1117/1.1469618.

[20] Boban Bondžulić, Nenad Stojanović, Vladimir Petrović, Boban Pavlović, and Zoran Miličević. Efficient Prediction of the First Just Noticeable Difference Point for JPEG Compressed Images. *Acta Polytechnica Hungarica*, 18(8):201–220, 2021. ISSN 17858860. doi: 10.12700/APH.18.8.2021.8.11. URL http://acta.uni-obuda.hu/Bondzulic_Stojanovic_Petrovic_Pavlovic_Milicevic_115.pdf.

[21] Sergey Krivenko, Vladimir Lukin, Olha Krylova, Liudmyla Kryvenko, and Karen Egiazarian. A Fast Method of Visually Lossless Compression of Dental Images. *Applied Sciences*, 11(1):135, dec 2020. ISSN 2076-3417. doi: 10.3390/app11010135. URL https://www.mdpi.com/2076-3417/11/1/135.

[22] N. N. Ponomarenko, V. V. Lukin, M. S. Zriakhov, A. Kaarna, and J. Astola. Automatic Approaches to On-Land/On-Board Filtering and Lossy Compression of AVIRIS Images. In *IGARSS 2008 - 2008 IEEE International Geoscience and Remote Sensing Symposium*, volume 3, pages III – 254–III – 257. IEEE, 2008. ISBN 978-1-4244-2807-6. doi: 10.1109/IGARSS.2008.4779331. URL http://ieeexplore.ieee.org/document/4779331/.

[23] F. Bellard. BPG Image Format. [Online]. Available:, 2014. URL https:%0A//bellard.org/bpg/.%0A.

[24] David Yee, Sara Soltaninejad, Deborsi Hazarika, Gaylord Mbuyi, Rishi Barnwal, and Anup Basu. Medical image compression based on region of interest using better portable graphics (BPG). *2017 IEEE International Conference on Systems, Man, and Cybernetics (SMC)*, 2017-Janua: 216–221, oct 2017. doi: 10.1109/SMC.2017.8122605. URL http://ieeexplore.ieee.org/document/8122605/.

[25] Lucana Santos, Sebastián Lopez, Gustavo M. Callico, José F. Lopez, and Roberto Sarmiento. Performance Evaluation of the H.264/AVC Video Coding Standard for Lossy Hyperspectral Image Compression. *IEEE Journal of Selected Topics in Applied Earth Observations and Remote Sensing*, 5(2):451–461, apr 2012. ISSN 1939-1404. doi: 10.1109/JSTARS.2011.2173906. URL `https://ieeexplore.ieee.org/document/6080751/`.

[26] Umar Albalawi, Saraju P. Mohanty, and Elias Kougianos. Energy-Efficient Design of the Secure Better Portable Graphics Compression Architecture for Trusted Image Communication in the IoT. In *2016 IEEE Computer Society Annual Symposium on VLSI (ISVLSI)*, volume 2016-Septe, pages 302–307, Pittsburgh, PA, USA, jul 2016. IEEE. ISBN 978-1-4673-9039-2. doi: 10.1109/ISVLSI.2016.21. URL `http://ieeexplore.ieee.org/document/7560214/`.

[27] Liudmyla Kryvenko, Olha Krylova, Sergey Krivenko, and Vladimir Lukin. A Fast Noniterative Visually Lossless Compression of Dental Images Using BPG Coder. In *2022 11th Mediterranean Conference on Embedded Computing (MECO)*, pages 1–6. IEEE, jun 2022. ISBN 978-1-6654-6828-2. doi: 10.1109/MECO55406.2022.9797114. URL `https://ieeexplore.ieee.org/document/9797114/`.

[28] Jong-Sen Lee and Eric Pottier. *Polarimetric Radar Imaging.* CRC Press, Boca Raton, dec 2017. ISBN 9781315219332. doi: 10.1201/9781420054989. URL `https://www.taylorfrancis.com/books/9781420054989`.

[29] Adugna G. Mullissa, Claudio Persello, and Valentyn Tolpekin. Fully Convolutional Networks for Multi-Temporal SAR Image Classification. In *IGARSS 2018 - 2018 IEEE International Geoscience and Remote Sensing Symposium*, volume 2018-July, pages 6635–6638, Valencia, jul 2018. IEEE. ISBN 978-1-5386-7150-4. doi: 10.1109/IGARSS.2018.8518780. URL `https://ieeexplore.ieee.org/document/8518780/`.

[30] Mohammed Hamzah Abed and C. Namrata Mahender. SAR Image Compression using SPIHT Algorithm. *International Journal of Computer Applications*, 22(8):14–18, may 2011. ISSN 09758887. doi: 10.5120/2605-3634. URL `http://www.ijcaonline.org/volume22/number8/pxc3873634.pdf`.

[31] Ruslan Kozhemiakin, Sergey Abramov, Vladimir Lukin, Blazo Djurovic, Igor Djurovic, and Marko Simeunovic. Strategies of SAR image lossy

compression by JPEG2000 and SPIHT, jun 2017. URL http://ieeexplore.ieee.org/document/7977176/.

[32] J.-L. Starck, F. Murtagh, and M. Louys. High quality astronomical image compression. *Vistas in Astronomy*, 41(3):439–445, jan 1997. ISSN 00836656. doi: 10.1016/S0083-6656(97)00049-4. URL https://linkinghub.elsevier.com/retrieve/pii/S0083665697000494.

[33] Vladimir Lukin, Sergey Krivenko, Ihor Kaluzhinov, Olha Krylova, and Liudmyla Kryvenko. Lossy Compression of Remote Sensing and Dental Images Corrupted by Spatially Correlated Noise. *Lecture Notes in Networks and Systems*, 367 LNNS:1003–1014, 2022. ISSN 23673389. doi: 10.1007/978-3-030-94259-5_77. URL https://link.springer.com/10.1007/978-3-030-94259-5_77.

[34] Nikolay N. Ponomarenko, Vladimir V. Lukin, Karen O. Egiazarian, and Leena Lepisto. Adaptive visually lossless JPEG-based color image compression, may 2013. ISSN 1863-1703. URL http://link.springer.com/10.1007/s11760-013-0446-1.

[35] Wang, Zhong, and Su. On-Orbit Signal-to-Noise Ratio Test Method for Night-Light Camera in Luojia 1-01 Satellite Based on Time-Sequence Imagery. *Sensors*, 19(19):4077, sep 2019. ISSN 1424-8220. doi: 10.3390/s19194077. URL https://www.mdpi.com/1424-8220/19/19/4077.

[36] Ping Zhong and Runsheng Wang. Multiple-Spectral-Band CRFs for Denoising Junk Bands of Hyperspectral Imagery. *IEEE Transactions on Geoscience and Remote Sensing*, 51(4):2260–2275, apr 2013. ISSN 0196-2892. doi: 10.1109/TGRS.2012.2209656. URL http://ieeexplore.ieee.org/document/6290356/.

[37] D. Wei, J.E. Odegard, H. Guo, M. Lang, and C.S. Burrus. Simultaneous noise reduction and SAR image data compression using best wavelet packet basis. In *Proceedings., International Conference on Image Processing*, volume 3, pages 200–203. IEEE Comput. Soc. Press, 1995. ISBN 0-7803-3122-2. doi: 10.1109/ICIP.1995.537615. URL http://ieeexplore.ieee.org/document/537615/.

[38] J E Odegard, H Guo, C S Burrus, and R G Baraniuk. Joint compression and speckle reduction of SAR images using embedded zerotree models. In *Workshop on Image and Multidimensional Digital Signal Processing, Mar*, pages 1996 80–81, 1996.

[39] R L White and J W Percival. Compression and progressive transmission of astronomical images. *Proceedings of SPIE*, 2199:703–713.

[40] O.K. Al-Shaykh and R.M. Mersereau. Lossy compression of noisy images. *IEEE Transactions on Image Processing*, 7(12):1641–1652, 1998. ISSN 10577149. doi: 10.1109/83.730376. URL http://ieeexplore.ieee.org/document/730376/.

[41] S.G. Chang, Bin Yu, and Martin Vetterli. Adaptive wavelet thresholding for image denoising and compression. *IEEE Transactions on Image Processing*, 9(9):1532–1546, 2000. ISSN 10577149. doi: 10.1109/83.862633. URL http://ieeexplore.ieee.org/document/862633/.

[42] Nikolay Ponomarenko, Sergey Krivenko, Vladimir Lukin, Karen Egiazarian, and Jaakko T. Astola. Lossy Compression of Noisy Images Based on Visual Quality: A Comprehensive Study. *EURASIP Journal on Advances in Signal Processing*, 2010(1): 976436, dec 2010. ISSN 1687-6180. doi: 10.1155/2010/976436. URL https://asp-eurasipjournals.springeropen.com/articles/10.1155/2010/976436.

[43] Guangtao Zhai and Xiongkuo Min. Perceptual image quality assessment: a survey. *Science China Information Sciences*, 63(11):211301, nov 2020. ISSN 1674-733X. doi: 10.1007/s11432-019-2757-1. URL https://link.springer.com/10.1007/s11432-019-2757-1.

[44] Nikolay Ponomarenko, Vladimir Lukin, Jaakko Astola, and Karen Egiazarian. Analysis of HVS-Metrics' Properties Using Color Image Database TID2013. In *Lecture Notes in Computer Science (including subseries Lecture Notes in Artificial Intelligence and Lecture Notes in Bioinformatics)*, volume 9386, pages 613–624, 2015. doi: 10.1007/978-3-319-25903-1_53. URL http://link.springer.com/10.1007/978-3-319-25903-1_53.

[45] N Ponomarenko, F Silvestri, K. Egiazarian, M. Carli, J. Astola, and V. Lukin. On Between-coefficient Contrast Masking of DCT Basis Functions. In *Proceedings of the Third International Workshop on Video Processing and Quality*, number X, pages 1–4, 2007. URL http://www.researchgate.net/publication/242309240_ON_BETWEEN-COEFFICIENT_CONTRAST_MASKING_OF_DCT_BASIS_FUNCTIONS/file/9c96052442be7c3176.pdf%0Ahttp://sp.cs.tut.fi/publications/archive/Ponomarenko2007-On.pdf.

[46] Zhou Wang, E.P. Simoncelli, and A.C. Bovik. Multiscale structural similarity for image quality assessment. In *The Thirty-Seventh Asilomar Conference on Signals, Systems and Computers, 2003*, volume 2, pages

1398–1402. IEEE, 2003. ISBN 0-7803-8104-1. doi: 10.1109/ACSSC.2003.1292216. URL http://ieeexplore.ieee.org/document/1292216/.

[47] Vladimir Lukin, Sergey Abramov, Sergey Krivenko, Andriy Kurekin, and Oleksiy Pogrebnyak. Analysis of classification accuracy for pre-filtered multichannel remote sensing data. *Expert Systems with Applications*, 40(16):6400–6411, nov 2013. ISSN 09574174. doi: 10.1016/j.eswa.2013.05.061. URL https://linkinghub.elsevier.com/retrieve/pii/S095741741300359X.

[48] Vladimir V. Lukin, Nikolay N. Ponomarenko, Alexander A. Zelensky, Andriy A. Kurekin, and Ken Lever. Compression and classification of noisy multichannel remote sensing images. In Lorenzo Bruzzone, Claudia Notarnicola, and Francesco Posa, editors, *Image and Signal Processing for Remote Sensing XIV*, volume 7109, page 71090W. International Society for Optics and Photonics, SPIE, 2008. doi: 10.1117/12.799497. URL https://doi.org/10.1117/12.799497.

[49] Nikolay Ponomarenko, Vladimir Lukin, Mikhail Zriakhov, Karen Egiazarian, and Jaakko Astola. Lossy Compression of Images with Additive Noise. In *Lecture Notes in Computer Science (including subseries Lecture Notes in Artificial Intelligence and Lecture Notes in Bioinformatics)*, volume 3708 LNCS, pages 381–386, 2005. ISBN 354029032X. doi: 10.1007/11558484_48. URL http://link.springer.com/10.1007/11558484_48.

[50] Nikolay Ponomarenko, Mikhail Zriakhov, Vladimir V. Lukin, Jaakko T. Astola, and Karen O. Egiazarian. Estimation of accessible quality in noisy image compression. *European Signal Processing Conference*, 4, 2006. ISSN 22195491.

[51] Sergey Krivenko, Olena Kolganova, and Vladimir Lukin. Lossy compression of images corrupted by spatially correlated noise. In *2016 13th International Conference on Modern Problems of Radio Engineering, Telecommunications and Computer Science (TCSET)*, pages 698–702. IEEE, feb 2016. ISBN 978-6-1760-7807-4. doi: 10.1109/TCSET.2016.7452157. URL http://ieeexplore.ieee.org/document/7452157/.

[52] V.V. Lukin, S.S. Krivenko, M.S. Zriakhov, N.N. Ponomarenko, S.K. Abramov, A. Kaarna, and K. Egiazarian. Lossy compression of images corrupted by mixed Poisson and additive Gaussian noise. In *2009 International Workshop on Local and Non-Local Approximation in Image Processing*, pages 33–40. IEEE, aug 2009. ISBN 9781424451678.

doi: 10.1109/LNLA.2009.5278407. URL http://ieeexplore.ieee.org/document/5278407/.

[53] Alexander N. Zemliachenko, Ruslan A. Kozhemiakin, Mikhail L. Uss, Sergey K. Abramov, Nikolay N. Ponomarenko, Vladimir V. Lukin, Benoît Vozel, and Kacem Chehdi. Lossy compression of hyperspectral images based on noise parameters estimation and variance stabilizing transform. *Journal of Applied Remote Sensing*, 8(1):083571, aug 2014. ISSN 1931-3195. doi: 10.1117/1.JRS.8.083571. URL http://remotesensing.spiedigitallibrary.org/article.aspx?doi=10.1117/1.JRS.8.083571.

[54] Vladimir Lukin, Alaxender Zemliachenko, Sergey Abramov, Benoit Vozel, and Kacem Chehdi. Automatic lossy compression of noisy images by spiht or jpeg2000 in optimal operation point neighborhood. In *2016 6th European Workshop on Visual Information Processing (EUVIP)*, pages 1–6, IEEE, oct 2016. IEEE. ISBN 978-1-5090-2781-1. doi: 10.1109/EUVIP.2016.7764581. URL http://ieeexplore.ieee.org/document/7764581/.

[55] V Naumenko, V Lukin, and S Krivenko. Lossy compression of single-channel images corrupted by additive white noise with performance prediction. *Integrated Computer Technologies in Mechanical Engineering*, 2021:911–923.

[56] Bogdan Kovalenko, Vladimir Lukin, Victoria Naumenko, and Sergey Krivenko. Analysis of noisy image lossy compression by bpg using visual quality metrics. In *2021 IEEE 3rd International Conference on Advanced Trends in Information Theory (ATIT)*, pages 20–25, Dec 2021. doi: 10.1109/ATIT54053.2021.9678575.

[57] Bogdan Kovalenko, Vladimir Lukin, Sergii Kryvenko, Victoriya Naumenko, and Benoit Vozel. BPG-Based Automatic Lossy Compression of Noisy Images with the Prediction of an Optimal Operation Existence and Its Parameters. *Applied Sciences*, 12(15):7555, jul 2022. ISSN 2076-3417. doi: 10.3390/app12157555. URL https://www.mdpi.com/2076-3417/12/15/7555.

[58] Alexander N. Zemliachenko, Sergey K. Abramov, Vladimir V. Lukin, Benoit Vozel, and Kacem Chehdi. Lossy compression of noisy remote sensing images with prediction of optimal operation point existence and parameters. *Journal of Applied Remote Sensing*, 9(1):095066, 2015. doi: 10.1117/1.JRS.9.095066. URL https://doi.org/10.1117/1.JRS.9.095066.

[59] Nikolay Ponomarenko, Vladimir Lukin, Karen Egiazarian, and Jaakko Astola. DCT Based High Quality Image Compression. In *Lecture Notes in Computer Science*, volume 3540, pages 1177–1185, 2005. doi: 10.1007/11499145_119. URL http://link.springer.com/10.1007/11499145_119.

[60] Nikolay N. Ponomarenko, Karen O. Egiazarian, Vladimir V. Lukin, and Jaakko T. Astola. High-quality dct-based image compression using partition schemes. *IEEE Signal Processing Letters*, 14(2):105–108, Feb 2007. ISSN 1558-2361. doi: 10.1109/LSP.2006.879861.

[61] Bogdan Kovalenko, Volodymyr Lukin, Sergii Kryvenko, Victoriia Naumenko, and Benoit Vozel. Prediction of Parameters in Optimal Operation Point for BPG-based Lossy Compression of Noisy Images. *Ukrainian Journal of Remote Sensing*, 9(2):4–12, 2022. doi: 10/36023/ujrs.2022.9.2.212. URL https://https://ujrs.org.ua/ujrs/article/view/212/230.

[62] Vladimir Lukin, Sergey Krivenko, Fangfang Li, Sergey Abramov, and Viktor Makarichev. On Image Complexity in Viewpoint of Image Processing Performance. In *CEUR Workshop Proceedings*, volume 3156, pages 458–473, Ukraine, 2022.

[63] Priyam Chatterjee and Peyman Milanfar. Is denoising dead? *IEEE Transactions on Image Processing*, 19(4):895–911, April 2010. ISSN 1941-0042. doi: 10.1109/TIP.2009.2037087.

[64] Sergey Abramov, Sergey Krivenko, Alexey Roenko, Vladimir Lukin, Igor Djurovic, and Mikhail Chobanu. *Prediction of filtering efficiency for DCT-based image denoising*. IEEE, Budva, Montenegro, jun 2013. ISBN 978-9940-9436-1-5. doi: 10.1109/MECO.2013.6601327. URL http://ieeexplore.ieee.org/document/6601327/.

[65] Miguel Colom, Antoni Buades, and Jean-Michel Morel. Nonparametric noise estimation method for raw images. *Journal of the Optical Society of America A*, 31(4):863, apr 2014. ISSN 1084-7529. doi: 10.1364/JOSAA.31.000863. URL https://opg.optica.org/abstract.cfm?URI=josaa-31-4-863.

[66] V. V. Lukin, S. K. Abramov, N. N. Ponomarenko, Benoit Vozel, and Kacem Chehdi. Methods for Blind Evaluation of Noise Variance in Multichannel Optical and Radar Images. *Telecommunications and Radio Engineering*, 65(6):527–556, 2006. ISSN 0040-2508. doi: 10.1615/TelecomRadEng.v65.i6.40. URL

http://www.dl.begellhouse.com/journals/0632a9d54950b268,00c8592e5cb80774,7b26a1d24e7a8379.html.

[67] B. Vozel, K. Chehdi, L. Klaine, V.V. Lukin, and S.K. Abramov. Noise Identification and Estimation of its Statistical Parameters by Using Unsupervised Variational Classification. In *2006 IEEE International Conference on Acoustics Speed and Signal Processing Proceedings*, volume 2, pages II–841–II–844. IEEE, 2006. ISBN 1-4244-0469-X. doi: 10.1109/ICASSP.2006.1660474. URL http://ieeexplore.ieee.org/document/1660474/.

[68] Stanislav Pyatykh, Lei Zheng, and Jürgen Hesser. Fast noise variance estimation by principal component analysis. In Karen O. Egiazarian, Sos S. Agaian, and Atanas P. Gotchev, editors, *Image Processing: Algorithms and Systems XI*, volume 8655, page 86550K, feb 2013. ISBN 9780819494283. doi: 10.1117/12.2000276. URL http://proceedings.spiedigitallibrary.org/proceeding.aspx?doi=10.1117/12.2000276.

[69] Oleg Ieremeiev, Vladimir Lukin, Krzysztof Okarma, and Karen Egiazarian. Full-reference quality metric based on neural network to assess the visual quality of remote sensing images. *Remote Sensing*, 12 (15), 2020. ISSN 2072-4292. doi: 10.3390/rs12152349. URL https://www.mdpi.com/2072-4292/12/15/2349.

[70] Sebastian Bosse, Dominique Maniry, Klaus-Robert Muller, Thomas Wiegand, and Wojciech Samek. Neural network-based full-reference image quality assessment. In *2016 Picture Coding Symposium (PCS)*, pages 1–5. IEEE, 2016. ISBN 978-1-5090-5966-9. doi: 10.1109/PCS.2016.7906376. URL https://ieeexplore.ieee.org/document/7906376/.

[71] Oleg Ieremeiev, Vladimir Lukin, Krzysztof Okarma, Karen Egiazarian, and Benoit Vozel. On properties of visual quality metrics in remote sensing applications. In *Electronic Imaging*, volume 34, pages 354–1–354–6, San Francisco, USA, jan 2022. doi: 10.2352/EI.2022.34.10.IPAS-354. URL https://library.imaging.org/ei/articles/34/10/IPAS-354.

[72] Hossein Ziaei Nafchi, Atena Shahkolaei, Rachid Hedjam, and Mohamed Cheriet. Mean Deviation Similarity Index: Efficient and Reliable Full-Reference Image Quality Evaluator. *IEEE Access*, 4:5579–5590, 2016. ISSN 2169-3536. doi: 10.1109/ACCESS.2016.2604042. URL http://ieeexplore.ieee.org/document/7556976/.

[73] Fangfang Li, Vladimir Lukin, Oleg Ieremeiev, and Krzysztof Okarma. Quality Control for the BPG Lossy Compression of Three-Channel Remote Sensing Images. *Remote Sensing*, 14(8):1824, apr 2022. ISSN 2072-4292. doi: 10.3390/rs14081824. URL https://www.mdpi.com/2072-4292/14/8/1824.

[74] F. Li and V Lukin. Providing a Desired Compression Ratio for Better Portable Graphics Encoder of Color Images: Design and Analysis. In *Proceedings of GeoRS 2022 Conference*, pages 633–640, China.

[75] Bogdan Kovalenko and Vladimir Lukin. Using visual metrics to analyze lossy compression of noisy images. *Aerospace Technic and Technology*, 0:83–91, 2021. doi: 10.32620/aktt.2021.6.09. URL http://nti.khai.edu/ojs/index.php/aktt/index.

[76] Oleksiy Pogrebnyak and Vladimir V. Lukin. Wiener discrete cosine transform-based image filtering. *Journal of Electronic Imaging*, 21 (4):043020, dec 2012. ISSN 1017-9909. doi: 10.1117/1.JEI.21.4.043020. URL http://electronicimaging.spiedigitallibrary.org/article.aspx?doi=10.1117/1.JEI.21.4.043020.

[77] Oleksii S. Rubel and Volodymyr V. Lukin. An Improved Prediction of DCT-based Image Filters Efficiency Using Regression Analysis. *Information and Telecommunication Sciences*, 0(1):30–41, oct 2014. ISSN 2411-2976. doi: 10.20535/2411-2976.12014.30-41. URL http://infotelesc.kpi.ua/article/view/28866.

[78] A. Colin Cameron and Frank A.G. Windmeijer. An R-squared measure of goodness of fit for some common nonlinear regression models. *Journal of Econometrics*, 77(2):329–342, apr 1997. ISSN 03044076. doi: 10.1016/S0304-4076(96)01818-0. URL https://linkinghub.elsevier.com/retrieve/pii/S0304407696018180.

[79] Peter J. Rousseeuw. Least Median of Squares Regression. *Journal of the American Statistical Association*, 79(388):871–880, dec 1984. ISSN 0162-1459. doi: 10.1080/01621459.1984.10477105. URL http://www.tandfonline.com/doi/abs/10.1080/01621459.1984.10477105.

[80] Oleksii Rubel, Vladimir Lukin, Sergey Abramov, Benoit Vozel, Oleksiy Pogrebnyak, and Karen Egiazarian. Is Texture Denoising Efficiency Predictable? *International Journal of Pattern Recognition and Artificial Intelligence*, 32(01):1860005, jan 2018. ISSN 0218-0014. doi: 10.1142/S0218001418600054. URL https://www.worldscientific.com/doi/abs/10.1142/S0218001418600054.

[81] Vladimir Lukin, Irina Vasilyeva, Sergey Krivenko, Fangfang Li, Sergey Abramov, Oleksii Rubel, Benoit Vozel, Kacem Chehdi, and Karen Egiazarian. Lossy Compression of Multichannel Remote Sensing Images with Quality Control. *Remote Sensing*, 12(22):3840, nov 2020. ISSN 2072-4292. doi: 10.3390/rs12223840. URL https://www.mdpi.com/2072-4292/12/22/3840.

[82] Isabel Rodrigues, Joao Sanches, and Jose Bioucas-Dias. Denoising of medical images corrupted by Poisson noise. In *2008 15th IEEE International Conference on Image Processing*, pages 1756–1759. IEEE, 2008. ISBN 978-1-4244-1765-0. doi: 10.1109/ICIP.2008.4712115. URL http://ieeexplore.ieee.org/document/4712115/.

[83] Benoit Vozel, Alexander N. Zemliachenko, Sergey K. Abramov, Vladimir V. Lukin, and Kacem Chehdi. Preliminary filtering and lossy compression of noisy remote sensing images. *Image and Signal Processing for Remote Sensing XXIV*, 10789(11):32, oct 2018. ISSN 1996756X. doi: 10.1117/12.2325102. URL https://www.spiedigitallibrary.org/conference-proceedings-of-spie/10789/2325102/Preliminary-filtering-and-lossy-compression-of-noisy-remote-sensing-sensingimages/10.1117/12.2325102.full.

5. Recommendation and Design of Personalized Garments based on Intelligent Human-Product Interaction

Zhujun Wang,[1,*] Xuyuan Tao,[2] Xianyi Zeng,[2] Yingmei Xing,[1] Zhenzhen Xu[1] and Pascal Bruniaux[2]

1 Introduction

Due to the rapid development and wide implementation of advanced technolo-gies, we are undergoing a revolution towards smart manufacturing, named Industry 4.0. A quick response to consumers' fast-shifting and individualized needs, is regarded as one of the core competitiveness for the success of en-terprises in many industry sectors[1]. In this context, as a hybrid production mode of made-to-measure and mass production, leveraging the personalized services and high productivity of the two modes, mass customization has been increasingly applied in the apparel industry recently[2].

Moreover, with the rapid development of innovative digital technologies such as AI,BD,VR and CC, e-MC has become a critical branding strategy of many fashion companies (garments, shoes, furniture, accessories, among oth-ers) for optimizing their manufacturing processes facing fierce international competition[3]. In practice, interactions between fashion products and humans (consumers and designers) usually play a pivotal role in implementing e-MC. However, the factors affecting these interactions are sophisticated[4]. Mean-while, fashion design is experience and knowledge-dependent work. A novice fashion designer will find it more difficult to deal with the implicit relation-ships among these factors efficiently and accurately, because they are full of uncertainty and nonlinearity. Therefore, advanced intelligent computational

[1] School of Textile and Garment, Anhui Polytechnic University, Wuhu 241000, China.
[2] GEMTEX Laboratory, Ecole Nationale Superieure des Arts et Industries Textiles, 59056 Roubaix, France.
* Corresponding author: ahpuwzj@ahpu.edu.cn

tools have been regarded as key to enhancing these interactions, and facilitating the fashion companies to have a good understanding of the relations between consumers' uncertain and vague requirements and product parameters. In this context, we propose a knowledge-based approach for personalized garment design and recommendation.

2 Research methodology

2.1 Literature review

Currently, many relevant computational tools, including recommendation systems[5;6], knowledge bases[7], decision-support systems[8], and 3D virtual try-on technologies[9] have been developed for solving various fashion design problems. For instance, to facilitate the general consumers in creating the preferred fashion design sketches, P.Y. Mok et al. developed an IGA-based knowledge model. They then presented a customized fashion sketch design support system[10]. Arzu et al. developed an intelligent system for fashion style selection for non-standard female figure types, combining a genetic algorithm and particle swarm optimization with a neural network classifier[11]. Zhou et al. proposed a recommendation model based on a hybrid method combining Kansei engineering with a traditional filtering algorithm to realize the personalized apparel recommendation services following users' emotional needs[12]. An open case-based color recommendation system was put forward by Hong et al., enabling the helping of designers and consumers in obtaining the most appropriate color range of consumer products[13]. Takatera et al. created a fabric retrieval system for apparel e-commerce for designers based on Kansei technology[14]. For realizing jeans recommendation for a specific consumer, Zhang et al. created a jeans knowledge base and then set up a recommendation system using sensory evaluation and an intelligent data fusion method[15]. Guan et al. explored an advanced apparel style learning and recommendation system that can recognize the deep design-associated features of clothes and learn the connotative meanings conveyed by these features as they relate to style and the body[16]. Dong et al. put forth a new interactive designer-oriented, knowledge-based recommender system, permitting the generation and optimization of a fashion design scheme for a specific consumer accurately and quickly[5]. These research results provide good tools for advanced computational decision support to the fashion industry. However, for dealing with the particular challenges in garment design in terms of garment fit accuracy and efficiency, as well as garment appearances[17], the techniques mentioned above still showed some drawbacks, which are summarized as follows:

(1)The techniques mentioned above are mainly oriented to consumers and are concentrated on fashion aesthetics, such as styles and colors. However, the garment recommendation systems integrating design elements for simultaneously dealing with fashion aesthetics and garment fit are still limited. Both garment fit and fashion style are the most crucial elements constituting the whole textile/apparel value chain[5].

(2)The previous works are limited to some specific garment styles, such as shirts and jeans. They lack generalized recommendation solutions for supporting designers' work.

In this context, we propose a knowledge-based approach (see Figure 1) for garment design recommendation, with a self-adjustment mechanism for predicting and controlling the style design and garment structural parameters from the consumers' personalized body measurements and requirements (consumer profile). Compared with the current systems, our approach will enable the recommendation and design of new garments in terms of accuracy, fit, and efficiency, fully taking colors, fabrics, styles, and patterns into account.

As one of the current computational tools for dealing with the uncertainties and ambiguities related to human factors, FL have been widely applied to fashion design. The concerned work includes design knowledge-based creation[5;7;15], fashion recommendation[13], human body classification[18], pattern design[8], product evaluation[19;20], garment size selection[21], and so forth. Meanwhile, GA has been regarded as a powerful tool for dealing with complex optimization problems due to their fast convergence and simple encoding, including human body clustering[22], clothing pressure prediction[23], thermal and moisture comfort estimation[24], production planning optimization[2], and design process optimization[10;25], and so on. Moreover, due to its easy implementation, simple network topology, and high generalized capability, SVR has gained much success in the textile and fashion industry, especially in the predictions of textile dyeing process parameters[26], yarn characteristics[27], and fabric qualities[28]. Therefore, the FL, GA, and SVR were employed to support the proposed approach in this study.

2.2 General research scheme

The proposed design recommendation system (see Figure 1) is mainly supported by a formalized, professional garment design knowledge base and a garment adaptation model.

The knowledge base in the system (see Figure 2) is composed of a series of basic databases (style, color, fabric, and pattern) and relational models on style, color, fabric, and garment fit. The major function of these relational models is to quantitatively characterize the relations between the formalized consumer profile provided by the user and the formalized garment profile generated from combinations of the basic databases. According to the results of the relational models, if no existing garment meets the consumer profile, a self-adjustment mechanism for design parameters will be activated by using an automatic garment adaptation model (see Figure 1) until the consumer's or designer's satisfaction is reached. A satisfactory design solution will be

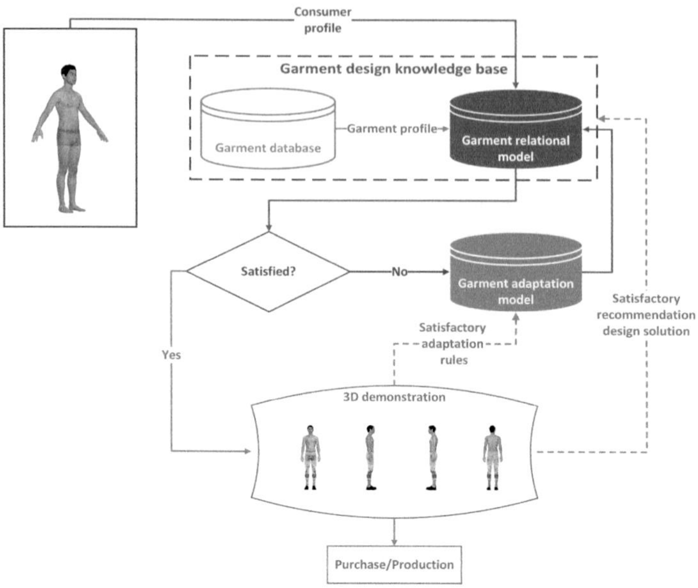

Figure 1: The proposed interactive design and recommendation approach for a personalized garment

recommended to the consumer by a 3D virtual demonstration. Meanwhile, the design knowledge base and adaptation model will be continuously updated by successful recommendation solutions recommended by the system. Finally, the performance of the proposed system will be improved progressively by generating personalized design solutions and learning the design knowledge for different business cases.

The significant contributions of our work are summarized as follows:

(1) Providing a feasible and efficient solution to realize accurate and efficient garment design recommendation and optimization towards garment e-MC;

(2) Supplying a friendly human–machine an interactive framework for representing, recommending, controlling, and optimizing fashion design dynamically and precisely;

(3) Providing efficient support to general consumers who have no professional design knowledge for designing their own products according to their personalized preferences;

(4) Dramatically promoting design quality by constructing an effective communication channel between the consumers and fashion designers;

(5) Effectively decreasing high recall rates and risks for fashion companies by offering more consumer-oriented fashion design solutions;

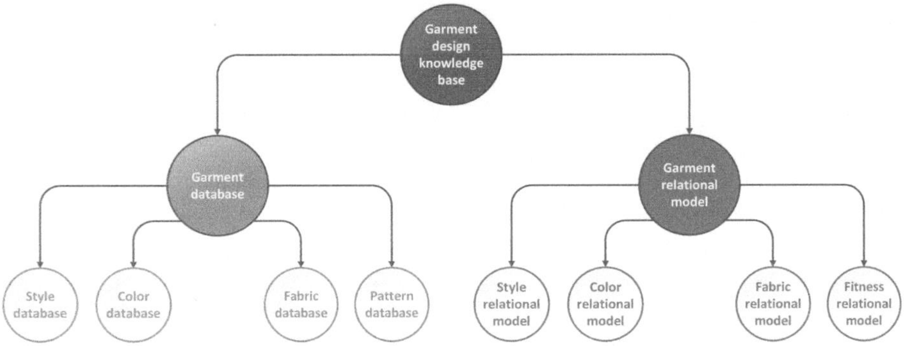

Figure 2: The structure of the proposed garment design knowledge base

(6) Facilitating the creation of new collaborative garment design processes by integrating the proposed recommendation system into a commercial 3D clothing design software, which can effectively upgrade the level of e-MC;

(7) Easily spreading the general principles of the proposed approach to other human-centered product design. A generalized human-centered product design platform can be constructed by modelling complex relations between human aesthetic preferences, body dimensions, and product parameters to provide the relevant decision support to designers.

The general principles of the proposed approach can be applied to various garment styles (e.g., pants, skirts, coats, shirt, besides other garments). However, for simplicity, we make it explicit using the specific case of pant designs. The involved pant samples, along with the corresponding patterns, were supported by a cooperative enterprise of our projects. Although the design solutions in this paper were demonstrated and validated using a commercial 3D garment CAD software named Clo 3D, the proposed approach and system can be generalized in other commercial garment CAD software and platforms.

2.3 General formalization

Let $CP = [SR \quad CR \quad MR \quad FR]$ be a vector for profiling the consumer's requirements, where *SR*, *CR*, *MR*, and *FR* refer to the requirements of style, color, fabric, and garment fit, respectively.

Let $GP = \{GP^1, \quad \ldots, \quad GP^k, \quad \ldots, \quad GP^n\}$ be a set of n garment profiles in the database.

Let $GP^k = [SC^k \quad CC^k \quad MC^k \quad FC^k]$ be the characteristics for garment profile GP^k in the database, where SC^k, CC^k, MC^k, and FC^k represent the characteristics of style, color, fabric, and garment fit, respectively. The elements of GP^k have a one-to-one correspondence with those of CP.

A complete garment style is composed of m categories of style design elements, such as the garment silhouette, length, pockets, and ornaments. In addition, each of the i-th category style elements can be further divided into u_i values. For example, the garment silhouette can be classified into the following types: H, X, T, O, and S according to the outline shapes. All the values of the style design element amount to v in total. Thus, the related symbols are defined as follows:

Let $SR = \{SR_1, \quad \ldots, \quad SR_i, \quad \ldots, \quad SR_m\}$ be a set of formalized vectors representing the requirements for m categories of style design elements, where $SR_i = [sr_{i_1} \quad \ldots \quad sr_{i_j} \quad \ldots \quad sr_{i_u}]$ is an i_u-dimensional normalized or one-hot vector expressing the requirements for the i-th category style design element. The sr_{i_j} is denoted as the nearness degree of the style requirement to the i_j-th style element.

We take the pants design, for example. The requirements for the pant silhouettes can be defined as a 5-dimensional one-hot vector, expressed by $SR_{silhouette} = [sr_H \quad sr_A \quad sr_X \quad sr_O \quad sr_T]$. If $SR_{silhouette} = [1 \quad 0 \quad 0 \quad 0 \quad 0]$, it means the silhouette of the required pants is the H type. Meanwhile, the waistline position requirements can be defined as a 3-dimensional normalized vector, expressed by $SR_{waist-line} = [sr_{lw} \quad sr_{nw} \quad sr_{hw}]$. The lw, nw, and hw refer to the lower, normal, and high waistline, respectively. If $SR_{waist-line} = [0.3 \quad 0.7 \quad 0]$, which can be defined by the consumer using a graphic interface, it indicates that the required pant waistlines are close to the normal waistline by 70% and the lower waistline by 30%.

Let $SC^k = \{SC_1^k, \quad \ldots, \quad SC_i^k, \quad \ldots, \quad SC_m^k\}$ be a set of formalized vectors representing the style characteristics of the garment G^k from the aspects of the m categories of style design elements, represented by the normalized vectors and the one-hot vectors; $SC_1^k = [sc_{i_1} \quad \ldots \quad sc_{i_j} \quad \ldots \quad sc_{i_u}]$ is an i_u-dimensional vector expressing the concrete style characteristics of a specific garment in the i-th category style design element. The structure of SC_m has a one-to-one correspondence with those of SR_m. Furthermore, the value of sc_{i_j} is defined by using the same method as that of sr_{i_j} mentioned above.

Considering the generalizability of the proposed approach and system, we introduced the RGB color space to describe the color characteristics as it can be intelligible for both consumers and fashion designers.

Let $CR = [cr_{red} \quad cr_{green} \quad cr_{blue}]$ be a vector expressing the color requirements for the target garment. The value in CR is from 0 to 255. For example, if $CR = [125 \quad 125 \quad 125]$, it means that the values corresponding to the characteristics of red, green, and blue for a specific color required by the consumer are 125, 125, and 125, respectively.

Let $CC^k = [cc_{red} \quad cr_{green} \quad cr_{blue}]$ be a vector expressing the color characteristics of a specific garment. The value in CC^k is from 0 to 255.

Let $MR = [sm^r \quad sf^r \quad wm^r]$ be a normalized vector expressing the fabric property requirements for the target garment, where sm, sf, and wm represent, respectively, the smoothness, softness, and warmth. The element value in MR is between 0 and 1.

Let $MC^k = [sm^c \quad sf^c \quad wm^c]$ be a vector expressing the fabric properties of a specific garment.

We suppose that there are n feature positions in a specific garment. The garment fit can be divided into five levels at each feature position, including too tight (tt), tight (ti), neutral (nl), loose (lo), and too loose (tl).

Let $FR = \{FR_1, \quad \ldots, \quad FR_i, \quad \ldots, \quad FR_n\}$ be a set of normalized vectors representing the requirements for various garment fit levels at n feature positions, where $FR_i = [tt_i^r \quad ti_i^r \quad nl_i^r \quad lo_i^r \quad tl_i^r]$ is a 5-dimensional normalized vector representing the requirements for five garment fit levels at the i feature position.

We take the pants, for example. The requirements for the fit level for the waist girth (wg) can be denoted as a 5-dimensional normalized vector $FR_{wg} = [tt_{wg}^r \quad ti_{wg}^r \quad nl_{wg}^r \quad lo_{wg}^r \quad tl_{wg}^r]$. For instance, $FR_{wg} = [0 \quad 0 \quad 0.8 \quad 0.2 \quad 0]$ represents the fact that the required fit level for the waist girth is close to the neutral level by 80% and the loose level by 20%.

Let $FC^k = \{FC_1^k, \quad \ldots, \quad FC_i^k, \quad \ldots, \quad FC_n^k\}$ be a set of normalized vectors representing the garment fit levels of a specific garment. The elements of FC^k are in a one-to-one correspondence with those of FR. The definition of FC_i^k is the same as that of FR_i.

Let $BD = \{BD_1, \quad \ldots, \quad BD_i, \quad \ldots, \quad BD_p\}$ be a set of human body dimension vectors for the p feature positions.

Let $GD^k = \{GD_1^k, \quad \ldots, \quad GD_i^k, \quad \ldots, \quad GD_p^k\}$ be a set of garment dimension vectors corresponding to the p feature positions for garment k.

Let $GE^k = \{GE_1^k, \quad \ldots, \quad GE_i^k, \quad \ldots, \quad GE_p^k\}$ be a set of garment ease allowance vectors to the p feature positions for garment k, where $GE_i^k = GD_i^k - BD_i$.

For one feature position, let $GE_i^k = [ge_1 \quad ge_2 \quad \ldots \quad ge_{i_u}]$ be an i_u-dimensional garment ease vector for i_u types of style characteristics. Take the waistline position for example. The garment ease for garment k in the waistline position can be defined by $GE_{waist-line}^k = [ge_{lw} \quad ge_{nw} \quad ge_{hw}]$.

For the garment fit levels, let $GF_i^k = [gf_{tt} \quad gf_{ti} \quad gf_{nl} \quad gf_{lo} \quad gf_{tl}]$ be a 5-dimensional garment ease vector corresponding to 5 types of garment fit levels at i feature positions for garment k.

3 General Scheme of the Design Knowledge Base Construction

3.1 Construction of the Garment Relational Model

In the garment relational model, we quantitatively define the similarity degree $r(CP, GP^k)$ between the consumer profile CP and the garment profile GP^k. During the recommendation, the most relevant garment profile is selected from the garment database and then adapted to meet the consumer's personalized requirements.

Formally, the $r(CP, GP^k)$ is defined as a linear combination of the similarity degrees of style $r(SR, SC^k)$, color $r(CR, CC^k)$, fabric $r(MR, MC^k)$, and garment fit $r(FR, FC^k)$ between the garment profile and the consumer profile (See Equation (1)), i.e.,

$$\begin{aligned} r\left(CP, GP^k\right) = \alpha_1 \times r\left(SR, SC^k\right) + \alpha_2 \times r\left(CR, CC^k\right) \\ + \alpha_3 \times r\left(MR, MC^k\right) + \alpha_4 \times r\left(FR, FC^k\right) \end{aligned} \tag{1}$$

where α_1, α_2, α_3, and α_4 refer to the weights of style, color, fabric, and garment fit, respectively. These weights, varying between 0 and 1, are selected by the user (designer or consumer) according to the importance of each component in its specific scenario.

Let ρ be a predefined threshold for evaluating the similarity degree $r(CP, GP^k)$ between the consumer profile CP and the garment profile GP^k. Thus, we define the recommendation rules as follows:

(1) Calculate a list of $r(CP, GP^k)$ for all the garment profiles in G and sort them in descending order;

(2) Recommend the garment profiles with $r\left(CP, GP^k\right) \geq \rho$ to the consumer or designer;

(3) If $r\left(CP, GP^k\right)$ of all the garment profiles GP is less than ρ, it means that all the existing garments cannot meet the requirements. Next, the self-adaptation mechanism will be activated based on the garment adaptation model.

3.1.1 Construction of the Style Relational Model

The main function of the style relational model was to evaluate the similarity degreer $r\left(SR, SC^k\right)$ between the style requirements SR of the consumer and the style characteristics SC^k of a garment profile GP^k. The $r\left(SR, SC^k\right)$ is defined by Equation (2):

$$r\left(SR, SC^k\right) = \frac{\sum_j \sum_i min\left(sr_i, sc_i\right)}{\sum_j \sum_i max\left(sr_i, sc_i\right)}, \left(sr_i \in SR_j \in SR, sc_i \in SC_j \in SC^k\right) \tag{2}$$

From Equation (2), we can find that values of $r\left(SR, SC^k\right)$ vary between 0 and 1. If all style elements of SR and SC^k are close to each other, the value of $r\left(SR, SC^k\right)$ is close to 1. Otherwise, it tends to 0.

3.1.2 Construction of the Color Relational Model

The color relational model is mainly built to represent the similarity degree $r\left(CR, CC^k\right)$ between the color requirements CR of the consumer and the color characteristics CC^k. Based on the Euclidean distance, to keep the consistency with the function monotonicity of Equation (2), the $r\left(CR, CC^k\right)$ is defined by Equation (3).

$$CR, CC^k = 1 - \frac{\sqrt{\left((cr_{red} - cc_{red})^2 + (cr_{green} - cc_{green})^2 + (cr_{blue} - cc_{blue})^2\right)/3}}{255} \tag{3}$$

Equation (3) shows that the larger the value is, the nearer the color requirements CR and the color characteristics CC^k could be.

3.1.3 Construction of the Fabric Relational Model

The fabric relational model was created to recommend the most relevant fabric in the database. The relationship $r\left(MR, MC^k\right)$ between the fabric requirements MR of the consumer and the fabric characteristics MC^k in the database is computed by Equation (4) using the same modeling method as Equation (2).

$$r\left(MR, MC^k\right) = \frac{min\left(sm^r, sm^c\right) + min\left(sf^r, sf^c\right) + min\left(wm^r, wm^c\right)}{max\left(sm^r, sm^c\right) + max\left(sf^r, sf^c\right) + max\left(wm^r, wm^c\right)} \tag{4}$$

According to Equation (4), the values of $r\left(MR, MC^k\right)$ are between 0 and 1. The larger value would indicate that the fabric characteristics MC^k are closer to the fabric requirements MR.

3.1.4 Construction of the Garment Fit Relational Model

The main objective of the garment fit relational model was established to express the relationship $r\left(FR, FC^k\right)$ between the garment fit requirements FR of the consumer and the characteristics of the garment fit in the system FC^k.

According to the same modeling method of $r\left(SR, SC^k\right)$ and $r\left(MR, MC^k\right)$, for each of the j feature positions of the garment, the local garment fit relational model $r(FR_i, FC_i)$ is denoted by Equation (5):

$$r\left(FR_i, FC_i^k\right) = \frac{min\left(tt_i^r, tt^c\right) + min\left(ti^r, ti^c\right) + min\left(nl^r, nl^c\right) + min\left(lo_i^r, lo_i^c\right) + min\left(tl_i^r, tl_i^c\right)}{max\left(tt_i^r, tt^c\right) + max\left(ti^r, ti^c\right) + max\left(nl^r, nl^c\right) + max\left(lo_i^r, lo_i^c\right) + max\left(tl_i^r, tl_i^c\right)} \quad (5)$$

Then, the global garment fit relational model $r\left(FR_i, FC_i^k\right)$ is calculated as an average value by Equation (6).

$$r(FR, FC^k) = avg(r(FR_i, FC_i^k)), (FR_i \in FR, FC_i^k \in FC^k) \quad (6)$$

Both the values of $r\left(FR_i, FC_i^k\right)$ and $r\left(FR, FC^k\right)$ are between 0 and 1. From Equation (5), if the local fit characteristics FC_i^k are close to the fit requirements FR_i, the value of $r\left(FR_i, FC_i^k\right)$ is close to 1. Otherwise, it tends to 0. Similarly, following Equation (6), if the value of $r\left(FR, FC^k\right)$ is close to 1, it means that the global fit characteristics FC^k are close to the fit requirements FR and vice versa.

3.2 Construction of Garment Design Adaptation Model

3.2.1 General Framework and Process of the Garment Design Adaptation Model

As shown in Figure 3, the complete garment design adaptation model is composed of three sequential linked parts, including the GA-based design solution generation model (sub-model 1), the linear model (sub-model 2) supported by a knowledge-based design database, and the SVR-based adaptation rule model (sub-model 3). The inputs of the adaptation model are the requirements for the style design elements SR expressed by the normalized vectors and the garment fit requirements FR for the feature positions. The outputs of the model are the movements of the controlling points on the garment patterns.

Concretely, the computation process of the adaptation model is composed of three sequential steps (see Figure 3). First, a new and normalized design solution with new style and garment fit levels is generated by the design solution generation model based on GA. Next, the general garment adaptation requirements are obtained by combining the outputs of the design solution generation model with the standard design and fitting solution from the knowledge-based design database. Eventually, the movements of the controlling points on the garment patterns are determined by a SVR-based adaptation rule model, following the adaptation requirements.

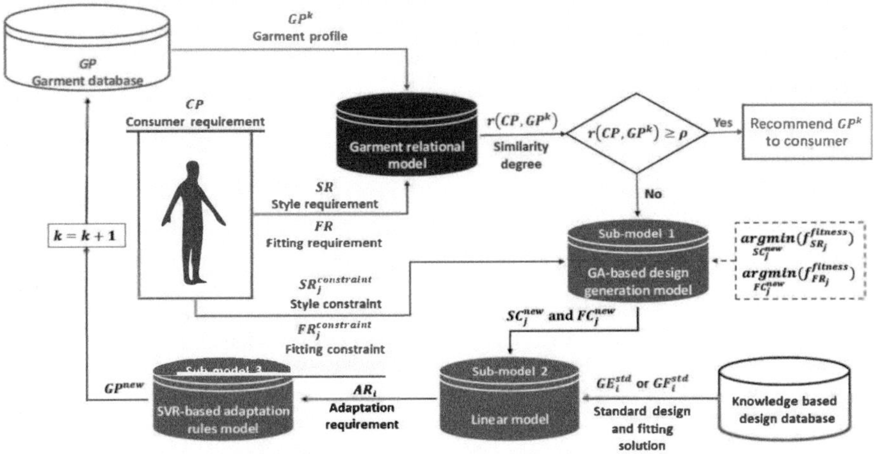

Figure 3: General calculation process of the garment design adaptation model

3.2.2 Construction of the Garment Design Solution Generation Model

As the GA has the advantages of fast convergence and simple encoding, it was employed in this study to rapidly generate multiple new solutions to meet the uncertain and vague requirements of the consumer.

Let the fit value $f_{SR_j}^{fitness}$ for the style design requirement SR_j be denoted as Equation (7), and the fit value $f_{FR_j}^{fitness}$ for the fit requirement FR_j be denoted as Equation (8).

$$f_{SR_j}^{fitness} = \left|SC_j^{new} - SR_j\right|, (SR_j \in SR) \tag{7}$$

where SC_j^{new} represents the new style characteristics.

$$f_{FR_j}^{fitness} = \left|FC_j^{new} - FR_j\right|, (FR_j \in FR) \tag{8}$$

where FC_j^{new} represents the new fit characteristics.

From Equations (7) and (8), a smaller value of the $f_{SR_j}^{fitness}$ or $f_{FR_j}^{fitness}$ indicates a more optimal fit degree.

The implementation process of the GA-based sub-model is described below. The termination criteria of the GA-based sub-model were determined as: (1) the predefined maximum iteration is reached; (2) the average variation in the fit function values with a specific number of generations is less than a pre-designed tolerance; (3) the consumer's requirements are satisfied. For each SR_j or FR_j, the consumer's requirement can be defined as two sets,

$SR_j^{constraint}$ and $FR_j^{constraint}$. If the terminal criteria were not satisfied, the process of "selection-crossover-mutation-calculation-evaluation" would be implemented repeatedly until the terminal criteria were reached. The objective functions of GA are denoted as Equations (9) and (10).

$$\arg min \left(f_{SR_j}^{fitness}\right), \left(SR_j \in SR, SC_j^{new} \in SR_j^{constraint}\right) \tag{9}$$

$$\arg min \left(f_{FR_j}^{fitness}\right), \left(FR_j \in SR, FC_j^{new} \in FR_j^{constraint}\right) \tag{10}$$

3.2.3 Calculation of the New Garment Ease Allowances

The ease allowance is denoted as a metric indicating the extra space difference between the human body and the garment to express the various characteristics of the style design elements and the garment fit levels.

The new garment ease allowance in this study is calculated by Equation (11):

$$GE_i^{new} = \begin{cases} OP_i \cdot GE_i^{std} & OP_i \quad is \quad SC_i^{new} \\ OP_i \cdot GF_i^{std} & OP_i \quad is \quad FC_i^{new} \end{cases} \tag{11}$$

where GE_i^{new} refers to the new garment ease allowance of the new garment design solution at the i feature position; OP_i represents the i-th output of the design solution generation model, which could be SC_i^{new} or FC_i^{new}; GE_i^{std} is denoted as a knowledge-based standard garment ease allowance vector at the i feature position; GF_i^{std} is denoted as a knowledge-based standard garment fitting vector at the i feature position (see Figure 3).

3.2.4 Determination of the General Garment Adaptation Requirements

The general garment adaptation requirements are defined by $AR = \{AR_1, \quad AR_2, \quad \ldots, \quad AR_i\}$, where AR_i is computed following Equation (12):

$$AR_i = GE_i^{new} - GE_i^k \tag{12}$$

where GE_i^k expresses the garment ease allowance of the most relevant existing garment in the proposed database corresponding to the $GE_i^n ew$. For instance, if the $GE_i^n ew$ represented the new ease allowance of the waistline position, then GE_i^k would express the ease allowance of the waistline position for the most relevant existing garment in the database.

3.2.5 Construction of the Garment Design Adaptation Model

Due to its easy implementation, simple network topology, and high generalized capability, for each structural line sl_i, the relations between the length variation of the line (dl_{sl_i}) and the movements of the corresponding points ($dx_{cp_i^{start}}$,$dy_{cp_i^{start}}$, $dx_{cp_i^{end}}$, $dy_{cp_i^{end}}$) are modeled based on SVR, as shown in Figure 4. The Bayesian approach presented by Jonas Mockus is a sequential design strategy for the global optimization of "black-box" functions. In this study, Bayesian optimization was utilized to determine the parameters of the SVR models in our work. Meanwhile, a 10-fold cross-validation approach was introduced to train the SVR models. The complete garment design adaptation model is composed of the SVR models of all the structural lines.

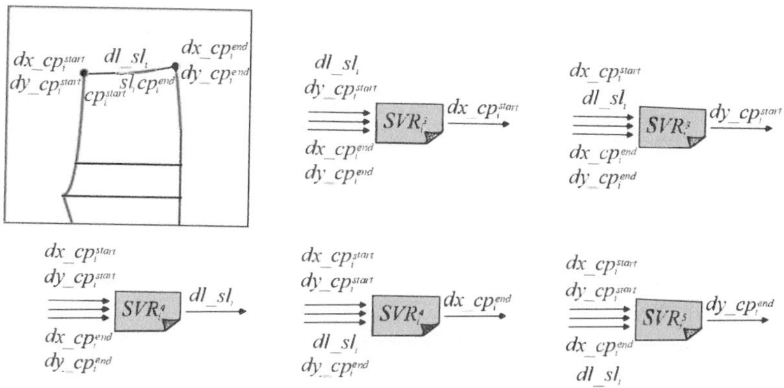

Figure 4: The SVR models for the structural line sl_i

4 Implementation and Application

Due to its simple structure with basic and necessary structure lines (e.g., waistline, outside seam, hemline, inseam, crotch line), the patterns of sports leggings were employed as the representative of the basic pattern of pants in this study. Therefore, in this section, to explicate our present approach with clarity, we implemented the proposed approach to design personalized pants (represented by sports legging) for a specific male consumer. In the sequential section (Section 5), the design of other styles of pants with more complicated structures and detailed design elements will be discussed.

As shown in Figure 5, the style design elements of the pants in the proposed system were classified into ten categories (i.e., silhouette, length, waistline position, and waistband) and 50 types (i.e., H type, A type, O type, T type, X type). The requirements and characteristics corresponding to the length and waistline position were defined by the normalized vectors, while those of the remaining style elements were defined by the one-hot vectors.

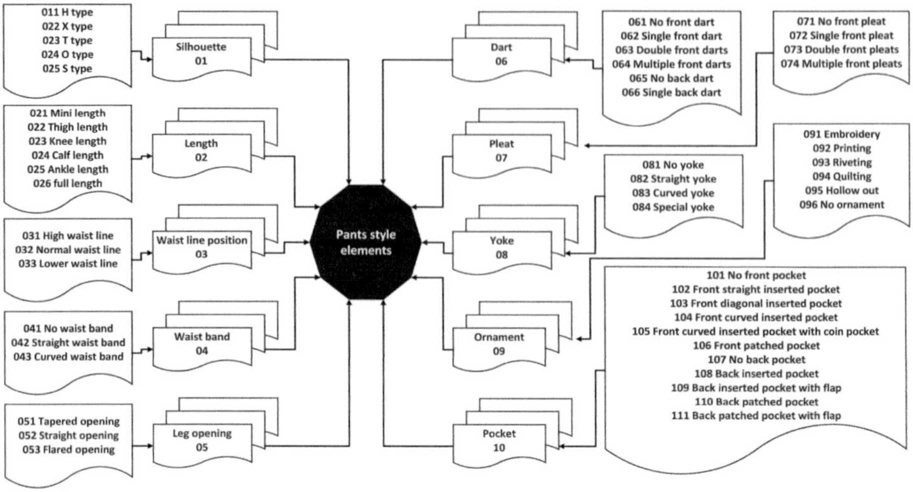

Figure 5: The classification of the pants style elements

4.1 Determination of the Consumer Profile

The consumer profile was constituted by the requirements for the style, color, fabric, and garment fit level. The requirements for the pant style elements and garment fit levels were shown in Table 1; Table 2. The color requirements were expressed by $CR = [77 \quad 73 \quad 72]$, meaning that the required value of red, green, and blue were 77, 73, and 72, respectively. The fabric requirements were represented by $MR = [0.4 \quad 0.45 \quad 0.4]$, indicating that the required values of fabric smoothness, softness, and warmth were 0.4, 0.45, and 0.4, respectively.

The requirements for pant style elements SR are as follows:

(1) Silhouette: $SR_1 = [0 \quad 0 \quad 0 \quad 0 \quad 1]$, indicating that the required pant silhouette is the skinny type;

(2) Length: $SR_2 = [0 \quad 0 \quad 0 \quad 0 \quad 0.2 \quad 0.8]$, indicating that the required pant's length is close to ankle length by 80% and full length by 20%;

(3) Waistline position: $SR_3 = [0.25 \quad 0.75 \quad 0]$, indicating that the required waistline position is close to a normal waistline by 75% and lower waistline by 25%;

(4) Waistband: $SR_4 = [0 \quad 1 \quad 0]$, indicating that the required waistband is straight type;

(5) Leg opening: $SR_5 = [1 \quad 0 \quad 0]$, indicating that the required leg opening is tapered;

(6) Dart: $SR_6 = [1 \quad 0 \quad 0 \quad 0 \quad 1 \quad 0]$, indicating that the required pant's style is without a front and back dart;

(7) Pleat: $SR_7 = [1 \quad 0 \quad 0 \quad 0]$, indicating that the required pant's style is without a front pleat;

(8) Yoke: $SR_8 = [1 \quad 0 \quad 0 \quad 0]$, indicating that the required pant's style is without a yoke;

(9) Ornament: $SR_9 = [0 \quad 0 \quad 0 \quad 0 \quad 0 \quad 1]$, indicating that the required pant's style is without an ornament;

(10) Pocket: $SR_{10} = [1 \quad 0 \quad 0 \quad 0 \quad 0 \quad 0 \quad 1 \quad 0 \quad 0 \quad 0 \quad 0]$, indicating that the required pant's style is without front and back pockets;

The requirements of garment fit levels at feature positions FR are as follows:

(1) Waist girth: $FR_1 = [0 \quad 1 \quad 0 \quad 0 \quad 0]$, indicating that the required garment fit level for waist girth is tight;

(2) Hip girth: $FR_2 = [0 \quad 1 \quad 0 \quad 0 \quad 0]$, indicating that the required garment fit level for hip girth is tight;

(3) Knee girth: $FR_3 = [0 \quad 1 \quad 0 \quad 0 \quad 0]$, indicating that the required garment fit level for knee girth is tight;

(4) Ankle girth: $FR_4 = [0 \quad 1 \quad 0 \quad 0 \quad 0]$, indicating that the required garment fit level for ankle girth is tight;

4.2 Determination of the Garment Profile

The garment profile was constituted by the characteristics of the style, color, fabric, and garment fit level. It is assumed that there are five kinds of legging patterns, five kinds of fabrics, and five kinds of colors involved in this study.

These were the two approaches for determining the style characteristics. For the style elements, whose requirements and characteristics were defined by one-hot vectors, including the silhouette (SR_1), the waistband (SR_4), the leg opening (SR_5), the dart (SR_6), the pleat (SR_7), the yoke (SR_8), the ornament (SR_9), and the pocket (SR_{10}).

For the remaining elements, involving the length (SR_2) and the waistline position (SR_3), their characteristics were calculated based on fuzzy classification. We take the calculation of the characteristics of the waistline position as an example by which to expound on the proposed method. The concrete procedures were as follows:

Step 1: The waistline position's fuzzy membership functions $MF_{SC_{wl}}$ are defined as below:

$$MF_{lower \quad waist \quad line}(ge^{wl}) = \begin{cases} 1, & ge^{wl} \in [-\infty, -3] \\ -ge^{wl}/3, & ge^{wl} \in [-3, 0] \\ 0, & otherwise \end{cases} \tag{13}$$

$$MF_{normal \quad waist \quad line}(ge^{wl}) = \begin{cases} ge^{wl}/3 + 1, & ge^{wl} \in [-3, 0] \\ -ge^{wl}/3 + 1, & ge^{wl} \in [0, 3] \\ 0, & otherwise \end{cases} \tag{14}$$

$$MF_{high \quad waist \quad line}(ge^{wl}) = \begin{cases} ge^{wl}/3, & ge^{wl} \in [0, 3] \\ 1, & ge^{wl} \in [3, \infty] \\ 0, & otherwise \end{cases} \tag{15}$$

where ge^{wl} refers to the garment ease allowance of the waistline position.

Step 2: For each pattern, we calculated the ge^{wl}, following the computation method illustrated and expounded in Figure 6.

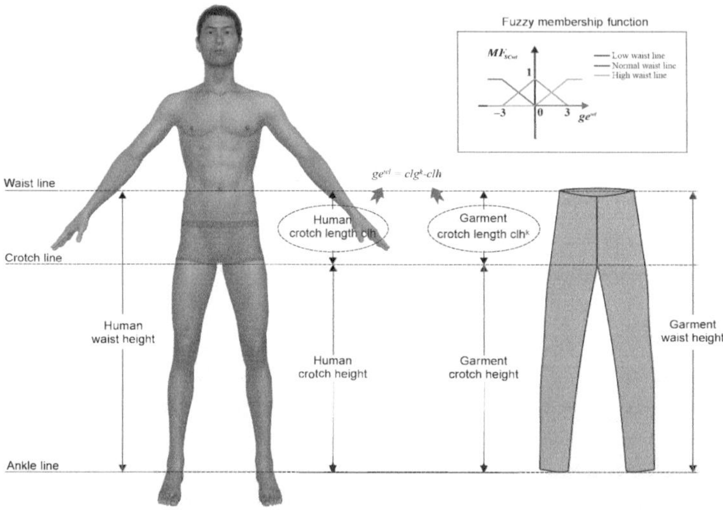

Figure 6: Fuzzy membership function of the waistline position

Meanwhile, the characteristics of the garment fit level were also determined by the fuzzy classification approach.

The characteristics of the length and waistline position of the five legging patterns are given below:

(1) Garment 1 (G_1): $SC^1_{length} = [0 \quad 0 \quad 0 \quad 0.58 \quad 0.42 \quad 0]$, indicating that the length of G_1 is the close to calf length by 58% and ankle length by 42%; $SC^1_{waist-line} = [1 \quad 0 \quad 0]$, indicating that the length of G_1 is lower waistline;

(2) Garment 2 (G_2): $SC^2_{length} = [0 \quad 0 \quad 0 \quad 0.47 \quad 0.53 \quad 0]$, indicating that the length of G_2 is close to calf length by 47% and ankle length by 53%; $SC^2_{waist-line} = [1 \quad 0 \quad 0]$, indicating that the length of G_2 is lower waistline;

(3) Garment 3 (G_3): $SC^3_{length} = [0 \quad 0 \quad 0 \quad 0.40 \quad 0.60 \quad 0]$, indicating that the length of G_3 is close to calf length by 40% and ankle length by 60%; $SC^3_{waist-line} = [1 \quad 0 \quad 0]$, indicating that the length of G_3 is the lower waistline;

(4) Garment 4 (G_4): $SC^4_{length} = [0 \quad 0 \quad 0 \quad 0.33 \quad 0.67 \quad 0]$, indicating that the length of G_4 is close to calf length by 40% and ankle length by 60%;

$SC^4_{waist-line} = [0.77 \quad 0.23 \quad 0]$, indicating that the length of G_4 is close to the lower waistline by 77% and normal waistline by 23%;

(5) Garment 5 (G_5): $SC^5_{length} = [0 \quad 0 \quad 0 \quad 0.26 \quad 0.74 \quad 0]$, indicating that the length of G_5 is close to the calf length by 26% and ankle length by 74%; $SC^5_{waist-line} = [0.50 \quad 0.50 \quad 0]$, indicating that the length of G_5 is close to the lower waistline by 50% and normal waistline by 50%.

The characteristics of the fit level of the five legging patterns are as below:

(1) G_1: $FC^1_{waist-girth} = [1 \quad 0 \quad 0 \quad 0 \quad 0]$, indicating that the fit level of waist girth for G_1 is too tight; $FC^1_{hip-girth} = [1 \quad 0 \quad 0 \quad 0 \quad 0]$, indicating that the fit level of hip girth for G_1 is too tight; $FC^1_{knee-girth} = [1 \quad 0 \quad 0 \quad 0 \quad 0]$, indicating that the fit level of knee girth for G_1 is too tight; $FC^1_{ankle-girth} = [1 \quad 0 \quad 0 \quad 0 \quad 0]$, indicating that the fit level of ankle girth for G_1 is too tight;

(2) G_2: $FC^2_{waist-girth} = [1 \quad 0 \quad 0 \quad 0 \quad 0]$, indicating that the fit level of waist girth for G_2 is too tight; $FC^2_{hip-girth} = [1 \quad 0 \quad 0 \quad 0 \quad 0]$, indicating that the fit level of hip girth for G_2 is too tight; $FC^2_{knee-girth} = [1 \quad 0 \quad 0 \quad 0 \quad 0]$, indicating that the fit level of knee girth for G_2 is too tight; $FC^2_{ankle-girth} = [1 \quad 0 \quad 0 \quad 0 \quad 0]$, indicating that the fit level of ankle girth for G_2 is too tight;

(3) G_3: $FC^3_{waist-girth} = [0 \quad 0.45 \quad 0.55 \quad 0 \quad 0]$, indicating that the fit level of waist girth for G_3 is close to tight by 45% and neutral by 55%; $FC^3_{hip-girth} = [1 \quad 0 \quad 0 \quad 0 \quad 0]$, indicating that the fit level of hip girth for G_3 is too tight; $FC^3_{knee-girth} = [1 \quad 0 \quad 0 \quad 0 \quad 0]$, indicating that the fit level of knee girth for G_3 is too tight; $FC^3_{ankle-girth} = [1 \quad 0 \quad 0 \quad 0 \quad 0]$, indicating that the fit level of ankle girth for G_3 is too tight;

(4) G_4: $FC^4_{waist-girth} = [0 \quad 0 \quad 0 \quad 0.45 \quad 0.55]$, indicating that the fit level of waist girth for G_4 is close to loose by 45% and too loose by 55%; $FC^4_{hip-girth} = [0.22 \quad 0.78 \quad 0 \quad 0 \quad 0]$, indicating that the fit level of hip girth for G_4 is close to too tight by 22% and tight by 78%; $FC^4_{knee-girth} = [1 \quad 0 \quad 0 \quad 0 \quad 0]$, indicating that the fit level of knee girth for G_4 is too tight; $FC^4_{ankle-girth} = [0.82 \quad 0.18 \quad 0 \quad 0 \quad 0]$, indicating that the fit level of ankle girth for G_4 is close to too tight by 82% and tight by 18%;

(5) G_5: $FC^5_{waist-girth} = [0 \quad 0 \quad 0 \quad 0 \quad 1]$, indicating that the fit level of waist girth for G_5 is too loose; $FC^5_{hip-girth} = [0 \quad 0.22 \quad 0.78 \quad 0 \quad 0]$, indicating that the fit level of hip girth for G_5 is close to tight by 22% and loose by 78%; $FC^5_{knee-girth} = [1 \quad 0 \quad 0 \quad 0 \quad 0]$, indicating that the fit level of knee girth for G_5 is too tight; $FC^5_{ankle-girth} = [0 \quad 0 \quad 0 \quad 0.42 \quad 0.58]$, indicating that the fit level of ankle girth for G_5 is close to loose by 42% and too loose by 58%.

The characteristics of the color are as follows:

(1) $CC^1 = [26 \quad 24 \quad 19]$, indicating that the values of red, green, and blue are 26, 24, and 19, respectively;

(2) $CC^2 = [24 \quad 27 \quad 38]$, indicating that the values of red, green, and blue are 24, 27, and 38, respectively;

(3) $CC^3 = [46 \quad 42 \quad 33]$, indicating that the values of red, green, and blue are 46, 42, and 33, respectively;

(4) $CC^4 = [51 \quad 47 \quad 36]$, indicating that the values of red, green, and blue are 51, 47, and 36, respectively;

(5) $CC^5 = [64 \quad 54 \quad 43]$, indicating that the values of red, green, and blue are 64, 54, and 43, respectively.

The characteristics of the fabric are as follows:

(1) $MC^1 = [0.12 \quad 0.12 \quad 0.15]$, indicating that the values of smoothness, softness, and warmth are 0.12, 0.12, and 0.15, respectively;

(2) $MC^2 = [0.10 \quad 0.10 \quad 0.25]$, indicating that the values of smoothness, softness, and warmth are 0.10, 0.10, and 0.25, respectively;

(3) $MC^3 = [0.90 \quad 0.10 \quad 0.45]$, indicating that the values of smoothness, softness, and warmth are 0.90, 0.10, and 0.45, respectively;

(4) $MC^4 = [0.85 \quad 0.65 \quad 0.15]$, indicating that the values of smoothness, softness, and warmth are 0.85, 0.65, and 0.15, respectively;

(5) $MC^4 = [0.70 \quad 0.20 \quad 0.55]$, indicating that the values of smoothness, softness, and warmth are 0.70, 0.20, and 0.55, respectively.

4.3 Evaluation of the Relationship between Consumer Profile and Garment Profile

According to Equation (1), α_1, α_2, α_3, and α_4, corresponding to the weights of the style, color, fabric, and garment fit, were set to 0.45, 0.05, 0.05, and 0.45, using the analytic hierarchy process (AHP). According to Equation (2), we obtained the relationships between the style requirements and the style characteristics of the five patterns: $r(SR, SC^1) = 0.6721$,$r(SR, SC^2) = 0.6727$, $r(SR, SC^3) = 0.6730$, $r(SR, SC^4) = 0.7148$, and $r(SR, SC^5) = 0.7665$.

According to Equation (3), we obtained the relationships between the color requirements and the five color characteristics in the database:
$r(CR, CC^1) = 0.011$,$r(CR, CC^2) = 0.013$, $r(CR, CC^3) = 0.017$,
$r(CR, CC^4) = 0.019$, and $r(CR, CC^5) = 0.027$.

According to Equation (4), we obtained the relationships between the fabric requirements and the three fabrics in the database: $r(MR, MC^1) = 0.312$,$r(MR, MC^2) = 0.360$, $r(MR, MC^3) = 0.500$, $r(MR, MC^4) = 0.530$, and $r(MR, MC^5) = 0.590$.

According to Equation (5) and (6), we got the relationships between the garment fit requirements and the garment fit characteristics of the pants in the database: $r(FR, FC^1) = 0$,$r(FR, FC^2) = 0$, $r(FR, FC^3) = 0.0726$, $r(FR, FC^4) = 0.1330$, and $r(FR, FC^5) = 0.1846$.

Then, with the combinations of 5 fabrics, five colors, and five sizes patterns, we created 125 pant profiles and evaluated them based on Equation (1). The top 5 $r(CP, GP^k)$ are 0.4588, 0.4584, 0.4583, 0.4581, and 0.4580, respectively.

If the tolerance ρ was set to 0.40, all the top five pant combinations would be recommended to the consumer based on 3D virtual try-on technology. Also, the Top 1 would be recommended priorly due to the highest $r(CP, GP^k)$ value. If the tolerance ρ was set to 0.80, none of the combinations at present would meet the requirements. A self-adaptation mechanism will be active until $r(CP, GP^k) \geq \rho$.

4.4 Adaptation of the Pants Style and Fit Level

According to $r(CP, GP)$, the combination with the highest value was selected as the most relevant style and pattern for further adaptation. The six parts, including the pants length, waist line position, waist girth, hip girth, knee girth, and ankle girth, were chosen for further adjustment until $r(CP, GP^{new}) \geq \rho$. GP^{new} refers to the new pants profile.

Based on the modeling method described in Section 3.2.2, we constructed six various GA-based models to generate new normalized style and garment fit level characteristic vectors, which were used as the inputs of the garment ease allowance database to generate the new ease allowances of the new garment design solution. In this study, the population sizes of the individuals were set to 50. To remove the effects of the raw fitness scores returned by the fitness function, rank scaling is employed. Then, the roulette method was applied to realize the selection of the individuals. And the single point crossover method was utilized to generate new individuals. A constraint dependent approach was used to perform the mutation.

The new normalized style and garment fit level characteristics can be expressed as follows:

$$SC^{new}_{length} = [0 \quad 0 \quad 0 \quad 0.196 \quad 0.804 \quad 0],$$
$$SC^{new}_{waistline-position} = [0.282 \quad 0.734 \quad 0],$$
$$FC^{new}_{waist-girth} = [0 \quad 0.963 \quad 0.037 \quad 0 \quad 0],$$
$$FC^{new}_{hip-girth} = [0 \quad 0.977 \quad 0.023 \quad 0 \quad 0],$$
$$FC^{new}_{knee-girth} = [0 \quad 0.983 \quad 0.017 \quad 0 \quad 0],$$
$$FC^{new}_{ankle-girth} = [0 \quad 0.991 \quad 0.009 \quad 0 \quad 0],$$

Then, according to Equation (1) to (6), we got the $r(CP, GP^{new})$=0.837, larger than $\rho = 0.80$, meaning that the objective of the adaptation was reached.

4.5 Adaptation of the pant patterns

Step 1: The new garment ease allowance for the six parts were calculated based on Equation (9);

Step 2: The general adaption requirements for each feature position were obtained. Further, the adaptation requirements were distributed for each pattern element at each feature position.

Step 3: Adaptation of the front panel (see Figure 7):

(1) Calculation of the movement of the controlling point cp_2 of the structure line sl_1;

(2) Calculation of the length deviation of the adjacent structure line sl_2 under the condition of the movement of the controlling point cp_2;

(3) Checking the length deviation of the structure line sl_2;

(4) Checking the adaptation results of the whole front panel;

(5) The process of "checking-adaptation" will be executed repeatedly until all the remaining adaptation requirements are met.

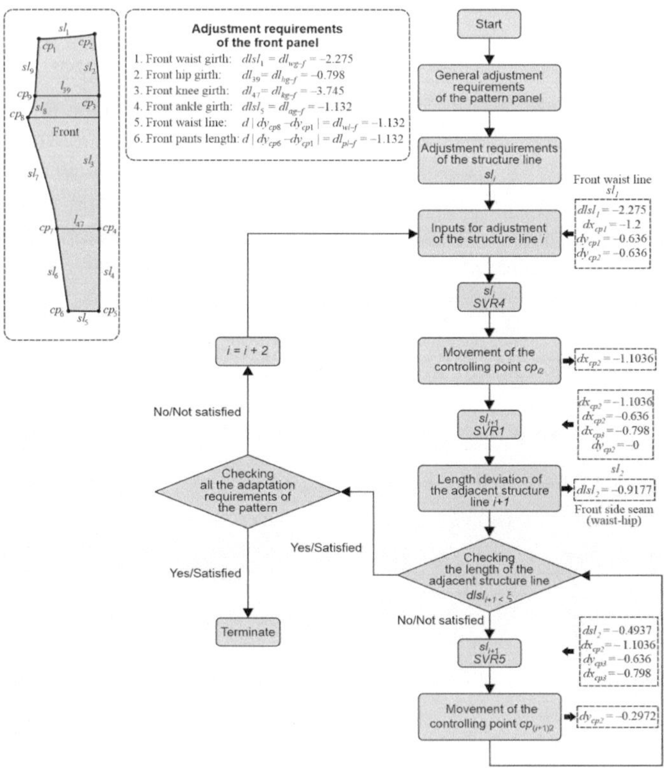

Figure 7: General adaptation flowchart of the structure lines and controlling points

Step 4: Adaptation of the back panel following the general principle described in Step 3. The final adaptation effects were illustrated in Figure 8. The hip part turned red from green for the adjusted pattern, meaning that the hip became tighter. Meanwhile, the knee part became blue, indicating that the part was looser than before.

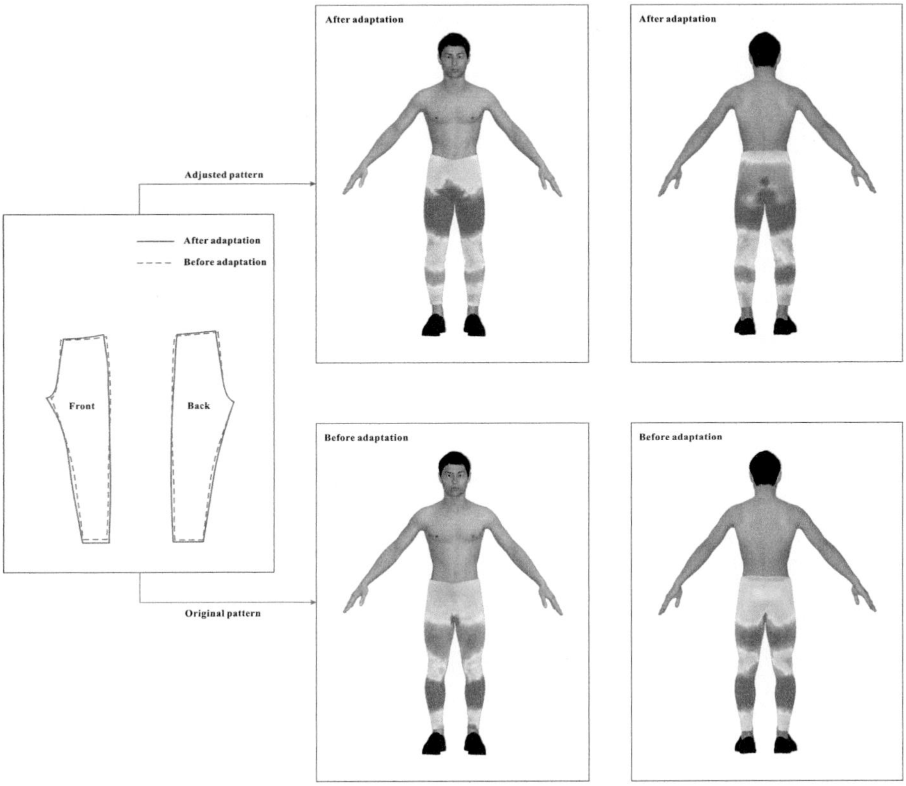

Figure 8: Comparison of the legging patterns before and after adaptation

5 Discussion

As mentioned in Section 4, the proposed system can be implemented in personalized pant design with a simple structure, such as that of a sports legging. More significantly, it can be extended to realize the personalized design of the pants with more complicated structures and detailed design elements (e.g., darts, pleats, split lines, pockets), by creating associate design rules.

The method and procedures of associate designs of detailed design elements will be concretely expounded in this section.

5.1 Associate adaptation of the darts/pleats

Darts and pleats are considered as the critical style design elements to promote the garment appearance and fitting. After adapting the outline of pant patterns, the associate adaptation of the darts/pleats can be realized by defining the relationships between the darts/pleats and the corresponding structure lines and controlling points. Take the realization of the associate adaptation of the front waist dart, for instance. The dart was positioned as shown in Figure 9. Then, the adaptation rules can be defined as follows:

(1) The length of the dart dl_{dp23} was set to a constant;

(2) $dx_{dp0} = (dx_{cp1} - dx_{cp2})/2, dy_{dp0} = (dy_{cp1} - dy_{cp2})/2$;

(3) $dx_{dp1} = dx_{dp0}, dy_{dp1} = dy_{cp1} = dy_{cp2}$;

(4) $dx_{dp2} = dx_{dp3} == dx_{dp0}, dy_{dp2} = dy_{dp3} = dy_{cp1} = dy_{cp2}$.

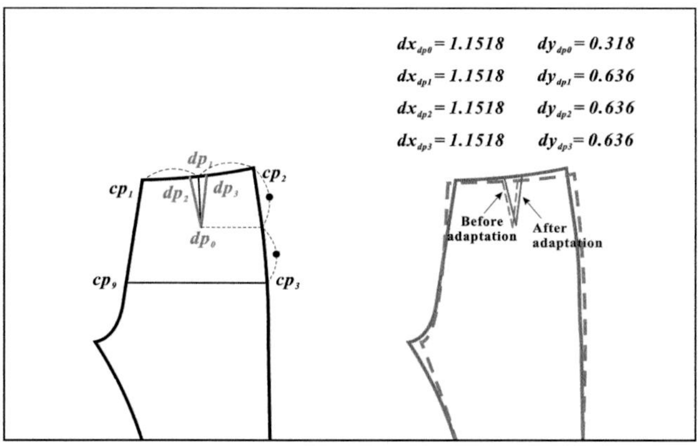

Figure 9: Associate adaptation of the waist dart

5.2 Associate design of the split lines

The split lines are the common design elements in fashion design. After the dart adaptation, the pattern can be further developed to meet the users' requirements on the split lines. Figure 10 shows a case of the development of the yoke pattern. From this case, it can be easily observed that the adjusted patterns based on the proposed system in this study can be very convenient to realize the design of any types of split lines by combining it with the traditional pattern design knowledge.

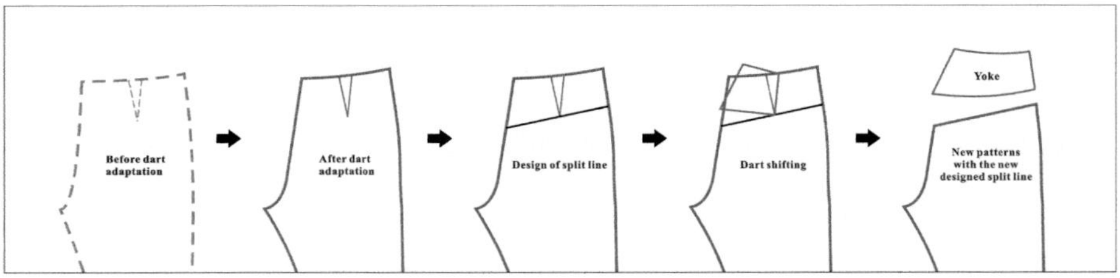

Figure 10: Associate design of the split lines

5.3 Adaptation effects comparison

The fitting effects were compared between two kinds of patterns. Pants A was developed based on the proposed system, and pants B was designed by a professional pattern designer. As shown in Figure 11 (a), the shapes of the two kinds of patterns are rather similar. Moreover, from Figure 11 (b), the garment pressure distribution of the two pants was also very closed. It indicates that the garment fitness of the two kinds of patterns is also similar. From the perspectives of pattern shape and garment fitness, it can be concluded that the performance of the patterns developed by the proposed system in this paper is close to that of patterns designed by the expert. Hence, it indicates that the proposed system offers a feasible solution for garment pat-tern-makers to realize automatic and intelligent design and adaptation of the garment patterns, particularly for the novices without sufficient expertise and experience.

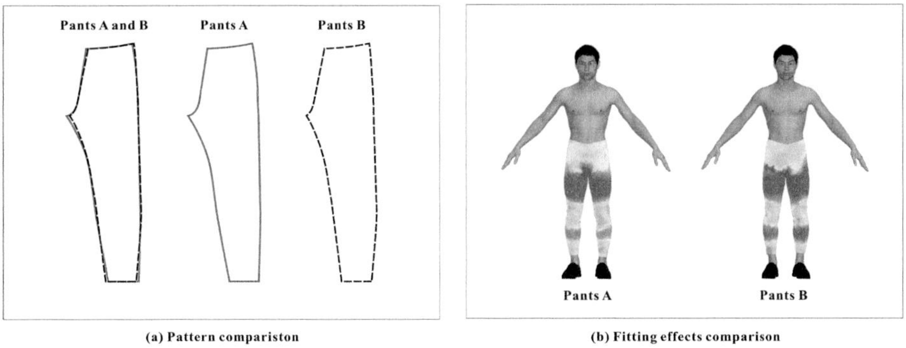

Figure 11: Comparison of the pants patterns and fitting effect

6 Conclusion

In this research, we put forth a new interactive personalized garment design recommendation approach based on a series of intelligent techniques, including fuzzy logic,genetic algorithms and support vector regression. This system can be easily combined with classical 3D clothing design software (e.g., Lectra Modaris 3D, Clo 3D, Vstitcher) to form a new interactive personalized garment design process, enhancing the level of mass customization for various fashion brand companies. The proposed personalized design recommendation system is composed of a formalized garment design knowledge base (databases and relational models) and a garment design adaptation model. It can effectively characterize relations between consumer profiles and garment profiles to find the most relevant garment design solution. It can also be used for the estimation of the garment design technical parameters. More significantly, the self-associate adaptation mechanism for the garment patterns provided in the pro-posed system is critical in garment mass customization. The main contributions of this study can be summarized as follows: (1) The proposed approach can supply personalized garments with high accuracy and velocity, facilitating the implementation of mass customization; (2) The proposed model not only can help general consumers without mastering professional design knowledge to select proper personalized garments, but also make the enterprises control the design qualities precisely; (3) A new collaborative garment design process can be formed by combining the proposed knowledge-based system with a commercial 3D garment design software, to satisfy the individualized demands of the consumer more intuitively, accurately, and promptly. Moreover, the performance of the knowledge bases and the system put forward can be enhanced progressively by supplementing new successful recommendation and design cases from the continuous implementation.

Due to the length limitation of the article, the implementation of the proposed system was validated in personalized pants' designs only. The performance of the pro-posed system should be verified further by various garment styles, such as shirts, skirts, coats, dresses, and so on. Apart from the garment styles, the factors affecting the decision making in garment recommendation processes, such as socio-cultural back-grounds, and user's emotions, should also be fully taken into account in future research. Furthermore, the development of a personalized garment design recommendation system using more advanced intelligent techniques for higher quality recommendations constitutes one of our future research focuses.

7 Acknowledgements

The authors wish to acknowledge the financial support of the Provincial Quality Engineering Project of Higher Education Institutions in Anhui (Grant No. 2022xxkc007), the Scientific Research Start-up Project for Recruited talents of Anhui Polytechnic University (Grant No. 2022YQQ062), the Teaching Quality Promotion Project for Undergraduate of Anhui Polytechnic University (Grant No. 2022jyxm95, 2022szyzk37), the Scientific Research Project of

Anhui Polytechnic University (Grant No. Xjky2022064), the Social Science Planning Project in Anhui (Grant No. AHSKQ2019D085), the Scientific Research Project of the Department of Education of Zhejiang Province (Grant No. Y202148250), the Open Project Program of Anhui Province College Key Laboratory of Textile Fabrics, the Anhui Engineering and Technology Research Center of Textile (Grant No. 2021AETKL04), the Key Teaching and Research project of Colleges and Universities in Anhui (Grant No. 2020jyxm0153), and the China National Arts Fund (named Traditional Chinese Garment Art Training Program on Digitalization).

References

[1] O. Rincon-Guevara, J. Samayoa, and A. Deshmukh. Product design and manufacturing system operations: An integrated approach for product customization. *Procedia Manufacturing*, 48:54–63, 2020.

[2] Y. Xu, S. Thomassey, and X. Zeng. Optimization of garment sizing and cutting order planning in the context of mass customization. *The International Journal of Advanced Manufacturing Technology*, 106:3485–3503, 2020.

[3] F. Turner, A. Merle, and D. Gotteland. Enhancing consumer value of the co-design experience in mass customization. *Journal of Business Research*, 117:473–483, 2020.

[4] J. Li and G. Lu. Modeling 3d garments by examples. *Computer-Aided Design*, 49:28–41, 2014.

[5] M. Dong, X. Zeng, L. Koehl, and J. Zhang. An interactive knowledge-based recommender system for fashion product design in the big data environment. *Information Sciences*, 540:469–488, 2020.

[6] A. Papachristou and N. Bilalis. Machine learning for clothing manufacture as a mean to respond quicker and better to the demands of clothing brands: a greek case study. *The International Journal of Advanced Manufacturing Technology*, 115(3):691–702, 2021.

[7] X. Ling, Y. Hong, and Z. Pan. Development of a dress design knowledge base (ddkb) based on sensory evaluation and fuzzy logic. *International Journal of Clothing Science and Technology*, 33(1):137–149, 2020.

[8] Z. Wang, Y. Xing, J. Wang, X. Zeng, and S. Xu. A knowledge-supported approach for garment pattern design using fuzzy logic and artificial neural networks. *Multimedia Tools and Applications*, 81:pages19013–19033, 2022.

[9] D. Song, T. Li, Z. Mao, and A. Liu. Sp-viton: shape-preserving image-based virtual try-on network. *Multimedia Tools and Applications*, 79: 33757–33769, 2020.

[10] P.Y. Mok, J. Xu, X. Wang, J. Fan, Y. Kwok, and J. Xin. An iga-based design support system for realistic and practical fashion designs. *Computer-Aided Design*, 45(11):1442–1458, 2013.

[11] A. Vuruskan, T. Ince, E. Bulgun, and C. Guzelis. Intelligent fashion styling using genetic search and neural classification. *International Journal of Clothing Science and Technology*, 27(2):283–301, 2015.

[12] X. Zhou, H. Liang, and Z. Dong. A personalized recommendation model for online apparel shopping based on kansei engineering. *International Journal of Clothing Science and Technology*, 29(1):2–13, 2017.

[13] H. Yan, X. Zeng, Y. Wang, P. Bruniaux, and C. Yan. Cbcrs: An open case-based color recommendation system. *Knowledge-Based Systems*, 141:113–128, 2018.

[14] M. Takatera, R. Yoshida, J. Peiffer, M. Yamazaki, K. Yashima, K. Kim, and K. Miyatake. Fabric retrieval system for apparel e-commerce considering kansei information. *International Journal of Clothing Science and Technology*, 32(1):148–159, 2019.

[15] J. Zhang, X. Zeng, K. Liu, H. Yan, and M. Dong. Jeans knowledge base development based on sensory evaluation technology for customers' personalized recommendation. *International Journal of Clothing Science and Technolog*, 30(1):34–55, 2018.

[16] C. Guan, S. Qin, and Y. Long. Apparel-based deep learning system design for apparel style recommendation. *International Journal of Clothing Science and Technology*, 31(3):376–389, 2019.

[17] K. Liu, C. Zhu, X. Tao, P. Bruniaux, and X. Zeng. Parametric design of garment pattern based on body dimensions. *International Journal of Industrial Ergonomics*, 72:212–221, 2019.

[18] K. Liu, J. Wang, X. Tao, X. Zeng, P. Bruniaux, and E. Kamalha. Fuzzy classification of young women's lower body based on anthropometric measurement. *International Journal of Industrial Ergonomics*, 55:428–434, 2016.

[19] Y. Hong, P. Bruniaux, X. Zeng, A. Curteza, and K. Liu. Design and evaluation of personalized garment block for atypical morphology using the knowledge-supported virtual simulation method. *Textile Research Journal*, 88(15):1721–1734, 2018.

[20] H. Yan, K. Liu, P. Bruniaux, M. Dong an J. Zhang, and Y. Chen. Application of 3d-to-2d garment design for atypical morphology: A design case for physically disabled people with scoliosis. *Industria Textila*, 69 (1), 2018.

[21] E. Nasibov, M. Demir, and A. Vahaplar. A fuzzy logic apparel size decision methodology for online marketing. *International Journal of Clothing Science and Technology*, 31(2):299–315, 2019.

[22] P. Cheng, D. Chen, and J. Wang. Clustering of the body shape of the adult male by using principal component analysis and genetic algorithm–bp neural network. *Soft Computing*, 24:13219–13237, 2020.

[23] P. Cheng, D. Chen, and J. Wang. Research on underwear pressure prediction based on improved ga-bp algorithm. *International Journal of Clothing Science and Technology*, 33(4):619–642, 2020.

[24] P. Cheng, D. Chen, and J. Wang. Research on prediction model of thermal and moisture comfort of underwear based on principal component analysis and genetic algorithm–back propagation neural network. *International Journal of Nonlinear Sciences and Numerical Simulation*, 22(6): 607–619, 2020.

[25] X. Tao, X. Chen, X. Zeng, and L. Koehl. A customized garment collaborative design process by using virtual reality and sensory evaluation on garment fit. *Computers Industrial Engineering*, 115:683–695, 2018.

[26] W.A. dricks and W.K. Robey. Modeling color fading ozonation of reactive-dyed cotton using the extreme learning machine, support vector regression and random forest. *Textile Research Journal*, 90(7-8):896–908, 2019.

[27] E. Doran and C. Sahin. The prediction of quality characteristics of cotton/elastane core yarn using artificial neural networks and support vector machines. *Textile Research Journal*, 90(13-14):1558–1580, 2019.

[28] S. Güneolu and M. Yüceer. A modeling study of micro-cracking processes of polyurethane coated cotton fabrics. *Textile Research Journal*, 88(24): 2766–2781, 2017.

6. A Probabilistic Neural Network-based Approach to Garment Fit Level Evaluation in 3D Digitalized Environment

Zhujun Wang,[1,*] Yingmei Xing,[1] Xianyi Zeng,[2] Pascal Bruniaux[2] and Xuyuan Tao[2]

1 Introduction

Nowadays, with the advances in novel digital technologies (big data, VR/AR, AI, etc.) instead of transacting with brick-and-mortar stores, people world-wide are more and more familiar with online purchasing of garments in e-shops, by taking advantage of convenience, wide variety, good discounts, and time-saving[1]. In this context, providing data-based intelligent services to help professional designers to interact with online consumers for their personalized designs provides powerful support for the consumers to make purchase deci-sions. One of the key fundamental factors influencing the purchase decision is the garment fit. No matter how amazing the appearance of the garments is, once the garment fit is poor, it hardly results in a final purchase.

Garment fit is defined as how the garment conforms to the body and how it appears on the body[2]. Compared with other design attributes, such as garment styles, silhouettes, fabrics, and colors, fit is one of the most significant purchasing criterion for consumers. Unfortunately, the return rate due to poor garment fit remains high in garment e-shopping[3]. This situation requires fashion brand companies to process returned garments with additional time and money[4].

[1] School of Textile and Garment, Anhui Polytechnic University, Wuhu 241000, China.
[2] GEMTEX Laboratory, Ecole Nationale Superieure des Arts et Industries Textiles, 59056 Roubaix, France.
*Corresponding author: ahpuwzj@ahpu.edu.cn

When shopping at brick-and-mortar stores, a majority of consumers aim at finding well-fitted garments through multiple real try-ons. However, in an e-shop, finding the garment that best fits for a specific consumer is very difficult due to the lack of real try-ons. Consumers can roughly estimate fitting effects using their previous experiences but it is not reliable because different brands are usually designed to meet the needs of their respective target populations with different sizing systems, which may lead to a misfit of the garment. To address this issue, 3D virtual try-on software systems, such as Lectra 3D prototype, Clo 3D, 3D runway creator, VStitcher, and Vidya, have been developed commercially available for garment design and fit evaluation over the past few years[5]. These systems permit the simulation of a real garment design and creation of new 3D garment products by integrating 2D pattern design, 3D parametric mannequin modeling, 3D virtual sewing, and fabric properties, based on sophisticated geometric and mechanical modeling and simulation techniques. They provide the opportunity for any remote consumers or professionals to visualize, communicate and evaluate the virtual garments through a web platform by showing both static and dynamic draping and fitting effects from a specific 3D digital mannequin and selected fabrics[6]. However, a convincing fashion product demonstration based on virtual try-ons that powerfully supports consumers' purchasing decision-making is still a challenging issue. Therefore, we put forth a ML-based approach to garment fit evaluation, aiming to enhance the reliability and precision of virtual reality-based garment design by acting on the interactions between the design parameters in a 3D digitalized environment and the real garment fitting data.

2 Research methodology

2.1 Literature review

2.1.1 The approach to garment fit evaluation

Currently, the visual analysis implemented by experts' know-how based on photographs or videos is the most used approach for garment fit evaluation[7]. The disadvantages in the accuracy and reliability of this subjective method are obvious. Thus, many researchers have attempted advanced technologies in this regard, such as 3D human body scanning. However, this method has three main shortcomings: (1) high cost with the support of expensive equipment; (2) complexity of operation; (3) low work efficiency. These three points have hindered its application in online apparel shopping. In e-shopping scenarios, consumers commonly make purchase decisions from numerous garments in a short time. Thus, developing an approach to predict the garment fit level fast, precisely, and automatically is indispensable. A digital pressure-based method of automatic garment fit evaluation was proposed by Liu et al.[8]. However, the indicator, namely digital clothing pressure, used for fit evaluation, cannot predict the fit level of loose garments correctly, since some parts of the loose garment are not in contact with the human body and the clothing pressure may be nearly zero. Conventionally, ease allowance, which refers to a metric indicating the dimensional difference between the human body and garment[9], is employed by fashion designers to evaluate garment fit. However, the traditional ease allowance can hardly reflect the fit levels influenced by different fabrics. The styles of the garments are diverse and ever-changing. Neither digital pressure nor traditional ease allowance can be suitable for assessing the fit levels of all kinds of garments. In this context, we propose a ML-based garment fit level prediction model (see Figure 1) aimed at predicting and controlling local garment and comprehensive fit levels by learning data from garment ease allowance, digital pressures, and involved fabric mechanical properties measured in a 3D digitalized environment.

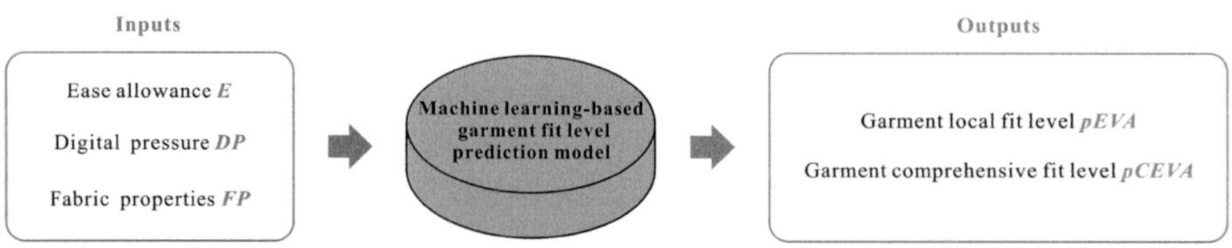

Figure 1: The proposed ML-based garment fit level prediction model

2.1.2 The application of PNN in the textile and apparel industry

Presently, AI-based ML techniques, such as ANN, FL, and GA, have been broadly implemented in the fashion industry[10], involving fashion product design[11], evaluation of garment and fabric products[12], garment comfort estimation[13], apparel manufacturing[14], fashion retailing[15], and garment supply chain management[16]. However, little attention is given to predicting the garment fit level in a 3D digitallized environment using the ANN technique.

PNN, proposed by Specht originally, is a particular type of ANN model based on the well-known Bayesian decision rules[17]. It has been treated as a powerful tool for tackling pattern recognition and classification problems in the textile and apparel sector, including textiles' classification and evaluation[18], textiles' defect detection[19], human body shape recognition[20], garment recommendation system[21], and manufacturing plant locations selection[22]. The advantages of PNN are as follows: (1) The PNN model has an inherently parallel structure, which can guarantee convergence to an optimal classifier with the increasing size of the representative training dataset; (2) The number of parameters for constructing and optimizing the PNN model is relatively fewer; (3) The PNN is not trained iteratively, which can shorten the learning process dramatically; (4) The learning dataset for PNN can be supplemented and removed without extensive retraining; (5) The PNN is less sensitive to noisy data with outliers, which can harm many other types of ANN. Contributing to these advantages, we adopted PNN to predict garment fit level in this work.

2.2 General research scheme

The general research process of this work is composed of the four sequential phases (See Figure 2). In Phase I, we collect the output learning data (i.e., sensory evaluations on garment fit) by performing a series of real try-on experiments, as well as the human body data. In Phase II, we acquire the input learning data, including the garment ease allowance values, digital pressures, and involved fabric properties measured in 3D digitalized environment by performing a series of virtual garment try-ons. In Phase III, a PNN-based garment fit level prediction model is set up and trained by the learning data. In Phase IV, we apply and verify the proposed model to predict the garment fit level.

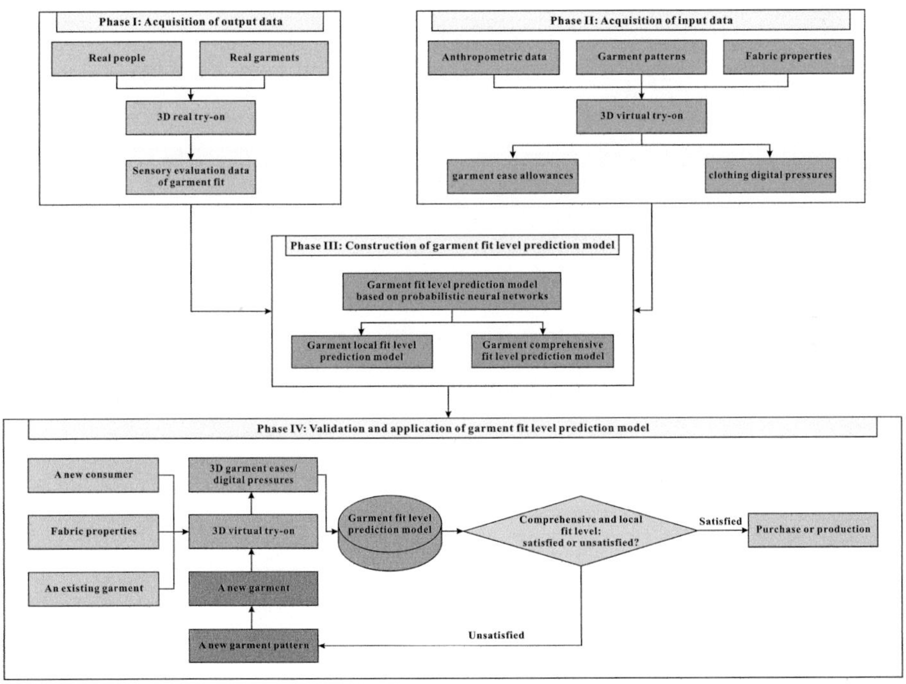

Figure 2: General research flowchart

3 Acquisition of the learning data

3.1 Subjects recruitment

In this study, a sensory experiment based on real try-ons was performed to collect the real garment fitting data. Considering the target markets of the companies involved in our research programs, we invited 17 male subjects with various figure types in China to participate in this experiment. The figure types of the subjects corresponded to 17 kinds of representative figure types demonstrated in the China National Standard (GB/T 1335.1-2008), which were 155/76A, 155/80A, 155/84A, 160/80A, 160/84A, 160/88A, 165/84A, 165/88A, 165/92A, 170/84A, 170/88A, 170/92A, 175/84A, 175/88A, 175/92A, 180/88A, and 185/92A, respectively.

3.2 Anthropometric data acquisition and 3D mannequin construction

To measure the anthropometric data precisely and expeditiously, we collected the anthropometric measurements of the subjects using Vitus Bodyscan. Next, the 3D mannequins (see Figure 3) corresponding to the subjects were generated using the software CLO 3D.

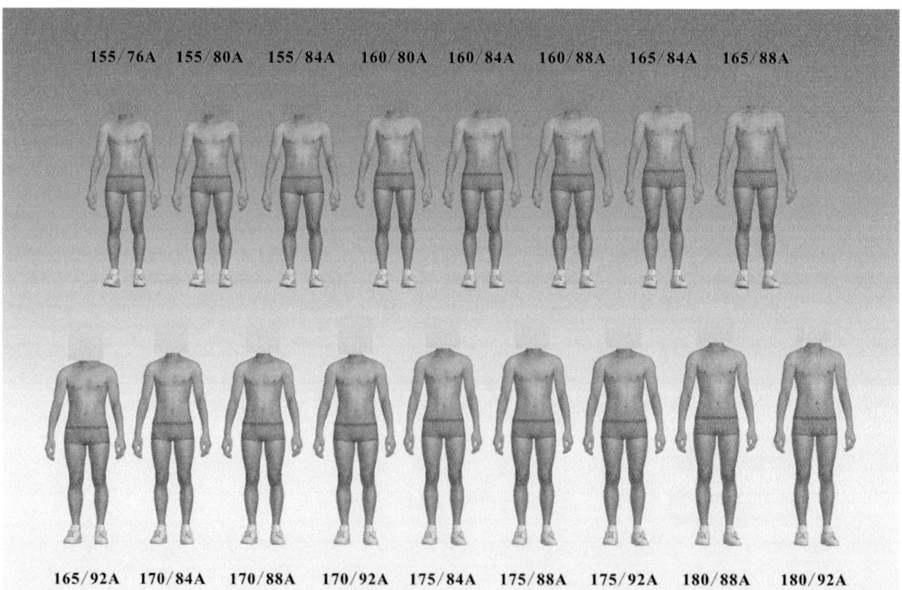

Figure 3: The created mannequins

3.3 Experimental garment style and fabrics

Both loose and tight garments were taken into account in this study. The general principles of the proposed approach can be suitable for all kinds of garment styles. Due to the length limitation, we elaborated on it using a real case of shirt and leggings from the companies involved in our research programs in this paper. The key dimensions of the shirts and leggings were present in Table 1 and 2. Parts of the mechanical properties' parameters of the fabrics for making the shirts and leggings were demonstrated in Table 3 and 4, where SW1 and SW2 refer to the stretch-weft and stretch-warp.

Table 1: Key dimensions of shirts (Unit: cm)

Shirt size	**Garment length**	**Bust girth**	**Collar girth**	**Shoulder width**	**Sleeve length**	**Waist girth**	**Hip girth**
38	70	98	38	43.4	60	92	94
40	72	102	39	44.6	60.5	96	98
42	74	106	40	45.8	60.5	100	102
44	76	110	41	47	60.5	104	106
46	78	114	42	48.2	61	108	110

Table 2: Key dimensions of leggings (Unit: cm)

Legging size	**Legging length**	**Waist girth**	**Hip girth**	**Knee girth**	**Calf girth**
XS	82.5	58	76	27	25
S	84	62	80	28	26
M	85.5	66	84	29	27
L	87	70	88	30	28
XL	88.5	74	92	31	29

Table 3: Parts of the involved fabric properties for making shirts

S.N.	**Structure**	**Composition**	**Thickness**	**SW1**	**SW2**	**Shear**
1	Woven	100% Cotton	0.23	55	58	14
2	Woven	97% Cotton, 3% Spandex	0.56	10	15	8
3	Woven	55% Linen, 45% Cotton	0.36	27	27	9
4	Woven	100% Cotton	0.6	43	40	7
5	Woven	100% Silk	0.14	3	3	0

Table 4: Parts of the involved fabric properties for making leggings

S.N.	Structure	Composition	Thickness	SW1	SW2	Shear
1	Knit	55% Cotton, 45% Rayon	0.23	11	12	10
2	Knit	100% Polyester	0.90	37	34	8
3	Knit	68% Viscose, 27% Nylon, 3% Elastane	0.86	15	15	15
4	Knit	100% Polyester	0.64	46	28	21
5	Knit	100% Polyester	0.50	27	27	9

3.4 Garment fitting data

A physical try-on experiment was conducted to collect the real garment fitting data. In a laboratory with constant temperature (20±2℃) and relative humidity (65±5%), each subject performed his try-on with the experimental garments and perceived the fit levels in static and dynamic scenarios, just like what he usually does in off-line stores. Meanwhile, they recorded the garment fitting data represented by {1,2,3,4,5}, corresponding to the linguistic values {"too tight/short(1)","tight/short(2)","perfect(3)","loose/long(4)", "too loose/long(5)"}. Finally, the data was used to form the output training dataset to construct the proposed model.

4 Construction of the garment fit level prediction model

4.1 General framework of the garment fit level prediction model

The proposed model (see Figure 4) is constituted by k sub-models and one comprehensive model. For a specific feature position k, the general framework of the PNN-based model for garment fit evaluation was illustrated in Figure 5. First of all, the local fit levels were predicted by the k sub-models respectively. After that, the comprehensive level was aggregated following Equation (1),

$$pCEVA\left(g_i^j, s_l\right) = \sum_{r=1}^{k} w^r peva^r\left(g_i^j, s_l\right) \tag{1}$$

where, $pCEVA\left(g_i^j, s_l\right)$ refers to the comprehensive fit level of the j-th size garment with the i-th fabric, given by the proposed model;$peva^r\left(g_i^j, s_l\right)$ is the predictive fit level value of the r-th feature position of the j-th size garment with the i-th fabric worn by the l-th subject;w^r are the weight values of the r-th feature position of the garments.

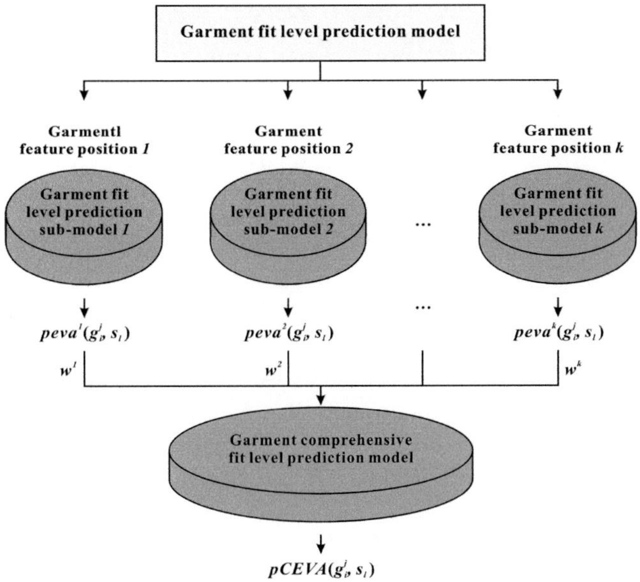

Figure 4: The general framework of the garment fit level prediction model

The sub-model based on PNN is a four-layer feed-forward network (see Figure 5) which can be constructed by the MATLAB software. The garment fit level prediction procedures for feature positions using the PNN model are given below:

Step1: The inputting vector x_i is distributed to the input layer. The number of input neurons is the same as the dimension of the inputting vector. For example, $x_i = \{1.5, 0.23, 55, 58, 14, 23, 29, 90, 90, 30, 39, 7\}$. The first element 1.5 represent the garment dimension at the k-th feature position. The remaining elements are defined as the fabric properties.

Step 2: In the pattern layer, the relationship between the input vector x_i and each pattern of the fit level $fl_l^{(j)}$ is calculated by the Euclidean distance following Equation (2). Then, the neuron output of the pattern layer is activated by a radial Gauss function like Equation (3):

$$Ed_{ij} = \left(x_i - fl_l^{(j)}\right)^T \left(x_i - fl_l^{(j)}\right) \tag{2}$$

$$Output_{ij}^{pattern} = exp\left[-\frac{\left(x_i - fl_l^{(j)}\right)^T \left(x_i - fl_l^{(j)}\right)}{2\sigma^2}\right] \tag{3}$$

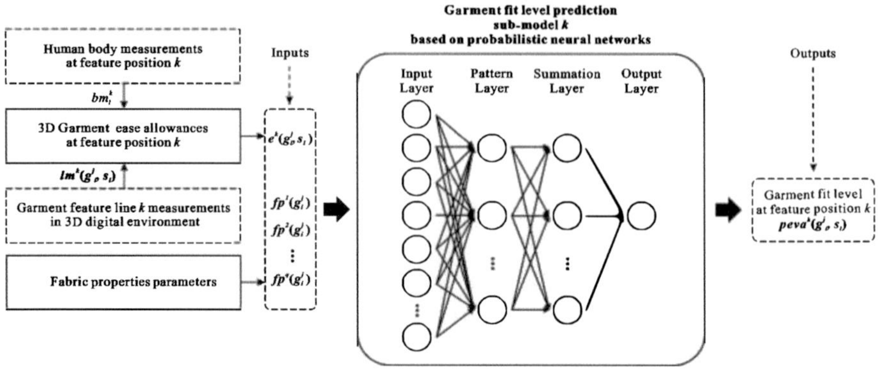

Figure 5: The structure of the PNN-based model for garment fit level evaluation at feature position k

where, $fl_l^{(j)}$ represent the l-th pattern of the fit level fl_j; Ed_{ij} is defined as the Euclidean distance between the input vector x_i and the $fl_l^{(j)}$.

Step 3: In the summation layer, one summation neuron corresponds to one specific fit level fl_j. For each summation neuron, the inputs from the pattern layer belonging to the fit level fl_j are aggregated following Equation (4) accordingly.

$$S_{ij} = \sum_{i=1}^{l} Output_{ij}^{pattern} \tag{4}$$

where, l is the number of samples belonging to the fit level fl_j.

Since the probability density function (PDF) of the fit level fl_j is usually unknown, the PNN algorithm introduces the Pazen window method to estimate the PDF. Based on this method, the class-conditional probability for the i-th sample belonging to the fit level fl_j, is computed following Equation (5).

$$Output_j^{summation} = P(x_i|fl_j) = \frac{S_{ij}}{l} \tag{5}$$

Step 4: Based on the Bayes theory, in the output layer, the fit level of the i-th sample is decided by Equation (6). It means that the largest $P(x_i|fl_j)$ will be identified and the corresponding j will be declared as the input vector x_i.

$$Category(x_i) = argmax(Output_j^{summation}) = argmax(P(x_i|fl_j) \tag{6}$$

4.2 Construction of the garment fit level prediction model for loose garments

4.2.1 Construction of the sub-models for predicting the local fit level

Step 1: Determining the indicators of predicting the garment fit level for loose garments

The garment ease allowances were utilized as the indicators for predicting the fit level of loose garments. The collected shirts' ease allowances are shown in Table 5, in which a negative value implies that the garment measurement is smaller than the corresponding body measurements.

Table 5: Shirts' ease allowances collected in the 3D digitalized environment (Unit: cm)

S.N.	Collar	Shoulder	Bust	Waist	Hip	Garment length	Sleeve length
1	4.54	3.51	21.94	28.58	13.80	29.02	9.02
2	3.52	2.32	17.86	24.58	10.65	28.78	7.60
3	2.45	1.33	13.95	20.60	7.42	26.60	5.97
4	1.42	0.51	10.00	16.62	4.68	26.15	4.59
5	0.42	-0.67	6.08	12.61	4.20	23.94	3.23
6	2.34	-0.40	6.18	12.58	3.7	23.24	1.66
...	...	...	...	...	...	...	...
425	3.48	6.98	21.37	28.64	16.92	27.78	1.95

Step 2: Determination of the input learning data and the output learning data

The inputs of the proposed model are constituted by the garment ease allowances combined with the involved fabric mechanical properties measured in a 3D digitalized environment (see Table 3), while the outputs are the predicted fit levels.

In this section, we take the construction of the sub-model for predicting collar fit levels for instance. The general principles of creating the sub-model can be applied to set up another sub-model for predicting local fit levels. For the collar fit levels prediction model, the input layer was composed of one ease data integrating with eleven fabric properties data. The output layer was the predicted collar fit level.

Step 3: Establishing the PNN model for predicting the garment fit level

To evaluate the proposed model close to reality, avoid the problem of over-fitting and guarantee the prediction accuracy of the proposed model, we utilized the K-fold (K=10) cross-validation approach in building the collar fit level prediction model. Concretely, the learning dataset was randomly split into K(=10) smaller sets initially. Next, for each of the K "folds", we created the collar fit level prediction model using the algorithm given in Section 4.1. Meanwhile, K-1(=9) small sets composed of 382 instances, were utilized as the training dataset, in turn, to train the model, with the remaining part of the original dataset as the validation dataset. Finally, the collar fit level prediction model was determined after 10-fold cross-validation.

4.2.2 Construction of the model for predicting the comprehensive fit level

After the local fit levels are obtained, the weights for every feature position should be determined. In this study, we perform the principal components analysis (PCA) on the ease of allowances to decide the weights using SPSS software. The general calculation procedure is introduced as follows:

Step 1: Extracting the principal components

Let GE be a matrix of garment ease data. In this section, $GE = \{Collar \quad ease, Shoulder \quad ease, Bust \quad ease, Waist \quad ease, Hip \quad ease, Garment \quad length \quad ease, Sleeve \quad length \quad ease\}$.

Let $PC = \{pc^1, pc^2, \ldots, pc^k\}^T$ be a vector of principal components.

The relationship between PC and GE can be expressed by Equation (7).

$$PC = VGE = \lambda GE \tag{7}$$

where, V is denoted as the covariance matrix; $\lambda = \{eiv^1, eiv^2, \ldots, eiv^k\}$ represents a set of the eigenvalues of the covariance matrix V.

Firstly, we calculated the standardized matrix **GES** following Equation (8).

$$GES_{ij} = GE_{ij} - \frac{1}{n}\sum_{i=1}^{n} GE_{ij} \tag{8}$$

Second, we formed the covariance matrix $V = GES^T GES$. In our study, the components whose eigenvalues were larger than 1 were selected. From Table 6, the first three components whose cumulative contribution rate came up to 87.527% meet the criterion. Thus, they were extracted for further research, and the corresponding eigenvalues eiv^1,eiv^2, and eiv^3 were 3.681, 1.270, and 1.175, respectively.

Step 2: Calculating the coefficients in linear combinations

Let $FLR_i = \{flr_i^1, flr_i^2, \ldots, flr_i^j\}$ be a set of rotated factor loadings of the principal components.

Let $CLC_i = \{clc_i^1, clc_i^2, \ldots, clc_i^j\}$ be a set of the coefficients in linear combinations, which can be calculated following Equation (9).

$$clc_i^j = \frac{flr_i^j}{\sqrt{eiv^i}} \tag{9}$$

Step 3: Calculating the un-normalized weights

Let $VCR^i = \{vcr^1, vcr^2, \ldots, vcr^i\}$ be a set of rotated variation contribution principle components' rates.

Let $UW = \{uw^1, uw^2, \ldots, uw^j\}$ be a set of un-normalized weights, which can be calculated following Equation (10).

$$uw^j = \frac{\sum_{i=1}^{3}\left(clc_i^j \times vcr^i\right)}{\sum_{i=1}^{3} vcr^i} \tag{10}$$

Step 4: Calculating the normalized weights

Let $W = \{w^1, w^2, \ldots, w^j\}$ be a set of normalized weights, which can be calculated following Equation (11).

$$w^j = \frac{uw^j}{\sum_{i=1}^{7} uw^j} \tag{11}$$

The weights for the shirts feature positions are 0.134, 0.105, 0.142, 0.163, 0.164, 0.159, and 0.133, corresponding to the garment length, sleeve length, collar, bust, waist, hip, and shoulder, respectively.

Therefore, we can obtain the comprehensive fit level of the shirt:

$$\begin{aligned} pCEVA(g_i^j, s_l) = {} & 0.134 \times peva^{GL}(g_i^j, s_l) + 0.105 \times peva^{SL}(g_i^j, s_l) \\ & +0.142 \times peva^{C}(g_i^j, s_l) + 0.163 \times peva^{B}(g_i^j, s_l) \\ & +0.164 \times peva^{W}(g_i^j, s_l) + 0.159 \times peva^{H}(g_i^j, s_l) \\ & +0.133 \times peva^{S}(g_i^j, s_l) \end{aligned} \tag{12}$$

4.3 Construction of the garment fit level prediction model for tight garments

We established two kinds of fit level prediction models for tight garments. One model is based on the ease allowances only. The other one is based on both the digital pressures and ease allowances.

4.3.1 Construction of the garment fit level prediction model for tight garments based on ease allowances

Step 1: Determining the number of sub-models

Since one sub-model corresponds to one feature position, and there are six feature positions highly related to the leggings fit, in terms of the waist, hip, thigh, knee, calf, and the legging length, the leggings' fit level prediction model is composed of six sub-models.

Step 2: Creating local fit level prediction models for leggings

First, the collected dataset (see Table 6) is split into the learning dataset and the testing dataset in a ratio of approximately 9:1. Next, the ease allowances (see Table 6) and the involved fabric properties (see Table 3) were utilized as the inputs of each sub-model, and the local fit level of the leggings was used as the outputs. Afterward, we established and trained the PNN-based leggings fit level prediction model using the 10-fold cross-validation approach described in Section 4.2.1. **Step 3:** Constructing the comprehensive fit level prediction model for the leggings

Table 6: Leggings' ease allowances collected in the 3D digitalized environment (Unit: cm)

S.N.	Waist	Hip	Thigh	Knee	Calf	Length
1	3.83	6.20	4.05	0.39	1.87	-13.50
2	3.40	5.40	3.42	0.42	1.84	-14.30
3	3.83	3.90	3.39	0.33	1.81	-15.00
4	3.60	3.30	3.29	0.35	1.76	-16.00
5	3.60	3.30	2.90	0.19	1.84	-16.57
6	3.60	3.30	3.71	0.13	1.82	-17.85
...	...	...	...	...	...	...
425	6.01	4.60	0.84	1.33	1.82	-15.75

The weights of the leggings' feature positions were computed by PCA. Concretely, they are 0.212, 0.176, 0.176, 0.158, 0.158, and 0.119, corresponding to legging length, waist, hip, thigh, knee, and calf, respectively.

The comprehensive fit level of the leggings can be acquired:

$$\begin{aligned} pCEVA(g_i^j, s_l) = {} & 0.212 \times peva^{LL}(g_i^j, s_l) + 0.176 \times peva^{W}(g_i^j, s_l) \\ & + 0.176 \times peva^{H}(g_i^j, s_l) + 0.176 \times peva^{T}(g_i^j, s_l) \\ & + 0.158 \times peva^{K}(g_i^j, s_l) + 0.119 \times peva^{C}(g_i^j, s_l) \end{aligned} \tag{13}$$

4.3.2 Construction of the garment fit level prediction model for tight garments based on digital pressures and ease allowances

Step 1:: Constructing the sub-models for predicting the local fit levels based on the digital pressures and garment eases

The inputs of each sub-model were the digital pressures, ease allowances of the legging length, and the involved fabric properties. The outputs were also the local fit of the leggings. Then, we established and trained a PNN-based model to predict the local fit level.

Step 2: Constructing the comprehensive fit level prediction model for the leggings based on the garment eases and digital pressures

Table 7: Leggings' digital pressures collected in the 3D digitalized environment (Unit: cm)

S.No.	Waist	Hip	Thigh	Knee	Calf
1	2.92	3.83	2.16	1.26	4.29
2	3.40	4.46	2.91	1.66	5.38
3	5.60	5.75	4.13	4.03	6.31
4	6.75	6.53	4.80	4.50	7.71
5	7.30	6.30	5.25	4.75	8.03
6	7.76	6.51	5.00	4.86	8.00
...	...	...	...	...	...
425	11.1	9.90	9.50	17.80	28.50

Based on the weights of the leggings' feature positions, the comprehensive fit level of the leggings can be obtained:

$$\begin{aligned} pCEVA(g_i^j, s_l) = {} & 0.063 \times peva^{LL}(g_i^j, s_l) + 0.171 \times peva^{W}(g_i^j, s_l) \\ & + 0.188 \times peva^{H}(g_i^j, s_l) + 0.184 \times peva^{T}(g_i^j, s_l) \\ & + 0.181 \times peva^{K}(g_i^j, s_l) + 0.213 \times peva^{C}(g_i^j, s_l) \end{aligned} \tag{14}$$

5 Validation and application of the garment fit level prediction model

5.1 Comparison of the performance of the proposed garment fit level prediction model

The performance of the proposed PNN model was compared with other typical classifiers, including the linear regression model, the SVM model, the RBF ANN model, and the BP ANN model. Tables 8 to 10 present the comparison results of different models.

For shirt fit level prediction (see Table 8), from the perspective of the average prediction accuracy, the proposed PNN model has the highest value (93.36%), followed by the BP ANN model (90.37%), LR model (69.77%), the SVM model (67.11%), and the RBF ANN model (60.13%), which indicates that the prediction performance of the proposed model is superior to other classifiers. Also seen from Table 8, the prediction accuracy of the specific model fluctuates at different feature positions. From the perspective of standard deviation (SD), the SD of the proposed PNN model (8.32%) is the lowest among the four classifiers, meaning that, the prediction accuracy fluctuations of the PNN-based model are minimal. Therefore, the PNN model has the optimal performance in predicting shirt fit levels among four classifiers relatively.

Table 8: Performance comparison for shirt fit level

S.N.	**Prediction accuracy(%)**				
	PNN	**SVM**	**RBF ANN**	**BP ANN**	**LR**
Garment length	86.05	74.42	53.49	67.44	100
Sleeve length	79.07	69.77	65.12	86.05	60.47
Collar	97.67	39.53	41.86	88.37	53.49
Bust	100	83.72	62.79	100	65.12
Waist	100	83.72	62.79	100	48.84
Hip	90.70	18.60	34.88	90.70	90.70
Shoulder	100	100	100	100	39.53
Average	93.36	67.11	60.13	90.37	69.77
Standard deviation	8.32	28.31	20.99	11.74	20.80

For legging fit level prediction based on ease allowances only (see Table 9), the BP ANN model has the highest average prediction accuracy (79.07%) and the minimal SD (9.24%). The prediction accuracy of the PNN model is 72.10%, followed closely by that of the BP ANN model. However, the SD of the PNN-based model comes up to 21.67%, which is the maximal among the five models. Thus, it can be easily conclude that the comprehensive performance of the BP ANN model is optimal to ease-based legging fit level estimation.

Table 9: Performance comparison of the ease-based model for leggings fit level

S.N.	**Prediction accuracy(%)**				
	PNN	**SVM**	**RBF ANN**	**BP ANN**	**LR**
Legging length	100	74.42	65.12	76.74	53.49
Waist	86.05	88.37	67.44	93.02	55.81
Hip	79.07	60.47	55.81	95.35	53.49
Thigh	74.42	69.77	79.07	100	67.44
Knee	44.19	62.79	67.44	88.37	67.44
Calf	48.84	48.84	46.51	79.07	69.77
Average	72.10	67.44	63.57	88.76	61.24
Standard deviation	21.67	13.48	11.17	9.24	7.73

For legging fit level prediction based on clothing digital pressures combined with ease allowance (see Table 10), the prediction accuracy and the SD of the PNN model achieve values of 97.67% and 1.47%, respectively, which demonstrates that the overall performance of the PNN model is outstanding among five predictive models.

Table 10: Performance comparison of the ease and pressure-based model for leggings fit level

S.N.	Prediction accuracy(%)				
	PNN	**SVM**	**RBF ANN**	**BP ANN**	**LR**
Legging length	100	74.42	65.12	76.74	53.49
Waist	97.67	95.35	95.35	95.35	100
Hip	95.35	76.74	93.02	93.02	100
Thigh	97.67	88.37	93.02	93.02	90.70
Knee	97.67	81.40	72.09	90.70	100
Calf	97.67	60.47	62.79	76.74	90.70
Average	97.67	79.46	80.23	89.30	89.15
Standard deviation	1.47	12.08	15.20	8.48	18.05

5.2 Influence of the indicators on prediction accuracy for predicting the fit level of tight garments

For legging fit level prediction, it can easily be seen that the average prediction accuracy of the ease and pressure-based sub-models is over 95% (see Table 10), while that of the ease-based sub-models is less than 75% (see Table 9). Moreover, The SD value of the ease and pressure-based model is only 1.47%, which is dramatically lower than that of the ease-based model. Therefore, it indicates that multiple indicators contribute to promoting the prediction performance in the scenario of predicting the fit level of tight garments.

5.3 Application of the garment fit level prediction model

To validate the proposed garment fit level prediction model, we applied it to the fit level prediction of a real shirt case for a specific male consumer. The application process of this case is as follows:

Step 1: We collected the body measurements of the consumer relevant to the shirt, expressed by a vector $BM = \{39.3, 42.3, 93.3, 72.6, 88.6, 56.8, 45.2\}$, corresponding to the feature human body positions $FP = \{Neck \quad girth, Shoulder \quad width, bust \quad girth, waist \quad girth, hip \quad girth, arm \quad length, back \quad length\}$.

Step 2: We obtained three shirts of different sizes and the same fabric. Then, the ease allowances for the three samples were collected and presented as follows:

$Ease_1 = \{0.3, 0.8, 5.8, 15.3, 5.6, 0.6, 23.8\}$

$Ease_2 = \{1.5, 2.6, 11.2, 20.2, 10.6, 2.3, 27.6\}$

$Ease_3 = \{2.8, 4.2, 17.8, 26.5, 16.8, 3.8, 31.2\}$

The elements of these sets represent the ease allowances of the collar, shoulder, bust, waist, hip, sleeve length, and shirt length, respectively.

Step 3: Combined with the fabric parameters, the ease data were imported to propose the fit level prediction model. Then we can get the garment fit levels for the samples that follow below:

$Fitlevel_1 = \{1, 2, 2, 2, 2, 2, 1.858\}$

$Fitlevel_2 = \{3, 3, 3, 3, 4, 4, 3.239\}$

$Fitlevel_3 = \{4, 4, 4, 4, 5, 5, 4.239\}$

The first six elements of these sets represent the predicted fit levels of the collar, shoulder, bust, waist, hip, sleeve length, and shirt length, respectively. The last one refers to the predicted comprehensive garment fit level.

From the results, the predicted comprehensive garment fit level of sample 2 equals 3.239 which is closer to 3 than the other two samples, indicating that sample 2 is more fit for the consumer. Then we can recommend sample 2 to the consumer.

5.4 Extension of the proposed garment fit level prediction approach

The general ideas and principles can be extended to various target markets and target populations, as well as the garment styles. The major process is described as follows:

(1) According to the characteristics of the target market and target population, as well as the specific garment styles, we must recruit the corresponding representative subjects and collect the corresponding learning data following the experimental steps described in our study.

(2) Based on the collected learning data, we must construct the corresponding garment fit level prediction model based on the method proposed in this study.

Through the steps mentioned above, the garment fit level prediction model toward various target markets, target population, and garment styles can be realized and applied.

6 Conclusion

In this study, we put forward a new garment fit level prediction approach using PNN technology. From the results, we can conclude that: (1) the garment ease allowances in the 3D digitalized environment is a good indicator for the loose garment fit level recognition; (2) for the tight garment, multiple indicators combining digital pressures with the ease allowances are superior to the single ease allowance indicator in predicting the fit level; (3) the PNN model performs better than the SVM, RBF ANN, and BP ANN models in the prediction of garment fit levels for both loose and tight garments. The model proposed in this study has the following advantages: (1) the proposed model can not only predict the comprehensive fit levels, but also the local fit levels of the garments fast, precisely, and automatically; (2) the model provides a feasible and efficient solution for perceiving the garment fit levels without real try-ons, which can be utilized in both online and in-store shopping to enhance the consumers' satisfaction; (3) the performance of the proposed model can be enhanced continuously by learning new datasets. The proposed model can be further improved and extended in the future. A personalized garment design system can be developed to offer consumers accurate garments for each specific fit level. From the viewpoint of the industrial application, more datasets should be considered and acquired in future research.

7 Acknowledgements

The authors wish to acknowledge the financial support of the Provincial Quality Engineering Project of Higher Education Institutions in Anhui (Grant No. 2022xxkc007), the Scientific Research Start-up Project for Recruited talents of Anhui Polytechnic University (Grant No. 2022YQQ062), the Teaching Quality Promotion Project for Undergraduate of Anhui Polytechnic University (Grant No. 2022jyxm95, 2022szyzk37), the Scientific Research Project of Anhui Polytechnic University (Grant No. Xjky2022064), the Social Science Planning Project in Anhui (Grant No. AHSKQ2019D085), the Scientific Research Project of the Department of Education of Zhejiang Province (Grant No. Y202148250), the Open Project Program of Anhui Province College Key Laboratory of Textile Fabrics, the Anhui Engineering and Technology Research Center of Textile (Grant No. 2021AETKL04), the Key Teaching and Research project of Colleges and Universities in Anhui (Grant No. 2020jyxm0153), and the China National Arts Fund (named Traditional Chinese Garment Art Training Program on Digitalization).

References

[1] Z. Hu, X. Li, C. Wei, and H. Zhou. Examining collaborative filtering algorithms for clothing recommendation in e-commerce. *Textile Research Journal*, 89(14):2821–2835, 2019.

[2] J. Pei, J. Fan, and S. P. Ashdown. A novel optimization approach to minimize aggregate-fit-loss for improved breast sizing. *Textile Research Journal*, 90(15-16):1823–1836, 2020.

[3] N. Hernández, H. Mattila, and L. Berglin. Can virtually trying on apparel help in selecting the correct size? *Clothing and Textiles Research Journal*, 37(4):249–264, 2019.

[4] E. Nasibov, M. Demir, and A. Vahaplar. A fuzzy logic apparel size decision methodology for online marketing. *International Journal of Clothing Science and Technology*, 31(2):299–315, 2019.

[5] Y. Hong, P. Bruniaux, X. Zeng, A. Curteza, and K. Liu. Design and evaluation of personalized garment block for atypical morphology using the knowledge-supported virtual simulation method. *Textile Research Journal*, 88(15):1721–1734, 2018.

[6] K. Liu, X. Zeng, P. Bruniaux, X. Tao, X. Yao, V. Li, and J. Wang. 3d interactive garment pattern-making technology. *Computer-Aided Design*, 104:113–124, 2018.

[7] Y. Lu, G. Song, and J. Li. A novel approach for fit analysis of thermal protective clothing using three-dimensional body scanning. *Applied Ergonomics*, 45(6):1439–1446, 2014.

[8] K. Liu, X. Zeng, P. Bruniaux, J. Wang, E. Kamalha, and X. Tao. Fit evaluation of virtual garment try-on by learning from digital pressure data. *Knowledge-Based Systems*, 133:171–182, 2017.

[9] Z. Liu, Q. He, F. Zou, Y. Ding, and B. Xu. Apparel ease distribution analysis using three-dimensional motion capture. *extile Research Journal*, 89(19-20):4323–4335, 2019.

[10] C.Giri, S. Jain, X. Zeng, and P. Bruniaux. A detailed review of artificial intelligence applied in the fashion and apparel industry. *IEEE Access*, 7: 95376–95396, 2019.

[11] H. Yan, K. Liu, P. Bruniaux, M. Dong an J. Zhang, and Y. Chen. Application of 3d-to-2d garment design for atypical morphology: A design case for physically disabled people with scoliosis. *Industria Textila*, 69 (1), 2018.

[12] X. Chen, X. Tao, X. Zeng, L. Koehl, and J. Boulenguez-Phippen. Control and optimization of human perception on virtual garment products by learning from experimental data. *Knowledge-Based Systems*, 87:92–101, 2018.

[13] R. Wang, H. Du, F. Zhou, D. Deng, and Y. Liu. An adaptive neural fuzzy network clothing comfort evaluation model and application in digital home. *Multimedia Tools and Applications*, 71:395–410, 2014.

[14] Y. Xu, S. Thomassey, and X. Zeng. Optimization of garment sizing and cutting order planning in the context of mass customization. *The International Journal of Advanced Manufacturing Technology*, 106:3485–3503, 2020.

[15] S. Ren, T. Choi, and N. Liu. Fashion sales forecasting with a panel data-based particle-filter model. *IEEE Transactions on Systems Man Cybernetics Systems*, 45(3):411–421, 2015.

[16] x. Wen, T. Choi, and S. Chung. Fashion retail supply chain management: A review of operational models. *International Journal of Production Economics*, 207:34–55, 2019.

[17] Specht. Probabilistic neural networks for classification, mapping, or associative memory. In *IEEE International Conference on Neural Networks*, 1988.

[18] S. Li, B. Xu, H. Fu, X. Tao, and Z. Chi. A two-scale attention model for intelligent evaluation of yarn surface qualities with computer vision. *Journal of the Textile Institute*, 109(6):798–812, 2018.

[19] A. Hasnat, A. Ghosh, A. Khatun, and S. Halder. Pattern classification of fabric defects using a probabilistic neural network and its hardware implementation using the field programmable gate array system. *Fibres Textiles in Eastern Europe*, 25(1):42–48, 2017.

[20] J. Jin, Y. Yang, and F. Zou. Developing an intelligent model for the construction a hip shape recognition system based on 3d body measurement. *Fibres textiles in Eastern Europe*, 24(5):110–118, 2016.

[21] J. Shahrabi, E. Hadavandi, and M. S. Esfandarani. Developing a hybrid intelligent model for constructing a size recommendation expert system in textile industries. *International Journal of Clothing Science Technology*, 25(5):338–349, 2013.

[22] W. K. Wong, K. F. Wu, and X. H. Zeng. Classification decision model for a hong kong clothing manufacturing plant locations using an artificial neural network. *International Journal of Advanced Manufacturing Technology*, 28(3-4):428–434, 2006.

7. Explainable Machine Learning based Control Charts for High-Dimensional Non-Stationary Time Series Data in IoT Systems: Challenges, Methods, and Future Directions

Aamir Saghir,[1,2] Kim Duc Tran[2] and Kim Phuc Tran[3,2]

1 Introduction

In recent years, the increasing cooperation between people and intelligent production systems and combining the best of both worlds (the speed and accuracy guaranteed by automation with humans' cognitive skills and critical thinking) have profoundly impacted industrial manufacturing, leading to the 5th Industrial Revolution, known as Industry 5.0. In Industry 5.0, rapid advancements in Digital Twin (DT) technologies and the pervasiveness of cost-effective sensor technologies have pushed the incor-poration of artificial intelligence, particularly data-driven machine learning models, for use in smart manufacturing (SM). However, the persistent issue with such mod-els is their high sensitivity to the training data and the lack of interpretability in the outcomes, at times generating unrealistic results. The Internet of Things (IoT) aspires to unite the physical and digital worlds into a single system, opening up significant business prospects for the manufacturing industry toward SM. SM are systems where the Internet in factory networks is connected to the physical world

[1] 1Mirpur University of Science and Technology (MUST), Mirpur-10250(AJK), Pakistan.
[2] International Chair in DS & XAI, International Research Institute for Artificial Intelligence and Data Science, Dong A University, Danang, Vietnam.
[3] Univ. Lille, ENSAIT, ULR 2461 - GEMTEX - Génie et Matériaux Textiles, F-59000 Lille, France.

through sensors and can be thought of as the management of a network of devices, home appliances, and vehicles of the IoT.

Quality control (QC) (an application of statistical computing) is extremely necessary for the industrial sector, ranging from manufacturing, and production activities to the results of production. QC is a guideline or set of procedures intended to ensure that a manufactured product or performed service adheres to a defined set of quality criteria or meets the requirements of the client or customer. The goal of quality control is to produce products that are free from defects and meet the expectations of customers. In traditional manufacturing, quality control was a manual process that involved inspectors checking each product for defects and ensuring that it met the required standards. The rapid development of technologies like the Internet of Things (IoT) and Artificial Intelligence (AI), has engendered the concept of smart manufacturing (SM) that adds intelligence into the manufacturing process to drive continuous improvements, knowledge transfer, and data-based decision-making using digital twin technology. Digital twin technology is a real-time simulation of physical objects and processes. The technology uses sensors to gather data from the physical objects and processes, which is then used to create a digital replica of the object or process. This digital replica can be used to monitor and analyze the physical object or process in real time, allowing for faster and more accurate quality control and inspection. With the advent of IoT, quality control has become much more sophisticated and efficient in IoT sensor data.

IoT-enabled quality control systems can use data and advanced analytics to monitor production in real time, identify potential quality problems, and provide feedback to manufacturers so that they can take corrective action. One of the key benefits of IoT-enabled quality control is that it can significantly improve product quality. In addition, IoT-enabled quality control can also help manufacturers reduce costs. IoT in industrial automation can help to automate the quality control process and help manufacturers eliminate the need for manual inspections, reduce waste, and increase production efficiency. By monitoring production in real-time, manufacturers can identify and fix quality problems before they become more significant, reducing the risk of faulty products and saving money on recalls and repairs. IoT solutions are the perfect option for quality control managers looking to maintain or improve their high standard levels while streamlining the processes through which these levels are reached. This is because the 24/7 real-time data collected through sensors and monitoring systems allow these managers to efficiently and precisely take note of any factors that may threaten their quality standard.

The following challenges should be considered in IoT quality control for efficiency and achieving the required standards. Companies must consider the security, scalability, data processing, and storage requirements when implementing an IoT system in their existing quality control process. Security is of utmost importance since any breach or hacking could result in the loss of confidential customer information or product details. Scalability needs to be considered since the number of connected devices can vary significantly based on production demand fluctuations. Data processing should also be considered carefully so that all relevant information from individual sensors can be collected, analyzed, and used for decision-making purposes. Finally, companies need to ensure they have adequate storage capacity for large datasets generated by networked IoT devices. By taking these implementation challenges into consideration beforehand, businesses can maximize their return on investment while ensuring that their customers receive high-quality products every time.

Although, smart quality control (SQC) systems are founded in SM to optimize the quality of manufacturing operations and manufactured products. However, this system is essentially based on a technology-centered (Industry 4.0) and lacks adequate interpretability. Industry 5.0 leverages humans' work alongside machines with better integration, enabling faster and better automation combined with the power of human brains, and the flexibility that allows for early system failure detection and system automation. In order to achieve that with a lower cost and high productivity, SQC requires more advanced human-machine interfaces using AI algorithms, in addition to an explainable module, with capabilities to deal with complex data from the continuous and simultaneous process, which brings fruitful business results in the long run. This has led to a growing interest in machine learning-based statistical control charts in SM[1].

This chapter discusses the challenges of applying anomaly detection and proposes a perspective about end-to-end efficient and explainable ML-based statistical control charts that overcome the difficulties of existing ML-based control charts. Additionally, this chapter provides a perspective of ML techniques-based control charts for AD in SM. The remainder of this chapter is organized as follows: Sections. 2 and 3 describe state-of-the-art approaches for anomaly detection and Control charts-based anomaly detection, respectively. Section 4 provides the ML-based control charts for AD in multivariate time series modeling. In Section 5, we analyze the difficulties and challenges of ML-based control charts and apply them to anomaly detection, with the goal of developing efficient, and explainable machine learning-based control charts for anomaly detection systems. Section 6 presents our end-to-end framework, which provides insights into explainable machine learning-based SQC methods for anomaly detection in the context of smart manufacturing. This framework addresses the aforementioned drawbacks by employing a comprehensive system based on blockchain machine learning and explainable AI, with a hybrid deep learning network. Section 7 outlines future directions for a comprehensive study, and finally, Sect. 8 briefly discusses the concluding remarks of this chapter.

2 Anomaly Detection methods and applications

Anomaly Detection (AD) is a process of identifying instances (anomalies, outliers, exceptions) in data that differ significantly from the expected norm. This is achieved by analyzing the characteristics of normal data. In today's world of distributed systems, effective management and monitoring of system performance are of paramount importance. With a multitude of components to monitor, AD can play a critical role in identifying errors, facilitating root cause analysis, and speeding up technical support. As a result, AD has gained significant traction across a variety of industries, including healthcare, fraud detection, network security, industrial damage detection, and military surveillance Chandola et al.[2],Huong et al.[3]. Anomaly detection (AD) can be further characterized into two methodologies, including a statistical approach and a machine-learning-based approach. The statistical approach relies on the assumption of data normality and posits that normal data objects are generated by a probabilistic and statistical model with a closed-form probability distribution. The parameters of the model are estimated, and data that deviates from the model is considered an outlier.

Statistical process control (SPC) monitors manufacturing processes with technology that measures and controls quality. SPC triggers various machines and instruments to provide quality data from product measurements and process readings. Once collected, the data is evaluated and monitored to control that process. SPC is an essential approach for AD that is widely applied in industry. The aim of this approach is to monitor and reduce variation in the process as soon as possible to guarantee high product quality at a minimal cost. In manufacturing, Control Charts, an effective tool of SPC, continuously monitor a process as well as detect process abnormalities in order to enhance and optimize the process Tran et al.[1]. For this aim, numerous forms of Control Charts have been produced in Tran[4]. On the other hand, the machine-learning-based approach employs machine-learning techniques to identify abnormalities.

Machine learning-based approaches for anomaly detection can be classified in a variety of ways, depending on the type of algorithm used, the presence of labels, and the form of input data. With respect to the availability of labels, anomaly detection can be further divided into three categories: supervised, semi-supervised, and unsupervised. The process of supervised anomaly detection involves training a supervised binary or multi-class classifier, utilizing the labels of both normal and anomalous data instances. In contrast, unsupervised anomaly detection techniques are based solely on the intrinsic properties of data instances and are used to identify outliers. The difficulty of obtaining labeled data, particularly for anomalous instances, makes the use of unsupervised techniques essential in the automatic labeling of unlabeled data. Semi-supervised anomaly detection techniques address this challenge by leveraging the available labels of the normal class to differentiate outliers from normal instances in a given dataset. This combination of both labeled and unlabeled data has resulted in the widespread adoption of semi-supervised anomaly detection techniques.

The use of time series data is prevalent in the area of Anomaly Detection, and it can be divided into two main categories: univariate and multivariate time series. Univariate time series is characterized by the variation of a single variable or feature over time. The recent progress in the field of deep learning has presented opportunities to extract more sophisticated features, which have the potential to improve the efficiency of anomaly detection in univariate time series data. In contrast, the detection of anomalies in multivariate time series data presents a complex and compelling challenge, attracting significant research attention, see, Chandola et al.[2],Xie and Qiu[5], etc.

Abnormalities in complex production systems are difficult to avoid and may result in serious degradations. Thus it is important to be able to detect potential anomalies in the monitored system as early as possible. Early AD can therefore provide significant improvements in the reduction of operational and maintenance costs, and system down-time, and lead to increased levels of safety, which is of ever-growing importance Tran et al.[6],Tran et al.[1].

3 Control Charts-based Anomaly Detection

Control charts, one of the essential tools in Statistical Process Control (SPC) first introduced by Shewhart [7], have been widely implemented in the manufacturing industry as an effective approach for anomaly detection Castagliola et al. [8],Afshari et al. [9], Nguyen et al. [10], Saghir et al. [11],Saghir et al. [12],Saghir et al. [13]. Through control charts, SPC helps manufacturing companies monitor product qualities (detection of abnormal products) and discover the defects in the production lines (assignable causes to fix). The application of control charts for AD has been studied in numerous different disciplines Tran et al. [1],Nguyen et al. [14],Kadri et al. [15],Münz and Carle [16], Tran [4]. However, traditional control charts now find difficulties in reality activities including designing, pattern recognition, and interpreting stages to deal with high-dimensional data from the continuous and simultaneous process. Multivariate control charts are the most popular process monitoring methods to use when overall quality depends on simultaneously taking many quality variables into consideration Pham [17]. Several multivariate control charts, such as Hotelling's T2 chart, multivariate exponentially weighted moving average (MEWMA) chart, and multivariate cumulative sum (MCUSUM) chart have been successfully implemented in the literature. But, a few challenges of these traditional multivariate control charts are that (i) they are ineffective with a large number of correlated quality variables, (ii) a large number of variables may degrade the ability to detect an anomaly and lead to a multicollinearity problem and (iii) only able to detect a shift in the process mean vector, i.e., out-of-control signals of the process. It is impossible to indicate which variable(s) or a group of variables is responsible for out-of-control signals of the process.

4 ML-based Control Charts for Anomaly Detection

ML is a domain of Artificial Intelligence (AI), which consists of programming algorithms to learn automatically from data and experiences or interactions with the environment. There are three main branches of ML: supervised learning (to find correlations between input data and output data), unsupervised learning (discovering the structure according to the data), and reinforcement learning (concerned with how to make a sequence of decisions). ML algorithms have emerged as powerful analytic tools and great assistance that can be integrated into control charts of SPC to solve these issues. Since, the goal of control charts is similar to that of one-class classification methods as only one class is represented in the training data, which can be used to learn its characteristics as well as to provide a measure of abnormal behavior of new observations (see, Truong et al. [18],Huong et al. [19]). The application of ML-based control charts emphasizes combining data-driven reliability models with sensor data collected from monitored systems. This is very useful for tackling (i) the pattern recognition [20;21;22] and (ii) the issues in the interpreting stage as discussed by Wang and Chen [23], Low et al. [24], Salehi et al. [25], Niaki and Abbasi [26].

Dimension reduction is a method of ML that studies the data in a smaller dimension instead of studying the data involved in a grand dimension. For monitoring high-dimensional data, various latent variable-based control charts have been proposed. The latent variable models can extract important information from the process by projecting the high-dimensional process data into a lower-dimensional space. A principal component analysis (PCA) is one of the most important methods of dimensionality reduction. Several ML-based control charts of SPC using the PCA and Kernel PCA (KPCA) methods have been developed to address monitoring problems with a large number of correlated variables in case of linear and non-linear processes Stefatos and Hamza [27],Lee et al. [28], Lee et al. [29]. More recently, deep learning (DL) models, which are essentially neural networks with multi-layers, have been used as feature extraction techniques for non-linear process monitoring. Among other DL models, autoencoder (AE) is an efficient and flexible dimensionality reduction algorithm for high-dimensional data and has a specific structure of neural networks in which the input is the same as the output. This is an unsupervised learning model. Moreover, in the real world, due to the rapid evolution of sensor technologies, the high-dimensional data generated from IoT sensors not only have autocorrelation but also have non-stationary behaviors; that is, the underlying mechanism producing a series of measurements, changes over time Zhao et al. [30].

Non-stationarity in processes' behavior frequently occurs due to several factors such as seasonal changes, processes that involve emptying and filling cycles, throughput changes, the presence of unmeasured disturbances, and also the nature of the process itself (see, Chen et al. [31]). Forecasting and SPC of non-stationary multivariate time series data are two very important tasks in SM. Although, recent research has proposed ML-based control charts using long short-term memory networks (LSTM) [32], LSTM Autoencoder [10] and variational autoencoder (VAE) [33] for monitoring multivariate non-stationary time series data control charts based only on VAE often fail in detecting long-term anomalies as VAE models cannot analyze information beyond a short local window. Also, these are like "black-box", with no explainability of out-of-control signals and need to include human decision-making in the operation of even the most sophisticated industrial automation solutions. XAI, discussed in Raza et al. [34], is one approach to this problem and it is a new advancement in ML that is designed to provide AI algorithms with narrative functionality; capable of communicating the major steps taken in arriving at a solution, to a human.

5 Challenges for ML-based control charts

Despite the impressive advancements of machine learning-based models, there remain significant limitations when deploying purely data-driven models in real-world applications. In particular, data-driven machine learning models can suffer from several limitations such as a lack of robustness, interpretability, and adherence to physical constraints or commonsense reasoning. In this section, the core challenges of ML-based control charts for AD are analyzed and presented, including high-dimensional Data, Multicollinearity of Data, Interpretability, and Hyperparameter Optimization, which are the main bottleneck problems that limit the expansion of smart manufacturing.

5.1 Non-stationary Data

Data from different users can also exhibit statistical challenges due to variations in data generation and collection processes, leading to non-stationary data. Non-stationarity in processes' behavior frequently occurs due to several factors such as seasonal changes, processes that involve emptying and filling cycles, throughput changes, the presence of unmeasured disturbances, and also the nature of the process itself[31].

5.2 Interpreting out-of-control

Nowadays in smart factories, the data collected from IoT sensors is high-dimensional with a complex structure (such as multivariate time series), which show non-stationary behaviors; that is, the underlying mechanism producing a series of measurement changes over time. In this case, interpreting out-of-control signals is a challenge as studies on the topic almost always make assumptions about the distribution. Although there are several studies in the field of ML-based control charts for multivariate non-stationary time series in AD these methods are often implemented in a "black box" format and have been criticized due to lack of interpretability[34;35].

5.3 Hyperparameter Optimization

Hyperparameter tuning plays a significant role in the overall performance of ML-based control charts and can be the main factor in deciding whether a trained model turns out to be state-of-the-art or simply moderate. The growing sizes of data sets and the complexities of the models result in an increased training time and require more resources, which makes hyperparameter optimization an even harder task. When automating hypermeter tuning, there is a need for practical solutions that must be robust, scalable, and maintain a low computational budget, while ensuring a strong final performance.

6 A Perspective about Explainable Machine Learning-based Control Charts for Anomaly Detection

In the developing context of IoT, new algorithms and approaches in SQC are required to improve the reliability and explainability of complex machine learning models. The methodologies used in our study related to DL models[34] and reinforcement learning[36] for anomaly detection models with high-dimensional non-stationary time series data. Also, variational autoencoder (VAE) and Transformer-based VAE are the core architecture of the deployment to achieve data reduction, high accuracy, convergence, efficiency, and reliability in the high-dimensional IoT environment. VAE is an efficient and flexible dimensionality reduction algorithm for high-dimensional data. Transformers are robust neural networks that have achieved tremendous success in many areas of machine learning[37] and have become state-of-the-art models for various applications across different data modalities. In addition to their excellent performance on supervised learning tasks, transformers can effectively transfer the learned knowledge from a pretraining task to new tasks with limited or no supervision[34;38]. Therefore, Transformer is the potential algorithm to address time-series data problems in the manner of online anomaly detection. The development of technologies entails the complexity of management models, which is the case of social networks, telecommunications and IoT networks, biological networks, or device networks. Besides, XAI, a promising technique that can help in the interpretability of "out-of-control" in SPC systems, also is integrated to improve performance and applied to different data tasks, e.g., anomaly detection, energy, and power system, health monitoring, and time-series prediction[35;34].

In this section, a new approach for efficient and explainable ML-based control charts for anomaly detection in high-dimensional non-stationary time series data is proposed by using a Variational Autoencoder and Transformer-based Variational Autoencoder with an explainable artificial intelligence module that could eliminate the aforementioned drawbacks. Then, consider reinforcement learning (RL) as a practical solution to optimize the hyperparameters of the proposed ML-based control charts.

6.1 Limitations and Possible Solutions

Although there are several studies in the field of ML-based control charts for multivariate non-stationary time series in AD these methods are often implemented in a "black box" format and have been criticized due to lack of interpretability. This section presents different potential techniques that can effectively address these challenges.

6.1.1 Non-stationarity Problem

Nowadays in smart factories, the data collected from IoT sensors is high dimensional data with complex structures (such as multivariate time series), which show non-stationary behaviors; that is, the underlying mechanism producing a series of measurement changes over time. In this case, interpreting out-of-control signals is a challenge as studies on the topic almost always make assumptions about the distribution. Forecasting and SPC of non-stationary multivariate time series data are two very important tasks in SM. Although, recent research has proposed ML-based control charts using long short-term memory networks (LSTM)[32], LSTM Autoencoder[10] and variational autoencoder (VAE)[33] for monitoring multivariate non-stationary time series data. However, control charts based only on VAE often fail to detect long-term anomalies as VAE models cannot analyze information beyond a short local window. Hybrid autoencoder models such as; LSTM-VAE and Transformer-based VAE models can be used to handle the non-stationarity problem of multivariate time series data.

6.1.2 Hyperparameter Optimization

The overall performance of ML models significantly depends on the hyperparameter tuning of the model and can be the main factor in deciding whether a trained model turns out to be state-of-the-art or simply moderate. The complexities of multivariate data sets and the models result in an increased training time and require more resources, which makes hyperparameter optimization an even harder task. Therefore, finding a scalable and efficient policy for hyperparameter tuning is critical, especially for high-dimensional function estimators, such as deep neural networks that require a longer time to evaluate a single hyperparameter response. Therefore, a robust and efficient reinforcement learning (RL)[36] agent, which explores the hyperparameter space of a fixed network topology agent, as a solution to optimize hyperparameters of the proposed ML-based control charts, is utilized as a state of the art solution.

6.1.3 Interpretability/ Explainable ML-based Control charts

Although there are several studies in the field of ML-based control charts for multivariate non-stationary time series in AD these methods are often implemented in a "black box" format and are been criticized due to lack of interpretability[34], [35]. To eliminate those drawbacks, XAI[34] module could be applied which provides explanations of black-box models to reveal the behavior and underlying decision-making mechanisms of the models through tools, techniques, and algorithms. In summary, the drawbacks and potential solutions are described in Figure 1.

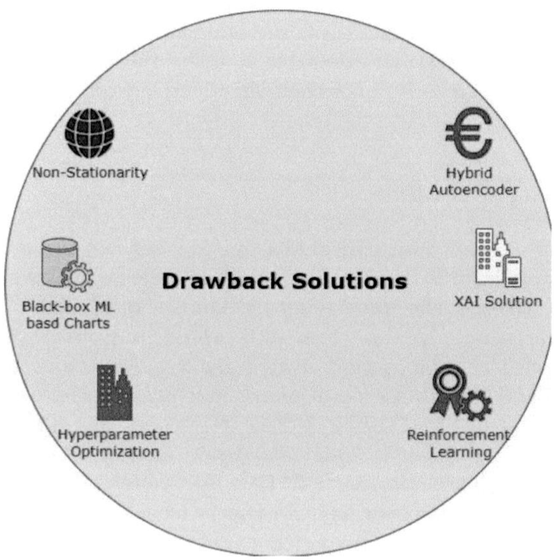

Figure 1: The existing drawbacks and proposed solutions in ML-based Control Charts

6.2 An End-to-End Efficient and Explainable SQC Anomaly Detection System

This section relies on the aforementioned solutions and formulates them into an end-to-end state-of-the-art method for the explainable ML-based Control charts in high-dimensional time series data of IoT systems. The end-to-end system described in Figure 2 consists of 3 layers as:

- Layer 1: The collected multivariate time series data is trained using DL models with VAE architecture. In this layer, turning and optimizing techniques are also applied for data reductions.
- Layer 2. At this layer, Hotelling T^2 and MEWMA control charts are used for data classification as anomaly/out-of-control or not anomaly/in-control.
- Layer 3. Finally, Explainable AI is utilized to enhance the transparency and interpretability of model decisions.

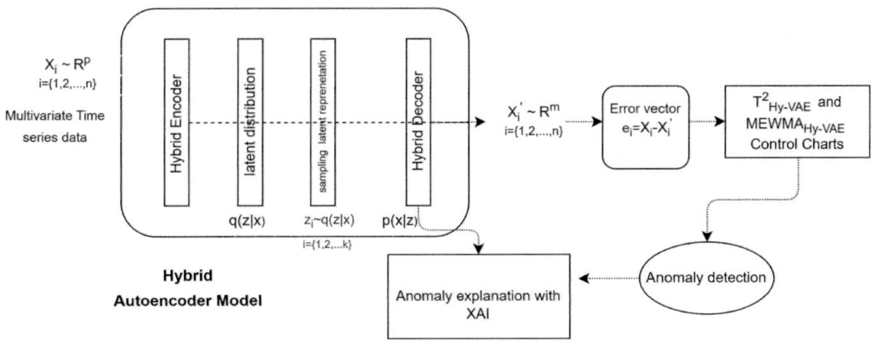

Figure 2: Explainable ML-based control charts for high-dimensional time series data.

7 Future Aspects

In this section, we aim to provide a comprehensive outlook by introducing the future perspectives of analog federated learning and its intersection with quantum aspects.

7.1 Multimodal Learning SQC Perspective

Multimodal learning (MML) in machine learning is a type of learning where the model is trained to understand and work with multiple forms of input data, such as text, images, and audio. *More specifically, MML aims to build models that can process and relate information from multiple modalities. It is a vibrant multi-disciplinary field of increasing importance and with extraordinary potential*[39]. Learning from multiple modalities is important because the information in the real world often involves more than one modality. In fact, in a dataset containing data of different modalities, one modality could carry information that is not available from the other modality. An example is an image of a city with its caption mentioning the name of the city Srivastava and Salakhutdinov [40]. Without the textual information, the name of the city could be hard or impossible to deduce from the image alone. Multimodal fusion has generated great interest in multimodal data processing and general concern among researchers [41;42]To solve the problem of the discrepancy between the information from the original data of each modality before learning, feature-level multimodal fusion is used to fuse the extracted features from each modality. regarding the decision-level fusion method, it integrates knowledge generated from multimodal information into different basic models, and then a balanced fusion decision is made for the results of these models. By combining feature-level fusion and decision-level fusion, hybrid multimedia fusion has a more flexible structure.

The challenges of AD in SQC are that higher performance is hard to achieve and multiple features are difficult to extract with single modal samples and individual deep learning models. Thus, it is necessary to combine different models and multimodal data, so as to achieve more robust and accurate anomaly detection results. Therefore, a promising future direction is multimodal learning AD employing multimodal machine learning. In this scheme, a hybrid scheme based on CNN for image database and LSTM for digital database can be used for high dimensional multivariate time series data for anomaly detection. A late multimodal fusion model with two networks, a convolutional neural network (CNN) and a recurrent neural network with long short-term memory layers (LSTM) can be built. The first network consists of the first branch with a convolutional network that receives the input images of IoT sensor data. In the second network, modeling is performed by the LSTM network to receive the digital input data of IoT sensors. This provides a framework to integrate information from two modalities to ensure anomaly detection in multivariate time series data of IoT sensors. This approach aims to provide high performance by extracting information from multiple features. The architecture of multimodal machine learning for anomaly detection is explained in Figure 3.

7.2 Physics-Informed Machine Learning SQC Perspective

Recent advances in data-driven machine learning have revolutionized fields like computer vision, reinforcement learning, and many scientific and engineering domains. In many real-world and scientific problems, systems that generate data are governed by physical laws. Recent work shows that it provides potential benefits for machine learning models by incorporating the physical prior and collected data, which makes the intersection of machine learning and physics to a prevailing paradigm. By integrating the data and mathematical physics models seamlessly, it can guide the machine learning model towards solutions that are physically plausible, improving accuracy and efficiency even in uncertain and high-dimensional contexts [43]. *The integration of physical laws or constraints into machine learning models, therefore, presents new opportunities for traditional scientific research, substantially advancing the discovery of new knowledge, and facilitating the research in persistent issues of machine learning, such as robustness, interpretability, and generalization* [44;45].

Physical-informed neural networks (PINNs), where the neural networks are trained to solve scientific problems leveraging underlying physics laws structured by generalizable nonlinear partial differential equations (PDEs). So far, PINNs have proven to be highly effective in solving many complex engineering problems (e.g., fluid mechanics, cyber-physical systems, power systems, molecular biology21, cuffless blood pressure estimation) with limited experimental data.

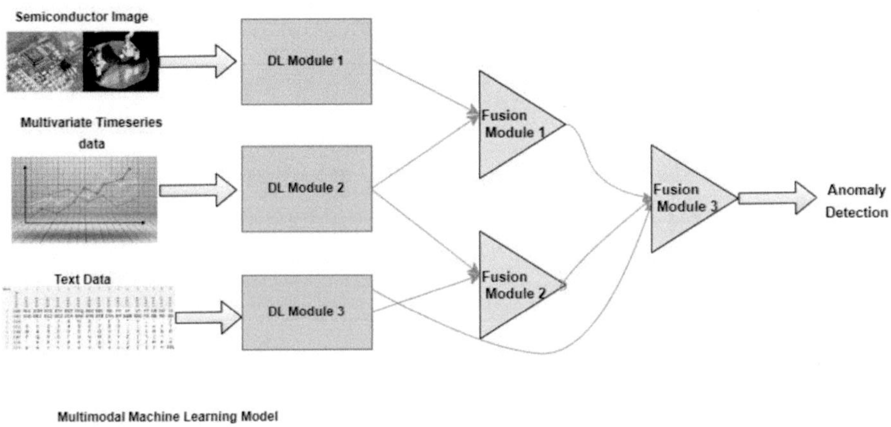

Figure 3: Multimodal Machine Learning structure for anomaly detection.

Another future direction is physics-informed AD employing physics-informed machine learning, which is a promising direction in AI. In this scheme, a framework for developing physics-informed neural network (PINN) models and PINN-based predictive control schemes for high-dimensional multivariate time series can be designed. These neural networks are trained to solve scientific problems leveraging underlying physics laws structured by generalizable nonlinear partial differential equations (PDEs). Additionally, the physical constraints on process states are embedded in the machine learning models to prevent physically unreasonable predictions. Subsequently, the PINN model that captures the dynamic behavior of the time series process is utilized in the design of a predictive model controller that optimizes the operation of the multivariate time series data. The PINN models used less training (i.e, labeled) data to achieve prediction accuracy and performance comparable to the purely data-driven model. The methodology is described in Figure 4. The overall process is analogous to conventional deep neural networks (DNN)leveraging underlying physics laws structured by generalized nonlinear PDEs. First, the DNN uses multivariate time series data to estimate the model. Taylor's approximation is defined for AD features from multivariate time series data. The parameters of the model are calculated with DNN auto-differentiation of predictions with respect to input features. This estimation is compared with the DNN predictors to estimate values for the physics-based loss functions. For PINN, the partial differentials of output data with regard to input features are calculated and fit into Taylor's approximation polynomial. A new error function that combines conventional error and physics-based NN error is used to train PINN

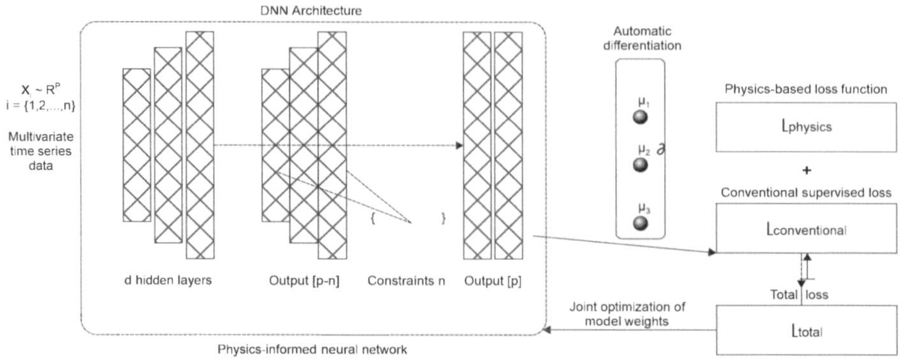

Figure 4: Physics-informed machine learning architecture for anomaly detection.

7.3 Self supervised learning

In many industrial environments, time series data is mainly dealt with to monitor machines, IT devices, spacecraft, or engines. Anomaly detection is one of the essential tasks for time series analysis, which can find defects in machines and prevent potential harm. In the literature, several deep learning-based approaches and Transformer architecture have been applied to treat multivariate data in the temporal order[46;47]. These approaches take account of the temporal characteristics of data and successfully adapt deep neural networks to the field of time series analysis. Most data sets do not provide ground truth labels for the training set in this area. In other words, it is unknown whether a point is anomalous or not in the training set. Therefore, previous studies have developed unsupervised learning methods for anomaly detection. Some of them insert an autoencoder or an adversarial network[48] into their model to understand data distribution efficiently. However, it is still a hard problem to detect anomalies from the temporal context without supervision. Moreover, abnormalities may display unexpected behavior and be related to multiple variables, which makes the detection task more difficult in real situations.

Therefore, an alternate direction for future research is the use of a self-supervised learning (SSL) approach in explainable ML-based control charts for multivariate time series data. SSL has been widely exploited in different domains, including computer vision[47], audio/speech processing[49], and time-series analysis[50;51]. SSL algorithm extracts useful features from large-scale unlabeled datasets to generate labels without the need for human annotation. Generally, the SSL process is separated into two successive tasks: pretext downstream task training. The model operates by training the unlabeled dataset through a self-supervised pretext task and, subsequently, transfers the learned parameters to downstream task training. In image data, the pretext task is used to learn representations with rotation prediction and distribution-augmented learning to extract useful features. Specifically, multiple research generates pseudo labels based on colorization, placing, corrupt, crop and resize, and so on. Those algorithms have demonstrated state-of-the-art performance in extracting useful features and efficiently applying semantic anomaly detection. The general framework of SSL is explained in Figure 5.

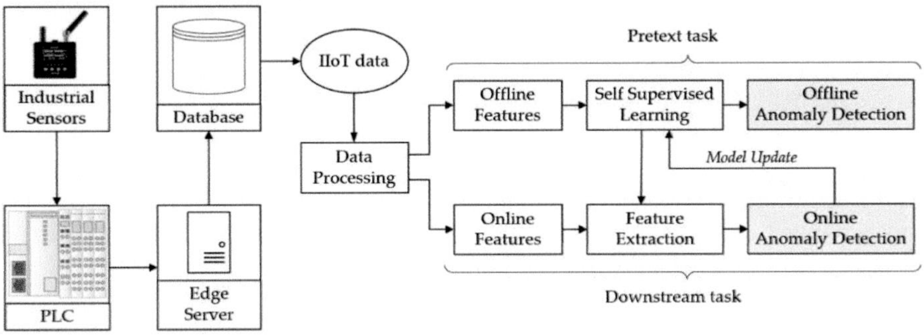

Figure 5: Anomaly detection framework in IIoT system using the SSL method.

7.4 Embedded AI

AI is profoundly changing human social life and the world at an unprecedented pace. AI technologies can create value for network devices across multiple aspects, such as parameter optimization, application identification, security, and fault diagnosis. In the current technological landscape, the convergence of artificial intelligence (AI) and electronic systems has paved the way for a novel paradigm: embedded AI. This trend has gained considerable traction recently, owing to its ability to facilitate the creation of low-power, cost-effective AI solutions that would be unfeasible otherwise. In a more technical sense, embedded AI is the application of AI algorithms and models at the device level that can function in isolation without the need for external intervention. Over the past several years, an important shift has occurred from cloud-level to device-level processing of artificial intelligence tasks, data, and results. Embedded AI is the direct result of this important shift. Traditionally, complex AI computations, such as producing search engine results, were performed at a data center in the cloud. With the implementation of AI models on graphics processing units (GPUs), session border controllers (SBCs), and systems on chips (SoCs), there is lesser dependence on the cloud for AI data processing. With embedded AI, devices have the ability to run AI models at the device level and directly use the results to perform an appropriate task or action. The cloud is still helpful from a data storage perspective, as data can be stored temporarily at the device level and eventually sent to a cloud server for safekeeping. While the uses and applications of embedded AI are vast, here's a short list of industries where this technology is automating processes, providing advanced analytics and business insights, and improving customer service, among numerous other benefits[52;53].

Machine learning applications or ML models are resource-intensive and require a system with a lot of computing power. Hence they are usually executed on resource-constrained devices, for example, on a PC or cloud servers, where processing data goes smoothly. With the recent advancements in data science, algorithms, and computing power of the processors, deployment of machine learning applications or

ML frameworks directly on embedded devices is now possible. This is known as Embedded Machine Learning (E-ML) or TinyML applications. Another promising future direction is employing embedded ML algorithms for anomaly detection. This pushes the processing towards the edge where the sensors gather data and help remove barriers such as interruptions in bandwidth and connection, security breaches due to the transfer of data over the internet, and power consumption to transmit data. This is especially important for deep learning as it facilitates autonomy and intelligence at the edge. Moreover, it also facilitates applications of neural networks, other ML frameworks, signal processing services, model development and gesture recognition.

Another future direction is embedded AI employing an embedded machine learning solution called the Isolation Forest technique to analyze machine information and detect anomalies. This way, the AI controller monitors machine status, like voltage and temperature, and alerts maintenance when this data shows anomalies that could lead to machine failure and unplanned downtime. It then uses the information learned to predict the best timing for maintenance services.

8 Summary and Conclusion

In this chapter, we discussed Machine learning-based control charts for Anomaly Detection, specifically for Smart Manufacturing in the new digital era in the future generation of industry, where systems become increasingly complex, producing massive data, necessitating more demand for transparency and performance. Machine learning has demonstrated its effectiveness in various applications; however, several challenges should be addressed, including stationarity problems, hyperparameter optimizations, resource restrictions, and interpretability. In addition, several state-of-the-art approaches and methods to address the limitations of ML-based anomaly detection are also presented. After that, perspectives of combining solutions toward efficient and explainable ML-based control charts for anomaly detection systems based on deep learning are shown through an end-to-end framework.

Finally, to provide a complete picture, various machine learning methods are introduced in AD of multivariate time series data as a potential research direction.

References

[1] Phuong Hanh Tran, Adel Ahmadi Nadi, Thi Hien Nguyen, Kim Duc Tran, and Kim Phuc Tran. Application of machine learning in statistical process control charts: A survey and perspective. In *Control charts and machine learning for anomaly detection in manufacturing*, pages 7–42. Springer, 2022.

[2] Varun Chandola, Arindam Banerjee, and Vipin Kumar. Anomaly detection: A survey. *ACM computing surveys (CSUR)*, 41(3):1–58, 2009.

[3] Truong Thu Huong, Nguyen Minh Dan, Nguyen Xuan Hoang, Kieu-Ha Phung, Kim Phuc Tran, et al. Anomaly detection enables cybersecurity with machine learning techniques. In *Machine Learning and Probabilistic Graphical Models for Decision Support Systems*, pages 124–183. CRC Press, 2022.

[4] Kim Phuc Tran. *Introduction to control charts and machine learning for anomaly detection in manufacturing.* Springer, 2022.

[5] Xiulin Xie and Peihua Qiu. Machine learning control charts for monitoring serially correlated data. In *Control Charts and Machine Learning for Anomaly Detection in Manufacturing*, pages 131–147. Springer, 2022.

[6] Kim Phuc Tran, Anh Tuan Mai, et al. Anomaly detection in wireless sensor networks via support vector data description with mahalanobis kernels and discriminative adjustment. In *2017 4th NAFOSTED Conference on Information and Computer Science*, pages 7–12. IEEE, 2017.

[7] Walter A Shewhart. Some applications of statistical methods to the analysis of physical and engineering data. *Bell System Technical Journal*, 3(1):43–87, 1924.

[8] P Castagliola, KP Tran, G Celano, AC Rakitzis, and PE Maravelakis. An ewma-type sign chart with exact run length properties. *Journal of quality technology*, 51(1):51–63, 2019.

[9] Robab Afshari, Adel Ahmadi Nadi, Arne Johannssen, Nataliya Chukhrova, and Kim Phuc Tran. The effects of measurement errors on estimating and assessing the multivariate process capability with imprecise characteristic. *Computers & Industrial Engineering*, 172:108563, 2022.

[10] HD Nguyen, Kim Phuc Tran, and Kim Duc Tran. The effect of measurement errors on the performance of the exponentially weighted moving average control charts for the ratio of two normally distributed variables. *European Journal of Operational Research*, 293(1):203–218, 2021.

[11] Aamir Saghir, Saddam Akber Abbasi, and Alireza Faraz. The exact method for designing the maxwell chart with estimated parameter. *Communications in Statistics-Simulation and Computation*, 50(1):270–281, 2021.

[12] Aamir Saghir, Muhammad Aslam, Alireza Faraz, Liaquat Ahmad, and Cedric Heuchenne. Monitoring process variation using modified ewma. *Quality and Reliability Engineering International*, 36(1):328–339, 2020.

[13] Aamir Saghir, Liaquat Ahmad, and Muhammad Aslam. Modified ewma control chart for transformed gamma data. *Communications in Statistics-Simulation and Computation*, 50(10):3046–3059, 2021.

[14] TH Nguyen, HD Nguyen, KD Tran, TH Truong, KH Phung, LH Nguyen, TTN Le, and KP Tran. One-sided synthetic-rz control charts: a new method for anomaly detection. In *2019 6th NAFOSTED Conference on Information and Computer Science (NICS)*, pages 262–267. IEEE, 2019.

[15] Farid Kadri, Fouzi Harrou, Sondès Chaabane, Ying Sun, and Christian Tahon. Seasonal arma-based spc charts for anomaly detection: Application to emergency department systems. *Neurocomputing*, 173:2102–2114, 2016.

[16] Gerhard Münz and Georg Carle. Application of forecasting techniques and control charts for traffic anomaly detection. In *Proc. 19th ITC Specialist Seminar on Network Usage and Traffic, Berlin, Germany*, 2008.

[17] Hoang Pham. Recent advances in reliability and quality engineering. 2001.

[18] Huong Thu Truong, Bac Phuong Ta, Quang Anh Le, Dan Minh Nguyen, Cong Thanh Le, Hoang Xuan Nguyen, Ha Thu Do, Hung Tai Nguyen, and Kim Phuc Tran. Light-weight federated learning-based anomaly detection for time-series data in industrial control systems. *Computers in Industry*, 140: 103692, 2022.

[19] Truong Thu Huong, Ta Phuong Bac, Dao Minh Long, Tran Duc Luong, Nguyen Minh Dan, Bui Doan Thang, Kim Phuc Tran, et al. Detecting cyberattacks using anomaly detection in industrial control systems: A federated learning approach. *Computers in Industry*, 132:103509, 2021.

[20] D Demircioglu Diren, S Boran, and I Cil. Integration of machine learning techniques and control charts in multivariate processes. 2020.

[21] Tai-fu Li, Sheng Hu, Zheng-yuan Wei, Zhi-qiang Liao, et al. A framework for diagnosing the out-of-control signals in multivariate process using optimized support vector machines. *Mathematical Problems in Engineering*, 2013, 2013.

[22] R-S Guh. Real-time recognition of control chart patterns in autocorrelated processes using a learning vector quantization network-based approach. *International Journal of Production Research*, 46(14):3959–3991, 2008.

[23] Tai-Yue Wang and Long-Hui Chen. Mean shifts detection and classification in multivariate process: a neural-fuzzy approach. *Journal of intelligent manufacturing*, 13:211–221, 2002.

[24] Chinyao Low, Chih-Ming Hsu, and Fong-Jung Yu. Analysis of variations in a multi-variate process using neural networks. *The International Journal of Advanced Manufacturing Technology*, 22:911–921, 2003.

[25] Mojtaba Salehi, Ardeshir Bahreininejad, and Isa Nakhai. On-line analysis of out-of-control signals in multivariate manufacturing processes using a hybrid learning-based model. *Neurocomputing*, 74(12-13):2083–2095, 2011.

[26] Seyed Taghi Akhavan Niaki and Babak Abbasi. Fault diagnosis in multivariate control charts using artificial neural networks. *Quality and reliability engineering international*, 21(8):825–840, 2005.

[27] George Stefatos and A Ben Hamza. Statistical process control using kernel pca. In *2007 Mediterranean Conference on Control & Automation*, pages 1–6. IEEE, 2007.

[28] Jong-Min Lee, ChangKyoo Yoo, Sang Wook Choi, Peter A Vanrolleghem, and In-Beum Lee. Nonlinear process monitoring using kernel principal component analysis. *Chemical engineering science*, 59(1):223–234, 2004.

[29] Wo Jae Lee, Gamini P Mendis, Matthew J Triebe, and John W Sutherland. Monitoring of a machining process using kernel principal component analysis and kernel density estimation. *Journal of Intelligent Manufacturing*, 31:1175–1189, 2020.

[30] Chunhui Zhao, He Sun, and Feng Tian. Total variable decomposition based on sparse cointegration analysis for distributed monitoring of nonstationary industrial processes. *IEEE Transactions on Control Systems Technology*, 28 (4):1542–1549, 2019.

[31] Qian Chen, Uwe Kruger, and Andrew YT Leung. Cointegration testing method for monitoring nonstationary processes. *Industrial & Engineering Chemistry Research*, 48(7):3533–3543, 2009.

[32] Kim Phuc Tran, Huu Du Nguyen, and Sébastien Thomassey. Anomaly detection using long short term memory networks and its applications in supply chain management. *IFAC-PapersOnLine*, 52(13):2408–2412, 2019.

[33] Seulki Lee, Mingu Kwak, Kwok-Leung Tsui, and Seoung Bum Kim. Process monitoring using variational autoencoder for high-dimensional nonlinear processes. *Engineering Applications of Artificial Intelligence*, 83:13–27, 2019.

[34] Ali Raza, Kim Phuc Tran, Ludovic Koehl, and Shujun Li. Designing ecg monitoring healthcare system with federated transfer learning and explainable ai. *Knowledge-Based Systems*, 236:107763, 2022.

[35] R Machlev, L Heistrene, M Perl, KY Levy, J Belikov, S Mannor, and Y Levron. Explainable artificial intelligence (xai) techniques for energy and power systems: Review, challenges and opportunities. *Energy and AI*, 9:100169, 2022.

[36] Hadi S Jomaa, Josif Grabocka, and Lars Schmidt-Thieme. Hyp-rl: Hyperparameter optimization by reinforcement learning. *arXiv preprint arXiv:1906.11527*, 2019.

[37] Alec Radford and Jeffrey Wu. Rewon child, david luan, dario amodei, and ilya sutskever. 2019. *Language models are unsupervised multitask learners. OpenAI blog*, 1(8):9, 2019.

[38] Danyang Liu and Gongshen Liu. A transformer-based variational autoencoder for sentence generation. In *2019 International Joint Conference on Neural Networks (IJCNN)*, pages 1–7. IEEE, 2019.

[39] Changchang Che, Huawei Wang, Xiaomei Ni, and Ruiguan Lin. Hybrid multimodal fusion with deep learning for rolling bearing fault diagnosis. *Measurement*, 173:108655, 2021.

[40] Nitish Srivastava and Russ R Salakhutdinov. Multimodal learning with deep boltzmann machines. *Advances in neural information processing systems*, 25, 2012.

[41] Khaled Bayoudh, Raja Knani, Fayçal Hamdaoui, and Abdellatif Mtibaa. A survey on deep multimodal learning for computer vision: advances, trends, applications, and datasets. *The Visual Computer*, pages 1–32, 2021.

[42] Tadas Baltrušaitis, Chaitanya Ahuja, and Louis-Philippe Morency. Multimodal machine learning: A survey and taxonomy. *IEEE transactions on pattern analysis and machine intelligence*, 41(2):423–443, 2018.

[43] Zhongkai Hao, Songming Liu, Yichi Zhang, Chengyang Ying, Yao Feng, Hang Su, and Jun Zhu. Physics-informed machine learning: A survey on problems, methods and applications. *arXiv preprint arXiv:2211.08064*, 2022.

[44] Nils Thuerey, Philipp Holl, Maximilian Mueller, Patrick Schnell, Felix Trost, and Kiwon Um. Physics-based deep learning. *arXiv preprint arXiv:2109.05237*, 2021.

[45] George Em Karniadakis, Ioannis G Kevrekidis, Lu Lu, Paris Perdikaris, Sifan Wang, and Liu Yang. Physics-informed machine learning. *Nature Reviews Physics*, 3(6):422–440, 2021.

[46] Ya Su, Youjian Zhao, Chenhao Niu, Rong Liu, Wei Sun, and Dan Pei. Robust anomaly detection for multivariate time series through stochastic recurrent neural network. In *Proceedings of the 25th ACM SIGKDD international conference on knowledge discovery & data mining*, pages 2828–2837, 2019.

[47] Jiehui Xu, Haixu Wu, Jianmin Wang, and Mingsheng Long. Anomaly transformer: Time series anomaly detection with association discrepancy. *arXiv preprint arXiv:2110.02642*, 2021.

[48] Ian Goodfellow, Jean Pouget-Abadie, Mehdi Mirza, Bing Xu, David Warde-Farley, Sherjil Ozair, Aaron Courville, and Yoshua Bengio. Generative adversarial nets. *Advances in neural information processing systems*, 27, 2014.

[49] Jean-Yves Franceschi, Aymeric Dieuleveut, and Martin Jaggi. Unsupervised scalable representation learning for multivariate time series. *Advances in neural information processing systems*, 32, 2019.

[50] Duc Hoang Tran, Van Linh Nguyen, Huy Nguyen, and Yeong Min Jang. Self-supervised learning for time-series anomaly detection in industrial internet of things. *Electronics*, 11(14):2146, 2022.

[51] Yungi Jeong, Eunseok Yang, Jung Hyun Ryu, Imseong Park, and Myungjoo Kang. Anomalybert: Self-supervised transformer for time series anomaly detection using data degradation scheme. *arXiv preprint arXiv:2305.04468*, 2023.

[52] Teresa Pelinski, Victor Shepardson, Steve Symons, Franco Santiago Caspe, Adan L Benito Temprano, Jack Armitage, Chris Kiefer, Rebecca Fiebrink, Thor Magnusson, and Andrew McPherson. Embedded ai for nime: Challenges and opportunities. In *International Conference on New Interfaces for Musical Expression*. PubPub, 2022.

[53] Farzam Farbiz, Mohd Salahuddin Habibullah, Brahim Hamadicharef, Tomasz Maszczyk, and Saurabh Aggarwal. Knowledge-embedded machine learning and its applications in smart manufacturing. *Journal of Intelligent Manufacturing*, 34(7):2889–2906, 2023.

8. Monitoring the Ratio of Two Normal Variables and Compositional Data: A Literature Review and Perspective

Thi Thuy Van Nguyen,[1,*] Thi Hien Nguyen,[2] Kim Duc Tran,[3,*] Cédric Heuchenne[1] and Kim Phuc Tran[4]

1 Introduction

In the evolving landscape of manufacturing, marked by the advent of Industry 5.0, the role of control charts in quality monitoring has become increasingly significant. Industry 5.0, characterized by a strong focus on personalization, sustainability, and the synergistic collaboration between humans and technology, demands innovative approaches in quality control to match its complex and dynamic nature. In this context, control charts, a fundamental tool in statistical process monitoring (SPC), are adapting to meet the nuanced needs of this new industrial era.

Control charts are efficient tools in providing a graphical representation of process data, essential for detecting changes in manufacturing processes. Implementing control charts allows manufacturing companies to monitor the quality of their products, thereby reducing waste, preventing defects, and improving production process efficiency. In Industry 5.0, where the manu-facturing processes are more intricate and tailored, the need for precise and adaptable quality monitoring tools is paramount. Control charts for monitoring the ratio of two normal random variables and Compositional Data (CoDa)

[1] HEC Liège - Management School, University of Liège, Belgium.
[2] Laboratoire AGM, UMR CNRS 8088, CY Cergy Paris, University, France.
[3] IAD, Dong A University, Da Nang, Vietnam.
[4] ENSAIT, GEMTEX, University of Lille, France.
* Corresponding author: ttv.nguyen@uliege.be; ductk@donga.edu.vn

have already received attention and continue to be an important point in this advanced manufacturing paradigm. The ratio of two normal random variables, a common measurement in numerous industrial applications, plays a critical role in ensuring process consistency and detecting production anomalies. This is especially relevant in Industry 5.0, where even minor variations in ratios, such as ingredient proportions in the food industry or dimensional ratios in manufacturing, can significantly impact the final product's quality and customization. CoDa, on the other hand, refers to data representing the proportions of different components that constitute a whole. CoDa is commonly encountered in various industrial applications, such as chemical research, econometrics, and food industries. The control chart for monitoring CoDa is used to monitor the proportions of the different components and detect any changes in their distribution. In the context of Industry 5.0, the control chart for monitoring CoDa becomes crucial for maintaining the integrity of products where the balance of components is key to quality and customer satisfaction. To obtain additional examples of the applications of the ratio of random variables and CoDa, refer to the works of Celano et al. [1] and Aitchison [2], respectively. While the ratio of two normal variables and CoDa represent distinct types of data, they both fundamentally involve ratios or proportions of different components or variables and share a common aspect of incorporating relative information between them. Specifically, when CoDa is limited to two components and follows a normal distribution within the simplex space, it may exhibit similarities to the ratio of two normal variables (even though they are still fundamentally different due to the constraints on CoDa). Given this shared conceptual foundation in the representation of relative information through ratios or proportions, our review strategically groups control charts designed for monitoring these two data types.

Consequently, in this chapter, we will provide a comprehensive overview of control charts for monitoring the ratio of two normal random variables and CoDa and provide intuitive outlines for papers in these fields. To achieve this, a comprehensive overview encompassing 87 studies was conducted to identify relevant studies for inclusion in this review. A total of 87 studies were selected based on their relevance, with 68 studies focusing on control charts for monitoring the ratio of two normal random variables and the remaining 19 studies focusing on control charts for monitoring CoDa. To facilitate the analysis and presentation of the selected studies, they were categorized based on data type into two groups: control charts for monitoring the ratio of two normal random variables and control charts for monitoring CoDa. The remainder of the paper is organized as follows: Sections 2 and 3 provide concerns about the control chart for monitoring the ratio of two normal variables and CoDa, together with some discussions on the advantages and disadvantages of these control charts. Section 4 offers some perspective views on the machine learning techniques in the stages of designing, pattern recognition, and interpreting control charts for monitoring Compositional Data (CoDa) and the ratio of two

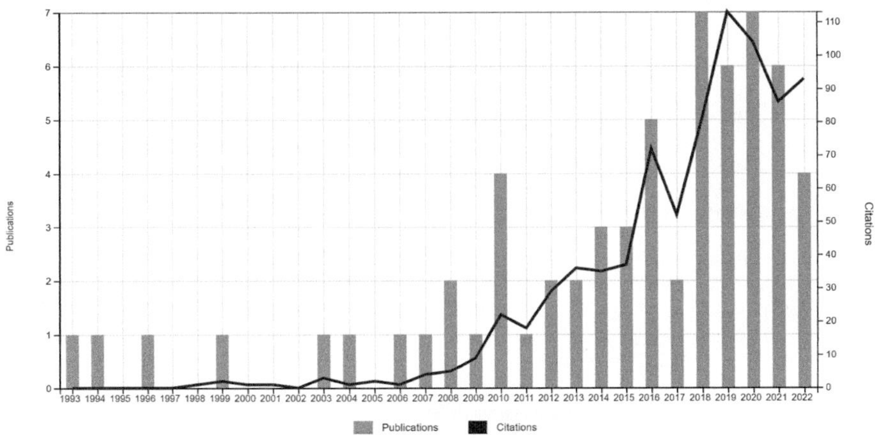

Figure 1: Revolution of publications and citations on Ratio control charts

normal random variables, enhancing the understanding of these essential tools in the context of Industry 5.0. Finally, the chapter ends with a summary of the contributions.

2 The Ratio Control Charts

In all production processes, the practitioners must monitor how well their products meet specifications. In more general terms, product quality has two "enemies": (1) deviations from target product specifications and (2) excessive dispersion around the target product specifications. During the early stages of the development of a production process, the design of experiments is often used to optimize these two quality characteristics, and control charts are effective tools in Statistical Process Control (SPC) for monitoring the stability of a process over time. Currently, most competitive manufacturing companies are implementing SPC in various applications: biology, genetics, medicine, finance, and other areas (see Montgomery[3]). The design and implementation of control charts monitoring the ratio of two variables are attractive from both theoretical and practical points of view. According to the data from *Web of Science*, interest in this type of chart has increased dramatically in recent years. In fact, Figure 1 shows the revolution of the number of publications and number of citations on the Ratio control charts. A large part of which is an *article* published in journals (89.71%), see Table 1 and is in the field of Industrial Engineering (28%) and Operations Research Management Science (26%) (see Fig. **??**). Two selection criteria are used to choose and accept the Ratio control chart research articles: 1) papers are found via a computerized search of the topic areas (Quality control, SPC). The search is narrowed using

Table 1: Document types of publications

Document Types	Number	Percentage
Article	61	89.71
Proceeding Paper	4	5.88
Early Access	2	2.94
Letter	1	1.47

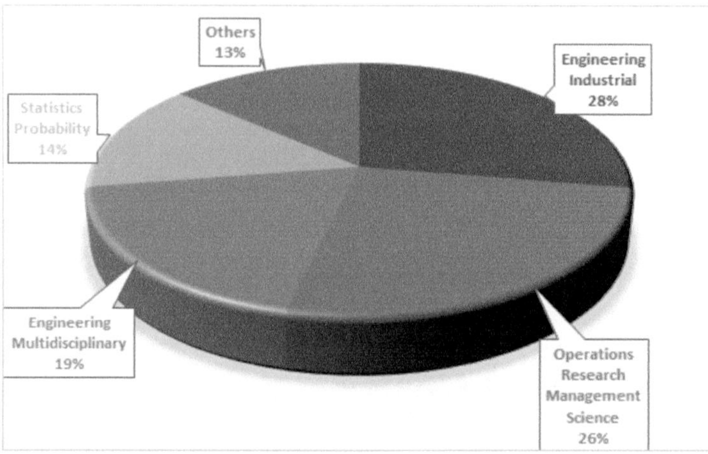

Figure 2: Document types of publications

the following terms: Ratio control chart, ratio distribution, and monitoring Ratio. This research considers the published research, including electronic literature sources such as Wiley, Taylor & Francis, Elsevier, Springer Nature and Scopus. Table 2 lists the number of published papers among 36 journals. *Quality and Reliability Engineering International* is ranked first (20 publications and $\sim 29.41\%$) in this area.

In this section, a brief review of the sample distribution of the ratio of two normal variables is recalled. Then, we summarize the differences between the various types of ratio charts and their performances.

2.1 Derivation of properties of the ratio distribution

Suppose that X and Y be the two normal random variables such that $\mathbf{W} = (X,Y)^T \sim N(\boldsymbol{\mu}_{\mathbf{W}}, \boldsymbol{\Sigma}_{\mathbf{W}})$, i.e. $\mathbf{W}$ is a bi-variate normal random vector with

Table 2: Number and percentage of papers in each journal

Journal name	Number	%
Quality And Reliability Engineering International	20	29.41%
International Journal Of Production Research	4	5.88%
Journal Of Quality Technology	3	4.41%
Communications In Statistics-Theory And Methods	3	4.41%
Journal Of Statistical Computation And Simulation	3	4.41%
Statistical Papers	2	2.94%
Computational Statistics & Data Analysis	2	2.94%
Technometrics	2	2.94%
Computers & Industrial Engineering	2	2.94%
Asqc 48th Annual Quality Congress Proceedings	1	1.47%
American Heart Journal	1	1.47%
Naval Research Logistics	1	1.47%
Journal Of Manufacturing Processes	1	1.47%
International Journal Of Reliability Quality & Safety Engineering	1	1.47%
Scientia Iranica	1	1.47%
Communications In Statistics-Simulation And Computation	1	1.47%
Journal Of Food Science	1	1.47%
Acta Scientiarum-Technology	1	1.47%
Symmetry-Basel	1	1.47%
European Journal Of Operational Research	1	1.47%
Biopharm International	1	1.47%
Arabian Journal For Science And Engineering	1	1.47%
Journal Of Probability And Statistics	1	1.47%
Structural Health Monitoring 2013, Vols 1 And 2	1	1.47%
IIE Transactions	1	1.47%
Engenharia Agricola	1	1.47%
Journal Of Mathematics	1	1.47%
Scientia Iranica Transaction E-Industrial Engineering	1	1.47%
Journal Of Testing And Evaluation	1	1.47%
Quality Innovation Prosperity-Kvalita Inovacia Prosperita	1	1.47%
British Medical Journal	1	1.47%
Science In China Series A-Mathematics	1	1.47%
Productivity & Quality Management Frontiers Iv, Vols 1 And 2	1	1.47%
Journal Of Industrial And Production Engineering	1	1.47%
Icim2012: Proceedings Of The 11th Inter. Conf. On Industrial Management	1	1.47%
Advances And Applications In Statistics	1	1.47%

mean vector and variance-covariance matrix as follows:

$$\boldsymbol{\mu}_{\mathbf{W}} = \begin{pmatrix} \mu_X \\ \mu_Y \end{pmatrix} \quad \text{and} \quad \boldsymbol{\Sigma}_{\mathbf{W}} = \begin{pmatrix} \sigma_X^2 & \rho\sigma_X\sigma_Y \\ \rho\sigma_X\sigma_Y & \sigma_Y^2 \end{pmatrix} \tag{1}$$

where μ_X and μ_Y are the means of two variables and ρ their correlation coefficient. Coefficients of variation (γ_X, γ_Y) and standard-deviation ratio (ω) of X and Y are denoted by $\gamma_X = \frac{\sigma_X}{\mu_Y}$, $\gamma_Y = \frac{\sigma_Y}{\mu_Y}$ and $\omega = \frac{\sigma_X}{\sigma_Y}$, respectively. Let Z be the ratio of X to Y $(Z = X/Y)$. Celano and Castagliola[4] derived an adequate approximation for the *c.d.f* (cumulative distribution function) of Z as a function of $\gamma_X, \gamma_Y, \omega$, and ρ as:

$$F_Z(z|\gamma_X, \gamma_Y, \omega, \rho) \simeq \Phi\left(\frac{A}{B}\right), \tag{2}$$

where

$$A = \frac{z}{\gamma_Y} - \frac{\omega}{\gamma_X}, \qquad \text{and} \qquad B = \sqrt{\omega^2 - 2\rho\omega z + z^2},$$

and Φ is the *c.d.f* of standard normal distribution (Sdn). After some tedious derivations, approximated *p.d.f* (probability density function) of the ratio Z is

$$f_Z(z|\gamma_X, \gamma_Y, \omega, \rho) \simeq \left(\frac{1}{B\gamma_Y} - \frac{(z-\rho\omega)\,A}{B^3}\right) \times \phi\left(\frac{A}{B}\right), \tag{3}$$

where $\phi(\cdot)$ is the p.d.f. of Sdn. Solving the equation $F_Z(z\,|\gamma_X, \gamma_Y, \omega, \rho) = p$ enables obtaining an approximate expression for the *i.d.f* (inverse distribution function) $F_Z^{-1}(p|\gamma_X, \gamma_Y, \omega, \rho)$. We have

$$F_Z^{-1}(p|\gamma_X, \gamma_Y, \omega, \rho) \simeq \begin{cases} \frac{-C_2 - \sqrt{C_2^2 - 4C_1C_3}}{2C_1}, & \text{if} \quad p \in (0, 0.5], \\ \frac{-C_2 + \sqrt{C_2^2 - 4C_1C_3}}{2C_1}, & \text{if} \quad p \in [0.5, 1), \end{cases} \tag{4}$$

where C_1, C_2, and C_3 are functions of $\omega, \rho, \gamma_X, \gamma_Y, p$ and they are:

$$C_1 = \frac{1}{\gamma_Y^2} - \left(\Phi^{-1}(p)\right)^2,$$

$$C_2 = 2\omega\left(\rho\left(\Phi^{-1}(p)\right)^2 - \frac{1}{\gamma_X\gamma_Y}\right),$$

$$C_3 = \omega^2\left(\frac{1}{\gamma_X^2} - \left(\Phi^{-1}(\mathrm{p})\right)^2\right),$$

and where $\Phi^{-1}(\cdot)$ is the *i.d.f.* of Sdn.

2.2 Various types of ratio control charts and their descriptions

In the SPC literature, the research on control charts for monitoring the ratio of two normal random distributions is mainly focused on some types of charts: the traditional Shewhart chart that we denote RZ-chart, overcome the problem of the Shewart chart in the case the small shift-size (Synthetic-RZ, Run-Rule RZ, cumulative sum chart (CUSUM -RZ) and the exponentially weighted moving average charts (EWMA-RZ). After that, to improve the performance of these charts, the researchers found some adaptive models. In other studies, they are looking for the effect of measurement errors on the performance of RZ-charts. In this subsection, we make a summary of these types and give some comments.

2.2.1 Shewhart-RZ charts

With this traditional type chart, the first author to mention is Spisak[5]. He investigated a quality control procedure for insurance against unemployment by estimating the bias in a ratio estimator. An illustrative example in his work is the Department of Labor's Unemployment Insurance Quality Control program. Other applications of ratio control charts are also mentioned. Oksoy et al.[6] discussed some useful guidelines to implement the ratio chart for supervising and controlling the ratio $Z = X/Y$ of glass oxide. Celano et al.[1] design the Shewhart chart based on individual measurements, (i.e. at fixed sampling times k, the sample size is $n = 1$). They are named the Shewhart-RZ control chart. This chart, monitoring the ratio $Z_k = X_k/Y_k$, is designed regarding a couple of probability action limits, upper control limit (UCL) and lower control limit (LCL) as follows

$$\mathrm{UCL}_{RZ} = F_Z^{-1}\left(1 - \frac{\alpha}{2}|\gamma_{X_0}, \gamma_{Y_0}, \omega_0, \rho\right)$$
$$\mathrm{LCL}_{RZ} = F_Z^{-1}\left(\frac{\alpha}{2}|\gamma_{X_0}, \gamma_{Y_0}, \omega_0, \rho\right)$$

where α is the type I error probability associated with the control chart. The chart issues an out-of-control when $Z_i > \mathrm{UCL}_{RZ}$ or $Z_i < \mathrm{LCL}_{RZ}$. The authors considered an illustrative example in food manufacturing; they check the stability of the ratio Z between the weight of almonds and raisins and the total weight of muesli (a mixture of whole grain, rolled oats, raisins, almonds, and several kinds of seeds). The result of the Shewhart-RZ chart showed that between samples #12 and #13, there is an error report (the process is out of control). Extending this work, Celano and Castagliola[4] investigated the RZ chart to the condition assuming that each subgroup consists of $n > 1$ sample units. This extension is very important because, in several manufacturing environments, sample units can be changed due to planning decisions or cannot be ensured with sufficient precision. The statistic they suggest monitoring is

$$\hat{Z}_i = \frac{\sum_{j=1}^{n} X_{i,j}}{\sum_{j=1}^{n} Y_{i,j}}$$

Therefore the control limits are now

$$\text{UCL}_{RZ} = F_{\hat{Z}_i}^{-1}\left(1 - \frac{\alpha}{2}\middle| n, \gamma_X, \gamma_Y, \omega_0, \rho_0\right)$$
$$\text{LCL}_{RZ} = F_{\hat{Z}_i}^{-1}\left(\frac{\alpha}{2}\middle| n, \gamma_X, \gamma_Y, \omega_0, \rho_0\right)$$

where

$$F_{\hat{Z}_i}^{-1}\left(p \middle| n, \gamma_X, \gamma_Y, \omega_0, \rho_0\right) = F_Z^{-1}\left(p \middle| n, \frac{\gamma_X}{\sqrt{n}}, \frac{\gamma_Y}{\sqrt{n}}, \frac{z_0\gamma_X}{\gamma_Y}, \rho_0\right)$$

The fact that the size of each sample unit is not strictly fixed when considering different subgroups simplifies the quality control procedure and makes the chart implementation more flexible. Always the same example in the food industry, the quality practitioner decides to sample $n = 5$ boxes every 30 minutes, and they compute the ratio of the sample average weights of pumpkin seeds and flax seeds from the muesli mixture. The RZ-Shewhart chart immediately signals the occurrence of the out-of-control condition by plotting to point #12 above UCL_{RZ}.

2.2.2 Synthetic RZ-charts

It is well known that Shewhart-type control charts are rather inefficient in some sensitive situations, for example, detecting small or moderate changes in a process. One of the advanced approaches is the synthetic control chart. They are a combination of a traditional chart and a conforming run-length (CRL) chart. These charts have been considered to maintain simplicity, are easy to implement and interpret by practitioners, and outperform traditional Shewhart charts. Furthermore, it supports the quality management strategy of many practitioners, who prefer waiting until the occurrence of a second point beyond the control limits before looking for an assignable cause. To monitor the ratio of two normal variables, these charts were denoted by Syn-RZ. Celano and Castagliola[7] designed and implemented the Zyn-RZ control charts. They also gave an example of the food industry to illustrate this model. Like the RZ-model, besides the control limits UCL_{RZ} and LCL_{RZ}, we consider the CRL sub-chart that monitors the CRL, which is defined as the number of inspected samples between two consecutive nonconforming samples. Therefore, the Syn-RZ control charts signal if and only if the plotting statistic of the RZ–Shewhart sub-chart plots outside the control limits ($\hat{Z}_j \notin$

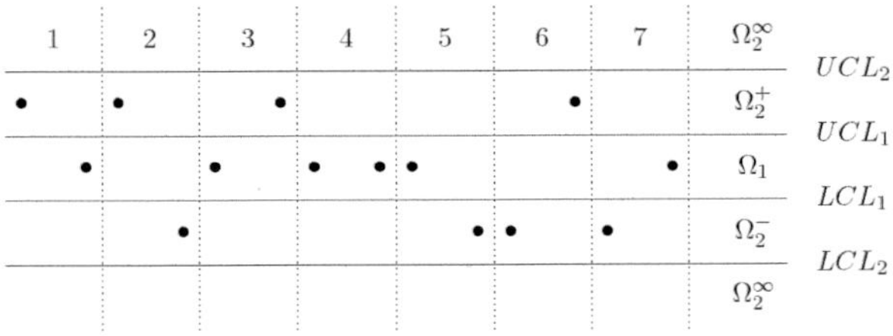

Figure 3: Intervals of $\Omega_1, \Omega_2^+, \Omega_2^-, \Omega_2^\infty$ and an example of transient states of the Markov chain $\mathrm{RR}_{2,3} - X$

$[\mathrm{LCL}_{RZ}, \mathrm{UCL}_{RZ}]$) and $\hat{Z}_k \in [\mathrm{LCL}_{RZ}, \mathrm{UCL}_{RZ}]$ for $k \in \{j+1, j+2, \cdots, i-1\}$. The resulting CRL is less than or equal to the H ($\mathrm{CRL} = i - j \leq H$ with some $H \in \mathbb{N}^*$ defined by the practitioners). The statistical design of the Syn-RZ charts is an optimization problem

$$(H^*, \alpha^*_{RZ}) =_{(H,\alpha_{RZ})} \left[(H, \alpha_{RZ}) \middle| n, \gamma_X, \gamma_Y, \rho_0, \rho_1, \tau\right] \tag{5}$$

subject to

$$\left[(H, \alpha_{RZ}) \middle| n, \gamma_X, \gamma_Y, \rho_0, \rho_1, \tau\right] =_0$$

Here τ is the shift size. The numeric results allow us to conclude that the optimal value H^* increases for moderate to small shift sizes. The Syn-RZ charts improve its detection performance for the smaller coefficient of variation (γ_X, γ_Y). The authors also gave a table showing that the Syn-RZ charts significantly outperform the RZ charts in Celano and Castagliola [4].

2.2.3 Runs Rules RZ charts

The statistical sensitivity of a Shewhart control chart can be improved by implementing supplementary Run Rules. Tran et al. [8] adopted the run rules for the Shewhart-RZ control charts, denoted as RR-RZ control charts, to increase its sensitivity to small shifts. A Run Rules chart (denoted by $\mathrm{RR}_{r,s}$) monitor the statistic X considering that the process is out-of-control, at instant i, if one of the following rules applies

- $X_i \in \Omega_2^\infty$
- $X_i \in \Omega_2^-$ and $\#\{X_j \in \Omega_2^-, j = i - s + 1, \cdots, i\} \geq r$ or $X_i \in \Omega_2^+$ and $\#\{X_j \in \Omega_2^+, j = i - s + 1, \cdots, i\} \geq r$

where the domain definitions of Ω_2^+, Ω_2^- and Ω_2^∞ are in Fig 3

In order to compute the statistical properties of the RR-RZ control charts, they defined

- a lower-sided r-out-of-s Run Rules control chart (denoted as RR-$\text{RZ}^-_{r,s}$). An out-of-control signal is given at time i if $\hat{Z}_i < \text{LCL}_{\text{RR-RZ}}$ (in this case, $\text{UCL}_{\text{RR-RZ}} = +\infty$).
- an upper-sided r-out-of-s Run Rules control chart (denoted as RR-$\text{RZ}^+_{r,s}$). An out-of-control signal is given at time i if $\hat{Z}_i > \text{UCL}_{\text{RR-RZ}}$ (in this case, $\text{LCL}_{\text{RR-RZ}} = -\infty$).

A comparison with the ARL values for the RZ control chart was established. As expected, the RR-RZ charts are more sensitive than RZ charts to small shifts of the nominal ratio z_0. Namely, it shows that in general, the shift-size $\tau \in [0.95, 1)$, the better option is RR-$\text{RZ}^-_{(r,s)}$ charts; when $\tau \in (1, 1.02]$, the better option is RR-$\text{RZ}^+_{(r,s)}$ chart and if $\tau < 0.95$ or $\tau > 1.02$, the better option is the RZ control chart (see Tran et al.[8]).

2.2.4 EWMA RZ charts

One of the control chart types that we cannot ignore is the EWMA type. In these types of charts, instead of averages or individual observations, a smoothed moving average is applied to make any trend or deviation (in averages away from the center line) more apparent. These charts take full advantage of the previous sample information, making charts faster at detecting relatively small shifts. For example, Tran et al.[9] proposed two one-sided EWMA-RZ charts, and it turned out that the EWMA-RZ chart is statistically more sensitive than the Shewhart-RZ chart on the whole. The reason for choosing two one-sided instead of one two-sided chart was explained by the authors: in general, the ratio distribution can be asymmetrical, and designing two one-sided charts will be more flexible for practitioners. They looked at two sub-charts:

- an upward EWMA-RZ chart (denoted as EWMA-RZ^+) with a single control limit $^+$ (here $^+ = z_0$)

$$Y_i^+ = \begin{cases} \max(z_0, (1-\lambda^+)Y_{i-1}^+ + \lambda^+ \hat{Z}_i & \text{if } i \geq 1 \\ z_0 & \text{if } i = 0 \end{cases}$$

- a downward EWMA-RZ chart (denoted as EWMA-RZ^-) with a single control limit $^-$ (here $^- = z_0$)

$$Y_i^- = \begin{cases} \min(z_0, (1-\lambda^-)Y_{i-1}^- + \lambda^- \hat{Z}_i & \text{if } i \geq 1 \\ z_0 & \text{if } i = 0 \end{cases}$$

here $\lambda^\pm \in (0, 1]$ is a smooth parameter

The design of two one-sided EWMA-RZ charts consists in optimizing a couple λ and the control limit such that the value of the out-of-control ARL is smallest.

If the ARL values for the out-of-control process are uniformly smaller than the in-control ARL, it is called an ARL-unbiased control chart. Otherwise, it takes a longer time on average to signal some assignable causes than when there is no assignable cause. In this case, control limits are chosen such that the ARL takes its maximum value at $\tau = 1$. Tran and Knoth[10] studied the long-term properties of ARL-unbiased EWMA-RZ control chart, they also investigated the steady-state ARL. Their results show that when $\Delta_\tau = |\tau - 1|$ increases, the values of optimal λ^* increase to 1; the performance of ARL-unbiased EWMA-RZ charts depends strongly on the sample size n and the value of (γ_X, γ_Y). Their report indicates that the unbiased ARL limits exhibit the most balanced performance in both the zero-state and steady-state ARL. Hence, it is an attractive version of a two-sided EWMA RZ control chart.

2.2.5 CUSUM-RZ charts

In this type of chart, instead of plotting individual observations, a cumulative sum of the deviations of the individual measurements from the central line or target specifications is plotted. These charts are particularly powerful for detecting small shifts: While in traditional Shewhart charts, such a deviation does not generate an out-of-control condition within a large number of samples, the CUSUM chart is very sensitive in this respect and quickly detects such slight changes. Tran et al.[11] proposed and investigated the statistical properties of two Phase II one-sided CUSUM control charts for monitoring the ratio of population means of a bi-variate normal distribution $\hat{Z}_i = \sum_{j=1}^{n} X_{i,j} / \sum_{j=1}^{n} Y_{i,j}$. More specifically, they suggest the following two separate one-sided CUSUM charts:

- upward CUSUM-RZ with the corresponding upper control limit $ULC^+ = H^+ \times z_0$

$$S_i^+ = \begin{cases} \max(0, S_{i-}^+ + \hat{Z}_i - z_0 + K^+) & \text{if } i \geq 1 \\ 0 & \text{if } i = 0 \end{cases}$$

- downward CUSUM-RZ with the corresponding lower control limit $LCL^- = H^- \times z_0$

$$S_i^- = \begin{cases} \max(0, S_{i-}^- - (\hat{Z}_i - z_0) - K^-) & \text{if } i \geq 1 \\ 0 & \text{if } i = 0 \end{cases}$$

where z_0 is the in-control ratio and K^+, K^- are the reference parameters that tune the sensitivity of the corresponding CUSUM-RZ charts.

The aim of their work is to find out the optimal couples $(K^{\pm *}, H^{\pm *})$ by minimizing the out-of-control ARL according to the target in-control average run length (ARL_0). The numeric results show a similar tendency to the results in Celano and Castagliola [4]. As expected, compared to the Shewhart-RZ charts, the one-sided CUSUM-RZ control charts have significantly better performances in the detection of out-of-control conditions.

2.3 Adaptive control charts types of ratio

An adaptive control chart involves varying at least one of the chart's parameters, such as the sampling interval, sample size, or the width constant of control limits. The purpose of this adaptive chart type is to improve the effectiveness of the control chart (i.e. detect out-of-control situations as quickly as possible). Varying the sampling interval (VSI) between samples is an alternative method adopted for quicker detection of an out-of-control process as compared with the conventional fixed sampling interval (FSI) Shewhart chart. The principle of this type is that the time until the next sample depends on what is being observed in the current sample. Sampling is less frequent when the process is at a high level of quality and vice versa. Nguyen et al. [12] suggest integrating the VSI feature into the Shewart chart for monitoring the ratio of population means of a bivariate normal distribution. It is demonstrated that the FSI RZ charts are slower than VSI-type charts in detecting process changes. The domain of in-control is divided into two: one is the "save region" where the time till the next sampling is h_L ("long") and the "warning region" where the time till the next sampling is h_S ("short"). Because of the asymmetric property of the ratio distribution the run length (ARL) is biased. In order to overcome this drawback, the researcher proposes designing two separate one-sided control charts, involving an upper-sided chart, which detects an increase in the ratio (denoted as "upward Shewhart-RZ chart"), and a lower-sided chart, which detects a decrease in the ratio (denoted as "downward Shewhart-RZ chart"). For example, to illustrate the upper situation, we can look at Figure 4. Instead of looking at the ARL, they are looking for the average time to signal (ATS) which counts the expected time before a control chart signals an out-of-control condition after the occurrence of an assignable cause or the issue of a false alarm.

$$\mathrm{ATS} = E(h) \times \mathrm{ARL} \tag{6}$$

where $E(h) = (h_S p_S + h_L p_L)/(1 - q)$ is the average sampling interval. Here p_S, p_L, and q are the probabilities that a sample point drops into the safe, warning, and out-of-control regions, respectively.

For the type EWMA and CUSUM, the readers can refer to Nguyen et al. [13] and Nguyen et al. [14].

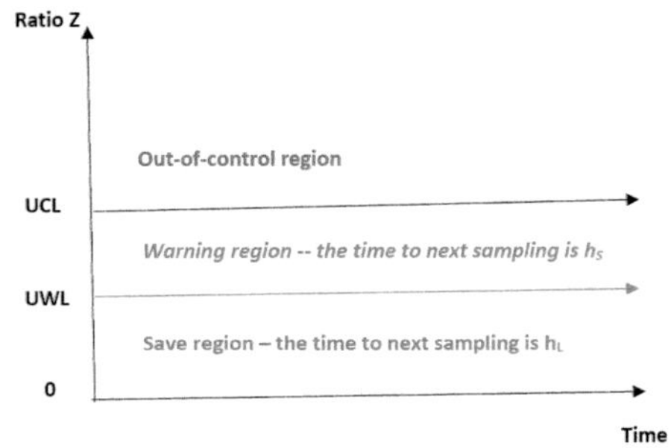

Figure 4: The upward VSI Shewhart-RZ chart

2.4 RZ - control charts under the effect of measurement errors

To enhance the practical use of the ratio chart, the effect precision measurement error has been considered in recent times. Most studies show that both the precision and accuracy errors have negative impacts on the performance of the control chart. In this model, under the presence of measurement error, the observations at time i will be

$$\mathbf{W}^*_{i,j,k} = \mathbf{A} + \mathbf{B}\mathbf{W}_{i,j} + \varepsilon_{i,j,k}, k = 1, \ldots, m, \tag{7}$$

where $\mathbf{W}_{i,j} = (X_{i,j}, Y_{i,j})^T \sim N(\boldsymbol{\mu}_\mathbf{W}, \boldsymbol{\Sigma}_\mathbf{W})$ is a bi-variate normal vector with mean $\boldsymbol{\mu}_\mathbf{W}$ and variance-covariance matrix $\boldsymbol{\Sigma}_\mathbf{W}$, $\mathbf{A} = (a_X, a_Y)^T$ is a (2×1) constants vector, $\mathbf{B}$ is a (2×2) matrix, $\varepsilon \sim N(\mathbf{0}, \boldsymbol{\Sigma}_M)$ is a centered bivariate normal random vector assumed to be independent of $\mathbf{W}$.

The monitoring statistics

$$\hat{Z}^*_i = \frac{\hat{\mu}_{X^*_i}}{\hat{\mu}_{Y^*_i}} = \frac{\sum_{j=1}^{n} \bar{X}^*_{i,j}}{\sum_{j=1}^{n} \bar{Y}^*_{i,j}}, \tag{8}$$

are recommended, where $\bar{X}^*_{i,j}$ and $\bar{Y}^*_{i,j}$ are two components of the bi-variate normal vector $\overline{\mathbf{W}}^*_{i,j}$.

In literature, we found that some models of RZ charts are considered with the effect of measurement errors like Shewhart-RZ charts (see Tran et al.[15]; Nguyen and Tran[16]), EWMA-RZ charts (see Nguyen et al.[17]), Run sum RZ charts (see Abubakar et al.[18]). Future research could investigate the effect of measurement errors on the control charts, monitoring the ratio of random normal variables like the CUSUM-RZ charts, Adaptive EWMA-RZ charts, Adaptive CUSUM-RZ charts, considering the case of short runs or considering Phase I implementation.

3 Monitoring the Compositional Data

CoDa and the ratio of two normal random variables are two data types that deal with ratios or proportions of different components or variables. CoDa represents proportions or percentages of different components that sum up to a constant, while the ratio of two normal random variables represents the ratio of two continuous variables having normal distributions. Although these data types are distinct, they share a common aspect of involving relative information of two or more components or variables. The ratio of two normal variables can be regarded as a special case of CoDa in which the number of components is limited to two, and the data has a normal distribution in a Simplex space. Control charts can be used to monitor the ratios involved in both data types and detect process changes over time. Due to their shared feature of involving ratios or proportions, we decided to group the control charts for monitoring CoDa and the ratio of two normal variables together in our analysis and review.

3.1 Modelling of Compositional Data

This section presents a brief overview of the CoDa and its simplex space, as well as several widely-used transformation techniques to transform CoDa into a vector representation in real space.

In statistics, CoDa refers to data that often describes the proportions, percentages, concentrations, or frequencies of some whole, and their individual values are strictly positive. An example of CoDa is the mineral composition in a rock sample, where a geologist can obtain a rock sample and perform an analysis to determine the proportions of its various mineral components. The applications of CoDa can be found in many domains, including but not limited to chemical research, econometrics and survey data analyses, geology and the food industry. where the composition of a sample is important for understanding the properties or behavior of the system, see Aitchison [2].

According to the definition provided by Aitchison [2], Pawlowsky-Glahn et al. [19], a (row) vector $\mathbf{x} = (x_1, \ldots, x_p)$ is considered a p-part composition if its components is strictly positive and carries only relative information. The term "relative information" here refers to the interrelationships among the composition's components, regardless of their numerical values. The sum of all the components of $\mathbf{x}$, $\sum_{i=1}^{p} x_i$, is a constant κ. Depending on the value of κ, the measurements can be interpreted as proportions (if $\kappa = 1$) or percentages (if $\kappa = 100$), for instance. Since the multiplication of a positive vector by a positive constant does not change the ratios between its components, each composition can be regarded as an equivalent class of proportional factors. Thus, if $\mathbf{x} = \lambda\mathbf{y}$ where $\mathbf{x}$ and $\mathbf{y}$ are compositions, and λ is a constant, we say that $\mathbf{x}$ and $\mathbf{y}$ are compositionally equivalent. Alternatively, the closure of a p-part composition $\mathbf{x} = (x_1, x_2, \ldots, x_p)$ to $\kappa > 0$ is defined by

$$\mathcal{C}(\mathbf{x}) = \left(\frac{\kappa x_1}{\sum_{i=1}^{p} x_i}, \frac{\kappa x_2}{\sum_{i=1}^{p} x_i}, \ldots, \frac{\kappa x_p}{\sum_{i=1}^{p} x_i} \right) \tag{9}$$

then the two p-part compositions are considered to be compositionally equivalent if their closures are equal, i.e., $\mathcal{C}(\mathbf{x}) = \mathcal{C}(\mathbf{y})$, for every positive constant κ.

The sample space of CoDa is represented by the simplex,

$$\mathcal{S}^p = \left\{ \mathbf{x} = (x_1, x_2, \ldots, x_p) \mid x_i > 0, i = 1, \ldots, D; \sum_{i=1}^{p} x_i = \kappa \right\} \tag{10}$$

In Euclidean space, vectors can be added or multiplied by scalars using Euclidean geometry to determine their properties or compute the distance between them. However, applying this geometry directly to CoDa in $\mathcal{S}^p$ is not feasible due to its special structure. To address this issue, Aitchison introduced the Aitchison geometry, with two fundamental operations required for a vector space structure of the simplex $\mathcal{S}^p$: perturbation and powering operators. These operators are equivalent to the addition and multiplication by scalar operations in real space, respectively. The perturbation operation, denoted by $\oplus$, adds a vector $\mathbf{y} \in \mathcal{S}^p$ to a vector $\mathbf{x} \in \mathcal{S}^p$, and is defined as follows:

$$\mathbf{x} \oplus \mathbf{y} = \mathcal{C}(x_1 y_1, \ldots, x_p y_p) \in \mathcal{S}^p$$

The powering operation, denoted by $\odot$, multiplies a vector $\mathbf{x} \in \mathcal{S}^p$ by a scalar constant $\alpha \in \mathbb{R}$, and is defined as follows

$$\alpha \odot \mathbf{x} = \mathcal{C}(x_1^{\alpha}, \ldots, x_p^{\alpha}) \in \mathcal{S}^p$$

The simplex $\mathcal{S}^p$ with the perturbation and the power operations result in a vector space structure, denoted by $(\mathcal{S}^p, \oplus, \odot)$.

In practical applications, it is common to transform CoDa into vectors in the Euclidean space rather than directly deploying the model in the Simplex space to eliminate the constraints of CoDa. One widely used transformation technique is the centered log-ratio transformation (clr) proposed by Aitchison [2]. This clr transformation is an isometry from the Simplex space $\mathcal{S}^p$ to a subspace $U \subset \mathbb{R}^p$, defined as follows

$$(\mathbf{x}) = \left(\ln \frac{x_1}{g_m(\mathbf{x})}, \ln \frac{x_2}{g_m(\mathbf{x})}, \ldots, \ln \frac{x_p}{g_m(\mathbf{x})} \right) = (\xi_1, \xi_2, \ldots, \xi_p) \tag{11}$$

Here, $g_m(\mathbf{x}) = (\prod_{i=1}^{p} x_i)^{\frac{1}{p}} = \exp\left(\frac{1}{p} \sum_{i=1}^{p} x_i\right)$ represents the component-wise geometric mean of the composition, and $\sum_{i=1}^{p} \xi_i = 0$. The inverse of the clr transformation retrieves $\mathbf{x}$ from $\xi = (\xi_1, \ldots, \xi_p)$ by

$$^{-1}(\xi) = \mathcal{C}(\exp(\xi)) = \mathcal{C}(\exp(\xi_1), \exp(\xi_2), \ldots, \exp(\xi_p)). \tag{12}$$

According to Egozcue et al.[20], the constraint in the component of $(\mathbf{x})$ results in a singular variance-covariance matrix for random composition. To address this issue, Egozcue et al.[20] introduced a new transformation called the isometric log-ratio (ilr) transformation, which is associated with an orthogonal basis in $\mathcal{S}^p$. Let $\mathbf{e}_1, \mathbf{e}_2, \ldots, \mathbf{e}_{p-1}$ be an orthonormal basis of $\mathcal{S}^p$, hence any composition $\mathbf{x} \in \mathcal{S}^p$ can be expressed as a linear combination of this basis, i.e,

$$\mathbf{x} = \bigoplus_{i=1}^{p-1} x_i^* \odot \mathbf{e}_i, \quad x_i^* = \langle (\mathbf{x}), (\mathbf{e}_i) \rangle$$

The ilr transformation of $\mathbf{x} \in \mathcal{S}^p$ is given by $(\mathbf{x}) = \mathbf{x}^* = (x_1^*, x_2^*, \ldots, x_{p-1}^*)$.

Let $\mathbf{B}$ be a $(p-1, p)$ matrix with the ith row defined as $(\mathbf{e}_i)$, for $i = 1, \ldots, p-1$. This matrix is commonly known as a contrast matrix associating with the orthonormal basis $\mathbf{e}_1, \mathbf{e}_2, \ldots, \mathbf{e}_{p-1}$. The ilr transformation of a composition $\mathbf{x}$ can be computed as follows:

$$\mathbf{x}^* = (\mathbf{x}) = (\mathbf{x}) \cdot \mathbf{B}^\intercal$$

There are many orthonormal basis candidates for $\mathcal{S}^p$. One such candidate was proposed by Egozcue and Pawlowsky-Glahn[21] using a sequential binary partition method. In this particular basis, $\mathbf{e}_i$ is defined as $\mathcal{C}(e_{i,1}, \ldots, e_{i,j}, \ldots, e_{i,p})$, where

$$e_{i,j} = \begin{cases} \exp\left(\sqrt{\frac{1}{i(i+1)}}\right) & \text{if} \quad j \le i \\ \exp\left(-\sqrt{\frac{i}{i+1}}\right) & \text{if} \quad j = i+1 \\ 1 & \text{otherwise} \end{cases}$$

Using this orthonormal basis to transform $\mathbf{x}$, the resulting coordinates x_i^* can be computed as

$$x_i^* = \sqrt{\frac{i}{i+1}} \ln \left(\frac{\left(\prod_{j=1}^{i} x_j\right)^{\frac{1}{i}}}{x_{i+1}} \right).$$

Given the ilr coordinate $\mathbf{x}^*$, the original composition $\mathbf{x}$ can be recovered by applying the inverse ilr transformation as follows

$$^{-1}(\mathbf{x}^*) =^{-1} (\mathbf{x}^*\mathbf{B}) = \mathcal{C}(\exp(\mathbf{x}^*\mathbf{B})).$$

tab:example1 illustrates the practical application of the ilr transformation for the case where the number of parts p is 4. The first five columns of the table present the components of ten compositions in $\mathcal{S}^5$, while the remaining three columns show their corresponding ilr coordinates in $\mathbb{R}^4$. It can be observed that the ilr coordinates x_i^* are no longer constrained by the constant sum. For a more comprehensive understanding of CoDa and its properties, see Pawlowsky-Glahn et al.[19].

Table 3: Example of ilr transformation in $\mathcal{S}^5$

x_1	x_2	x_3	x_4	x_5	x_1^*	x_2^*	x_3^*	x_4^*
0.02	0.3	0.14	0.25	0.29	-1.91	-0.48	-0.84	-0.82
0.15	0.08	0.22	0.3	0.25	0.44	-0.57	-0.67	-0.32
0.11	0.17	0.23	0.11	0.38	-0.31	-0.42	0.34	-0.82
0.01	0.08	0.22	0.08	0.61	-1.47	-1.67	-0.31	-2.04
0.39	0.1	0.33	0.07	0.11	0.96	-0.42	1.05	0.33
0.23	0.08	0.42	0.05	0.22	0.75	-0.92	1.19	-0.4
0.31	0.03	0.46	0.04	0.16	1.65	-1.28	1.21	-0.35
0.04	0.06	0.33	0.2	0.37	-0.29	-1.56	-0.67	-1.07
0.22	0.35	0.13	0.26	0.04	-0.33	0.62	-0.16	1.55
0.24	0.38	0.11	0.25	0.02	-0.32	0.82	-0.13	2.78

3.2 Control charts for monitoring CoDa

The first control chart for monitoring this special data was proposed by Boyles [22], in which the author explored the application of a chi-square control chart for monitoring data (CoDa) derived from the Dirichlet distribution. The degrees of freedom in the chi-square distribution were adjusted slightly to accommodate this approach. Subsequently, other types of control charts, such as Shewhart-type and memory-type charts, were also developed and examined. These control charts adhere to the same principles as conventional control charts but have been adapted to accommodate the special structure of CoDa, including its non-negative components and constant-sum constraints. The log-ratio transformation method is commonly employed in this context, as mentioned in Section 3. In this section, a brief review of research related to control charts for monitoring CoDa in Phase II MSPC will be provided.

3.2.1 Chi-square control chart

The chi-square control chart was initially proposed by Boyles [22] when studying CoDa derived from a Dirichlet distribution. Given a sample of size n independent observations $\mathbf{x}_1, \mathbf{x}_2, \ldots, \mathbf{x}_n$ where $\mathbf{x}_i = (x_{i1}, x_{i2}, \ldots, x_{ip})$ follows a

Dirichlet distribution $\mathcal{D}(\alpha)$ with $\alpha = (\alpha_1, \alpha_2, \ldots, \alpha_p)$ is a positive parameter, the statistic of interest to be monitored is given by

$$X^2 = \frac{(x_{i1} - \pi_1)^2}{\pi_1} + \frac{(x_{i2} - \pi_2)^2}{\pi_2} + \ldots + \frac{(x_{iD} - \pi_p)^2}{\pi_p}, \tag{13}$$

where $\pi_1, \pi_2, \ldots, \pi_p$ denote the process averages. Boyles [22] demonstrated that when the CoDa originate from a Dirichlet distribution, X^2 is asymptotically distributed as a multiple of χ^2_{p-1}, thereby the X^2 chart based on χ^2_{p-1} is valid as a certain type of asymptotic approximation. To achieve an overall false alarm rate of 0.0027, the upper control limit has been established based on the "3σ" threshold.

3.2.2 Shewhart-type control charts

Guevara et al. [23] pioneered the application of Hotelling T^2 control charts in Dirichlet regression to examine the connection between a response variable and one or more explanatory variables. Subsequently, Vives-Mestres et al. [24] studied a method for interpreting out-of-control signals in the T^2_C control chart used for monitoring three-part CoDa. Building upon this work, Vives-Mestres et al. [25] proposed an extension of the T^2_C chart to monitor individual CoDa observations based on isometric log-ratio (ilr) transformation. In these works, the authors conducted an investigation on the control charts to detect shifts in the mean vector and assumed that the covariance matrix of the distribution would remain constant throughout the study. Furthermore, they presumed that the in-control (IC) and out-of-control (OC) distributions of the dataset followed a normal distribution. The general design of the T^2_C chart is as follows: Let $\mathbf{x}_1, \mathbf{x}_2, \ldots,$ be Phase II p-parts compositions with the in-control (IC) data following normal distribution $\mathcal{N}(\mu_0, \Sigma_0)$ on the simplex space $\mathcal{S}^p$. Let $\mathbf{x}^*_t = (x^*_{t1}, x^*_{t2}, \ldots, x^*_{t,p-1})$ denote its ilr coordinates (as defined in Section 3) then $\mathbf{x}^*_t$ follows a normal distribution $\mathcal{N}(\mu^*_0, \Sigma^*_0)$ in the real space $\mathbf{R}^{p-1}$. In case μ^*_0, Σ^*_0 are known, the monitored statistic of T^2_C at the ith time point is defined by

$$T^2_C = (\mathbf{x}^*_i - \mu^*_0)'\Sigma^{*-1}_0(\mathbf{x}^*_i - \mu^*_0) \tag{14}$$

and the chart would give a signal of mean shift if $T^2_C > \chi^2_{1-\alpha,p-1}$ where $\alpha \in [0, 1]$ is the given significant level, $\chi^2_{1-\alpha,p-1}$ is the $100(1-\alpha)$-th quantile of χ^2_{p-1} distribution. In scenarios where both μ^*_0 and Σ^*_0 are unknown, as is frequently encountered in practical applications, they can be estimated using the sample mean vector $\bar{\mathbf{x}}$ and the sample covariance matrix $\mathbf{S}$ of the ilr transformed data of the IC sample (obtained from Phase I). Suppose that an IC sample of size n is available, the T^2_C statistic at the ith time point becomes

$$T^2_C = (\mathbf{x}^*_i - \bar{\mathbf{x}})'\mathbf{S}^{-1}(\mathbf{x}^*_i - \bar{\mathbf{x}}) \tag{15}$$

and the chart would give a signal of mean shift when $T_C^2 > \frac{(p-1)(n-1)(n+1)}{n(n-p-1)} \mathcal{F}_{1-\alpha,p-1,n-p-1}$ where $\mathcal{F}_{1-\alpha,p-1,n-p-1}$ is the $100(1-\alpha)$-th quantile of $\mathcal{F}$ distribution $\mathcal{F}_{p-1,n-p-1}$.

Vives-Mestres et al.[25] compared this T_C^2 chart using ilr transformation method to transform CoDa with the T^2 control chart after removing one component of the composition and has shown its out-performance with the other in terms of average run length (ARL) values. However, it (and Hotelling's T^2 control chart for monitoring other types of data) has a significant drawback in that it assumes that the variance-covariance matrix of the process is constant over time, which may not always hold in practical situations. Changes in raw materials, process parameters, equipment, environmental conditions, personnel, and other factors can lead to variability in the covariance matrix of the process, making the chart less effective in detecting changes in the process over time. Additionally, the control chart makes decisions about process performance solely based on the observed data at the current time point, which can be less sensitive to small shifts in the process distribution. This can lead to a higher probability of false alarms and missed opportunities to identify and correct process issues.

3.2.3 Memory-type control charts

As in unconstrained cases and as mentioned before, multivariate Shewhart charts used for monitoring CoDa assume that the variance-covariance matrix of the process is constant over time and relies only on the observed data at the current time point to make decisions about process performance. While they effectively detect relatively large and transient shifts, they are less sensitive to relatively small and persistent shifts in the process distribution. To address these limitations, researchers have suggested using multivariate control charts such as multivariate exponentially weighted moving average (MEWMA) and multivariate cumulative sum (MCUSUM), which incorporate the information from previous observations along with the current observation, improving their ability to detect small shifts. These memory-type control charts use both past and present observations to make decisions and update their estimates over time.

Suppose that at each sampling period $i = 1, 2, \ldots$, a sample of size m independent p-part compositions $\mathbf{x}_{i,1}, \ldots, \mathbf{x}_{i,m}$ is collected and each $\mathbf{x}_{i,j}$ follows a multivariate normal distribution $N_{\mathcal{S}^p}(\mu, \Sigma)$, on the simplex $\mathcal{S}^p$. Assume also that when the process is in control, the composition center is μ_0, and when the process is out-of-control, the composition center is μ_1. Let $\mathbf{x}_{i,1}^*, \ldots, \mathbf{x}_{i,m}^*$ represent the coordinates of $\mathbf{x}_{i,1}, \ldots, \mathbf{x}_{i,m}$ hence $\mathbf{x}_{i,j}^*$ follows a multivariate normal distribution, $\mathcal{N}_{\mathbb{R}^{p-1}}(\mu_0^*, \Sigma_0^*)$, on $\mathbb{R}^{p-1}$.

- **Multivariate EWMA-CoDa control charts**

Tran et al.[26] proposed to use a MEWMA scheme to monitor CoDa. The statistic to monitor the process suggested by Tran et al.[26] is

$$Q_i = \mathbf{w}_i^{\mathsf{T}} \Sigma_{\mathbf{w}_i}^{-1} \mathbf{w}_i, i = 1, 2, \ldots \tag{16}$$

where

$$\mathbf{w}_i = r(\bar{\mathbf{x}}_i^* - \mu_0^*) + (1 - r)\mathbf{w}_{i-1},\ i = 1, 2, \ldots \tag{17}$$

is the EWMA vector, $\bar{\mathbf{x}}_i^*$ is the sample mean of ith ilr-transformed sample, Σ_{w_i} is the variance-covariance matrix of $\mathbf{w}_i$. The smoothing parameter r in the EWMA vector determines the weights given to the current and past observations. A higher value of r results in a faster decrease in the weights given to past observations, meaning that past observations have less influence on the MEWMA statistic and vice versa. The MEWMA chart issues an out-of-control signal when the value of the statistic Q_i exceeds the upper control limit H, where H is chosen to achieve a specific in-control average run length ($_0$).

- **Multivariate CUSUM-CoDa control chart**

Imran et al.[27] considered the MCUSUM-CoDa control chart in both known and unknown parameters cases. In case μ_0^*, Σ_0^* are known, the monitoring statistic of the MCUSUM-CoDa will be

$$C_i = \left[m(\mathbf{s}_i^T \Sigma_0^{*-1} \mathbf{s}_{i-1})\right]^{1/2}, \quad i = 1, 2, \ldots \tag{18}$$

where

$$\mathbf{s}_i = \begin{cases} 0 & \text{if} \quad i = 0 \quad \text{or} \quad Q_i \leq k \\ (\mathbf{s}_{i-1} + \bar{\mathbf{x}}_i^* - \mu_0^*)(1 - k/Q_i) & \text{otherwise} \end{cases}$$

$$Q_i = \left(m(\mathbf{s}_{i-1} + \bar{\mathbf{x}}_i^* - \mu_0^*)^T \Sigma_0^{*-1} (\mathbf{s}_{i-1} + \bar{\mathbf{x}}_i^* - \mu_0^*)\right)^{1/2}$$

This MCUSUM-CoDa chart issues a signal when $C_i > h$, where h is chosen to achieve a specific value of in-control . In case μ_0^*, Σ_0^* are unknown, they can be estimated by the sample mean and sample covariance matrix of the available IC dataset

$$\hat{\mu}_0^* = \frac{1}{n} \sum_{j=1}^{n} \bar{\mathbf{x}}_j^*, \qquad \hat{\Sigma}_0^* = \frac{1}{n} \sum_{j=1}^{n} \mathbf{S}_j^* \tag{19}$$

where n is the number of samples of size m IC data, $\mathbf{S}_j^*$ is the sample covariance matrix of jth sample. The MCUSUM-CoDa chart has been demonstrated to exhibit similar performance to the MEWMA-CoDa chart and to outperform the T^2-CoDa chart in terms of ARL values.

3.2.4 Adaptive control charts

- **VSI MEWMA-CoDa control chart:**

  As an extension of the MEWMA-CoDa chart, Nguyen et al.[28] designed a Phase II MEWMA-CoDa control chart with variable sampling intervals to monitor CoDa based on an isometric log-ratio transformation. In the FSI MEWMA-CoDa control chart, the sampling interval is set at a consistent value denoted as h_F. Conversely, in the VSI MEWMA-CoDa control chart, the interval between consecutive samples $\bar{\mathbf{X}}_i, \bar{\mathbf{X}}_{i+1}$ changes based on the present value of Q_i. This chart maintains the same UCL as the FSI chart but adds an extra warning limit, represented as = where 0 <<. This warning limit helps switch between longer (h_L) and shorter (h_S) sampling intervals. The longer interval h_L is used when the control statistic $Q_i^2 \leq^2$ indicates a safe zone, and the shorter interval h_S is applied when $^2 < Q_i^2 \leq^2$, marking a warning zone. A signal indicating the process is out of control is generated when $Q_i^2 >^2$. As the sampling interval in the VSI chart is variable, the relationship between Average Time to Signal (ATS) and Average Run Length (ARL) in this chart can be expressed as:

  $$\text{ATS}^{\text{VSI}} = \text{E(h)} \times^{\text{VSI}} \tag{20}$$

  Here, E(h) signifies the average sampling interval. The proposed FSI MEWMA-CoDa control chart has shown a better performance compared with the original version of MEWMA-CoDa in terms of Average Time to Signal (ATS) values.

- **VSS MEWMA-CoDa control chart:**

  Traditionally, standard process monitoring control charts have primarily relied on a fixed sample size (FSS) approach. As another extension to the MEWMA-CoDa chart,[29] proposed a control chart that leverages a variable sample size (VSS), namely VSS-MEWMA-CoDa, to enhance the efficiency and effectiveness of the MEWMA control chart when applied to CoDa. In this work, the MEWMA vectors are calculated by

  $$\mathbf{w}_i = r\hat{\mathbf{z}}_i + (1-r)\mathbf{w}_{i-1}, \tag{21}$$

  where $\hat{\mathbf{z}}_i$ is the standardized sample mean of ith ilr-transformed sample with

  $$\hat{\mathbf{z}}_i = \frac{\hat{\mathbf{x}}_i^* - \mu_0^*}{\sigma_0^*/\sqrt{n}}.$$

  The asymptotic variance-covariance matrix proposed by Lowry et al.[30] is commonly used to estimate the variance-covariance matrix Σ_{W_i}

  $$\Sigma_{W_i} = \frac{r}{n(2-r)}\Sigma^* \tag{22}$$

The VSS-MEWMA-CoDa CC issues a signal when the statistic Q_i with

$$Q_i = \mathbf{w}_i^{\intercal} \Sigma_{\mathbf{w}_i}^{-1} \mathbf{w}_i, i = 1, 2, \ldots \tag{23}$$

is greater than the upper control limit (UCL). The sample size n is varied based on the value of statistic Q_i. The transition from a small to a large sample size, or vice versa, is based on the computed value of Q_i. Specifically, a smaller sample size is employed, denoted as $n = n_1$, when the value of Q_i falls within the range $0 < Q_i <$. Conversely, a larger sample size, indicated by $n = n_2$, is utilized when Q_i satisfies the condition $< Q_i <$. It is important to note that n_1, n_0, and n_2 are related such that $n_1 < n_0 < n_2$, where n_0 represents the predetermined fixed sample size.

3.2.5 Control charts under the effect of measurement error or autocorrelation between samples

- **Control charts under the effect of measurement error**

The impact of measurement error on control chart performance has been a focal point of research in statistical process monitoring, drawing attention to the need for addressing such errors to maintain the integrity of control charts, particularly those dealing with CoDa. The groundwork in this area was set by Zaidi et al. [31], who investigated the effect of measurement errors on the performance of the Hotelling T^2 control chart for CoDa. In this work, the authors defined the sample mean composition at time i by

$$\bar{\mathbf{x}}_i = \frac{1}{m}(\mathbf{x}_{i1} \oplus \cdots \oplus \mathbf{x}_{im}) = \mathbf{a} \oplus (b \cdot \mathbf{y}_i) \oplus \frac{1}{m}(\epsilon_{i1} \oplus \cdots \oplus \epsilon_{im}).$$

where $\mathbf{a} \in \mathcal{S}^p$ and $b \in \mathbb{R}$ are known constants, ϵ_{ij} is a multivariate normal random error term $\text{MNOR}_{\mathcal{S}^p}(\mathbf{0}, \Sigma_M^*)$ that accounts for measurement inaccuracies, and Σ_M^* is the known measurement error variance-covariance matrix. The Hotelling T^2 control chart accounts for measurement errors through the statistic:

$$Z_i = (\bar{\mathbf{x}}_i^* - \mathbf{a}^* - b\mu_0^*) \left(b^2 \Sigma^* + \frac{1}{m} \Sigma_M^* \right)^{-1} (\bar{\mathbf{x}}_i^* - \mathbf{a}^* - b\mu_0^*)^T.$$

Building upon this, Zaidi et al. [32] furthered the discourse by incorporating measurement errors into the MEWMA-CoDa control chart analysis, which provided insights into the implications for production processes. In a continued effort to unravel the effect of measurement errors, Imran et al. [33] examined their influence on the multivariate CUSUM control charts. Most recently, Imran et al.[34] emphasized the critical role of measurement error in the VSI Hotelling T^2 charting scheme. These studies highlight the evolving recognition of measurement error as a significant factor that affects the efficacy of control charts in ensuring quality and reliability in manufacturing processes.

- CoDa control chart for autocorrelated data

  In the SPC, it is a standard presumption that data is independent and lacks autocorrelation. However, this assumption does not hold true in many practical industrial applications, especially when dealing with CoDa. This prevalent autocorrelation within CoDa vectors poses a challenge, as it can significantly impair the interpretative accuracy of control chart signals and diminish their capacity for detecting anomalies. Addressing this issue, Zaidi et al.[35] introduced the application of a time series autoregressive moving average (ARMA) model to manage autocorrelated CoDa within the simplex space $\mathcal{S}^p$. The ARMA model for CoDa is written as:

$$\mathbf{x}_t = \alpha_1 \otimes \mathbf{x}_{t-1} \oplus \cdots \oplus \alpha_p \otimes \mathbf{x}_{t-p} \oplus \epsilon_t \ominus \beta_1 \otimes \epsilon_{t-1} \oplus \cdots \oplus \beta_q \otimes \epsilon_{t-q},$$

  where $\alpha = (\alpha_1, \ldots, \alpha_p)$ and $\beta = (\beta_1, \ldots, \beta_p)$ are the unknown parameters, p, q are the orders of autoregressive and moving average polynomials, respectively, and ϵ_t is the residual term. The residual control chart for CoDa is applied by assuming a multivariate normal distribution within the simplex space. Specifically, where the residual vector ϵ follows a multivariate normal distribution $\text{MNOR}_{\mathcal{S}^p}(\mu_\epsilon, \Sigma_\epsilon)$ in the simplex space $\mathcal{S}^p$, the Hotelling T^2 control chart for residuals is formulated as:

$$T^2 = (\epsilon^* - \bar{\epsilon}^*)(\Sigma_\epsilon^*)^{-1}(\epsilon^* - \bar{\epsilon}^*)^T,$$

  where ϵ^* is the ilr-transformed residual vector, $\bar{\epsilon}^*$ is the estimator of the mean vectorμ_ϵ, and Σ_ϵ^* is the estimator of variance-covariance matrix Σ_ϵ. The Hotelling T^2 control chart that utilizes residuals from the CoDa ARMA model demonstrated superior performance in detecting shifts in the mean vector when compared to the Hotelling T^2 based on isometric log-ratio transformed original data. This enhancement underscores the critical importance of accounting for autocorrelation in SPC to ensure the reliability of quality monitoring techniques.

3.3 Analysis

In this section, we present a comprehensive analysis of the papers collected for reviewing the topic of monitoring CoDa. Through a series of visualizations and summary tables, we aim to provide a clear and insightful overview of the research landscape in this field. The analysis includes the distribution of papers across various research categories, the publication trends over time, and a detailed breakdown of the journals that have featured in this research, as well as the key contributions of leading authors in this field.

Table 4 provides a quick reference to key aspects across the 19 collected papers on monitoring CoDa, spanning from 1997 to 2023. It delineates the progression and scope of research through various facets such as control chart categorizations, the number of p-parts of the CoDa vector, the transformation/analyzing method, and comparison between the proposed control chart (if available).

In order to provide a clear overview of the research focus in the field of monitoring CoDa, Fig. 5 provides a pie chart category of 19 papers in distinct research categories: Shewhart type, Memory type, Adaptive type, Control charts for monitoring CoDa under the influence of Measurement Error and Autocorrelation, and Others type. This visual representation highlights the proportions of research efforts dedicated to each category, offering insights into the most heavily researched areas and those that may require further exploration. According to the pie chart, Shewhart-type control charts, a classic in the field, account for 26% of the papers, indicating their importance in monitoring CoDa. Similarly, control charts considering the measurement errors and autocorrelation also make up 26%, highlighting the importance of considering the effect of measurement error and the intricacies of data dependencies in monitoring. Adaptive control charts represent 21%, reflecting their popularity of consideration in this field. Notably, 11% of the papers integrate artificial intelligence with control charts, pointing towards an innovative direction that could redefine future methodologies.

Figure 6 depicts the cumulative number of papers published from 1997 to 2023. It illustrates a significant upward trajectory in research related to monitoring CoDa, with a notably steep increase observed in 2023.

Table 5 presents the distribution of journals for the 19 papers reviewed in the field of monitoring CoDa. It is evident that "*Quality and Reliability Engineering International*" is at the forefront, with five papers (26.32%) of the total corpus, signifying its prominence as a platform for scholarly discourse in this field. Close behind is the "*Journal of Applied Statistics*", hosting four papers (21.05%), which underscores its influential role in disseminating research findings. *Computers & Industrial Engineering* has two papers (10.53%) and other journals, each contributing one paper (5.26%), illustrating the field's interdisciplinary breadth. This spread across various journals indicates a wide interest in applications of CoDa analysis. The data denotes a vibrant academic ecosystem where multiple specialized journals serve as conduits for advancing the frontiers of monitoring CoDa research.

Table 4: Sumary of control chart for monitoring CoDa papers

Papers	Control chart	Type	N^0 of p-part	Transformation/ Analyzed method	Comparison with other charts	Best chart
Boyles [22]	Chi-square	Shewhart-type	Arbitrary	No transformation	-	-
Guevara et al. [23]	T^2	Shewhart-type	Arbitrary	Dirichlet Regression	-	-
Vives-Mestres et al. [24]	T^2	Shewhart-type	p = 3	ilr	-	-
Vives-Mestres et al. [25]	T^2	Shewhart-type	p = 3	ilr	-	-
Vives-Mestres et al. [36]	T^2	Shewhart-type	Arbitrary	clr	-	-
Tran et al. [26]	MEWMA	Memory-type	Arbitrary	ilr	T^2	MEWMA
Zaidi et al. [31]	T^2	ME/Autocorrelation	Arbitrary	ilr	-	-
Zaidi et al. [32]	MEWMA	ME/Autocorrelation	Arbitrary	ilr	T^2	MEWMA
Nguyen et al. [28]	VSI-MEWMA	Addaptive	Arbitrary	ilr	MEWMA	VSI-MEWMA
Imran et al. [27]	MCUSUM	Memory-type	Arbitrary	ilr	T^2, MEWMA	CUSUM, MEWMA
Imran et al. [33]	CUSUM	ME/Autocorrelation	Arbitrary	ilr	-	-
Imran et al. [29]	VSS-MEWMA	Addaptive	Arbitrary	ilr	FSS-MEWMA	VSS-MEWMA
Imran et al. [37]	VSI-MEWMA	Addaptive	Arbitrary	ilr	FSI-MEWMA	VSI-MEWMA
Weiß [38]	Ordinal	Others	Arbitrary	alr	-	-
Zaidi et al. [35]	T^2_{ARMA}	ME/Autocorrelation	Arbitrary	T^2, CoDa ARMA	T^2	T^2_{ARMA}
Imran et al. [34]	VSI T^2	ME/Autocorrelation	Arbitrary	ilr	-	-
Zaidi et al. [39]	T^2	AI with CC	Arbitrary	MLFFNN	-	-
Imran et al. [40]	T^2	AI with CC	p =3, p = 5	MLPFNN	-	-
Imran et al. [37]	VSI MEWMA	Addaptive	Arbitrary	ilr	FSI-MEWMA	VSI-MEWMA

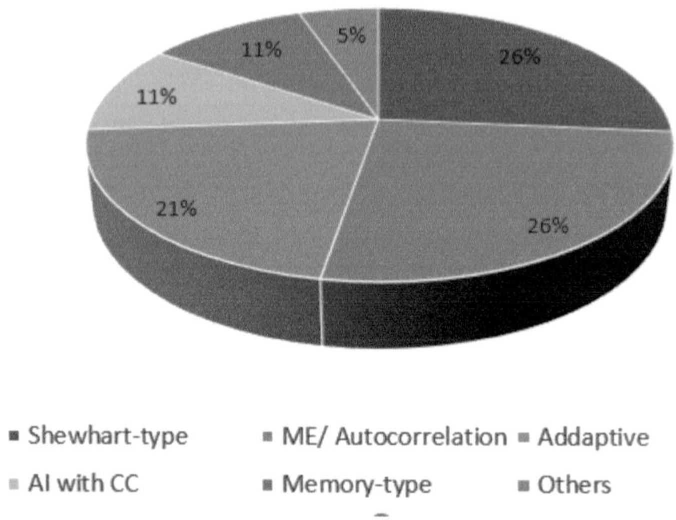

Figure 5: Cumulative number of papers over time

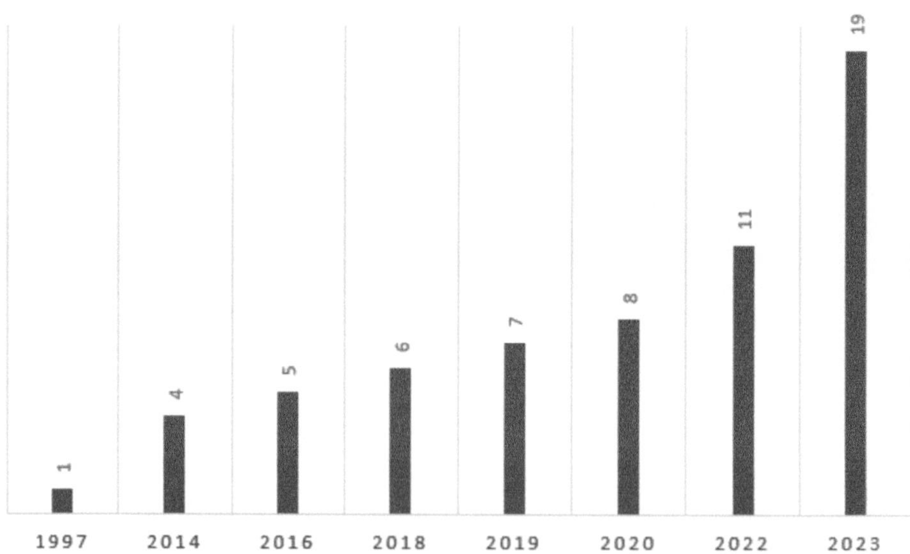

Figure 6: Cumulative number of papers over time

Table 5: Number and percentage of papers in each journal

Journal	Number	Percentage
Quality and Reliability Engineering International	5	26.32%
Journal of Applied Statistics	4	21.05%
Computers & Industrial Engineering	2	10.53%
Revista Colombiana de Estadística	1	5.26%
Journal of Quality Technology	1	5.26%
IIE Transactions	1	5.26%
Journal of Statistical Computation and Simulation	1	5.26%
CMES-Computer Modeling in Engineering & Sciences	1	5.26%
Statistical Modelling	1	5.26%
IFAC-PapersOnLine	1	5.26%
Expert Systems with Applications	1	5.26%

Table 6 provides a snapshot of the contributions made by various authors, presenting a clear disparity in the number of contributions among them. Fatima Sehar Zaidi emerges as the most prolific contributor, with a total of 11 contributions. Following closely is Muhammad Imran, with 9 contributions, and Kim Phuc Tran along with Jinsheng Sun, each with 7 contributions, showcasing their active participation. Other authors have made moderate contributions, ranging from 1 to 4 papers, indicating steady engagement in the field.

4 Perspectives for Monitoring the Ratio of Two Normal Variables and Compositional Data in the Industry 5.0

In this section, we will offer our perspective on applying existing ML methodologies in monitoring CoDa and the ratio of two normal random variables, and highlight potential areas for future research within the dynamic and evolving landscape of Industry 5.0. We place a special emphasis on the integration

Table 6: Authors' contribution to the area of monitoring CoDa

Author	**Number of contributions**
Fatima Sehar Zaidi	11
Muhammad Imran	9
Kim Phuc Tran	7
Jinsheng Sun	7
Zameer Abbas	4
Hafiz Zafar Nazir	4
Marina Vives-Mestres	3
Josep Daunis-I-Estadella	3
Josep-Antoni Martín-Fernández	3
Philippe Castagliola	3
Michael Boon Chong Khoo	3
Xuelong Hu	3
Hong-Liang Dai	3
Anan Tang	2
Russell Boyles	1
Rubén Darío Guevara-González	1
José Alberto Vargas-Navas	1
Dorian Luis Linero-Segrera	1
Giovanni Celano	1
Thi Thuy Van Nguyen	1
Cédric Heuchenne	1
Christian H. Weiß	1

of ML techniques at various stages – from the design to pattern recognition, and the interpretation of control charts. By dissecting the current trends and forecasting future developments, this section endeavors to provide valuable insights for researchers and practitioners alike, navigating the intricate intersection of statistical control processes and advanced technological paradigms.

4.1 Intergrating ML techniques in designing control charts

As stated before, autocorrelation is a common occurrence in real-life data, especially in industrial and manufacturing processes, where it significantly affects the integrity and interpretation of time-series data, making it a critical factor to be considered in the design of control charts for monitoring CoDa. In response, advanced ML techniques present a formidable solution. For example, long short-term memory (LSTM) networks, as introduced by Schmidhuber et al.[41], excel in capturing temporal dependencies, crucial for accurate data analysis in autocorrelated environments. Gaussian process regression (GPR), detailed by Williams and Rasmussen[42], provides a powerful probabilistic approach, suitable for handling complex, interdependent variables in such data. Additionally, transformer models introduced in the paper "Attention is all you need" by Vaswani et al.[43] with their novel multihead attention mechanisms, are highly effective in identifying relevant sequential data patterns. The attention mechanisms in transformer models allow them to focus on specific parts of a data sequence that are most relevant for predictions. This feature is particularly useful in scenarios with autocorrelation, as it helps in identifying significant changes or trends within the data. Autoencoders (AE), recognized for anomaly detection and feature extraction, can distinguish between normal and aberrant patterns, especially valuable in cases of complex deviations due to autocorrelation. By integrating these advanced ML techniques with existing control chart methodologies, a more robust, accurate, and adaptive control chart can be designed. This integration not only enhances the accuracy and reliability of control charts but also ensures they are well-equipped to handle the complexities of modern industrial data, leading to improved monitoring, decision-making, and operational efficiency in various industrial applications.

As an example of such integration, in our forthcoming paper, titled "A novel transformer-based anomaly detection approach for ECG monitoring healthcare system", we explored the integration of a Transformer-based variational AE with a MEWMA-Support vector data description (SVDD) control chart for the purpose of monitoring electrocardiogram (ECG) data. This innovative approach has demonstrated significant potential, yielding impressive results in terms of both accuracy and control of false alarm rates. The effectiveness of this method is particularly critical in healthcare settings, where precise and reliable monitoring is paramount, but it also holds substantial promise for various other fields that require precise data analysis and anomaly detection. This work exemplifies the advancements in leveraging sophisticated ML techniques to enhance the efficacy and reliability of monitoring systems in critical applications. Another example that can be mentioned is the work of Wang and Liu [44]. In this work, the authors proposed a new control chart utilizing the isolation forest (iForest) algorithm, offering advantages such as lower computational complexity, superior performance in detecting small shifts in various distributions, and effectiveness in handling challenging distributions with heavy tails, suggesting its potential applicability in designing efficient control charts for monitoring CoDa.

Toward the control chart for monitoring the ratio of two normal variables, considering autocorrelation is also important. In this context, incorporating a bivariate first-order vector autoregressive (VAR(1)) model is a prudent choice. This model accounts for both autocorrelation and cross-correlation, which are prevalent in scenarios with high-frequency data collection. In Nguyen et al. [45], the authors exemplified this approach and showed that autocorrelation among observations adversely affects the performance of the Shewhart control chart for monitoring the ratio of two normal variables. For future research, it is important to consider the performance of other types of control charts, such as EWMA and CUSUM, for monitoring ratios in the presence of autocorrelation. Additionally, conducting research to design advanced control charts that can alleviate the adverse effects of autocorrelation is crucial for enhancing process monitoring effectiveness.

4.2 ML-based control charts pattern recognition for monitoring CoDa/Ratio

This perspective will be applied to both control charts used for monitoring CoDa and the ratio of two normal random variables, which, for brevity and clarity in our discussion, will henceforth be referred to as "CoDa/Ratio".

The integration of ML techniques in Control Chart Pattern Recognition (CCPR) represents a transformative development in the field of SPC, including monitoring of CoDa/Ratio. In the 21st century, the rapid advancement of information technology and the proliferation of big data have catalyzed a significant shift in how control charts are interpreted and utilized. Historically, the interpretation of control charts relied heavily on heuristic rules, drawing largely on the experience and judgment of operators. This approach, while valuable, had its limitations, particularly in its dependence on human expertise and the potential for variability in interpretations. The emergence of expert systems marked a significant stride in addressing these limitations. These systems, which combine a set of expert-derived rules with data flows from the processes being controlled, represented an early form of automation in process monitoring. However, at the end of the 1980s, Neural Networks (NNs) began to be utilized for automating the reading and interpretation of control charts, as noted by Pugh [46]. NN, as a form of ML, offers a more dynamic and adaptable approach compared to expert systems. They could learn directly from data, identify patterns, and make predictions or decisions without being explicitly programmed with the rules. This marked the beginning of the dominance of ML in pattern recognition, including CCPR.

ML algorithms have been shown to outperform traditional models in various practical situations. For instance, Guh [47] highlighted the capabilities of NN models in learning, self-organizing, and recognizing patterns from noisy or incomplete data representations – tasks that are challenging for human operators, even with the support of expert systems. Li et al. [48] proposed a Support Vector Machine (SVM)-based CCPR framework, demonstrating its superior accuracy in classifying the sources of out-of-control signals compared to conventional multivariate control schemes. Similarly, Diren et al. [49] reported that traditional CCPR models often fall short in predicting unexpected new situations, a gap effectively bridged by ML techniques. By learning from historical data, ML models can anticipate and adapt to new scenarios, enhancing the predictive capabilities of control charts. The evolving landscape of ML-based techniques CCPR brings a promising development for monitoring complex scenarios. For instance, SVM, since its introduction by Vapnik et al. [50], has been instrumental in CCPR. Their ability to classify data in a higher-dimensional space for linear separability enhances their applicability in recognizing complex control chart patterns, see, for example, Lin et al. [51]. Neural Networks (NNs) and Deep Learning have brought a new dimension to CCPR. The pioneering work of Pugh [46] and the subsequent advancements by researchers like Cheng [52] and Addeh et al. [53] have highlighted the robust pattern recognition capabilities of these models. NNs' adaptability in various conditions, including their effectiveness in environments with noise or incomplete data, positions them as powerful tools for intricate monitoring tasks. These ML techniques collectively represent a significant advancement in the field of CCPR, offering refined and accurate tools for analyzing complex data patterns. Their effectiveness across various monitoring scenarios underscores their potential to revolutionize process control and monitoring in industries requiring high precision and reliability.

The application of advanced ML techniques for CCPR, particularly for CoDa/Ratio, remains a largely unexplored area. There is a significant need for further research to develop methods using ML to effectively monitor these types of data. While existing studies have made strides in reducing assumptions about data distributions when designing ML-based control charts, there are still challenges, such as autocorrelation in data, which is often encountered in time series data collected from Internet of Things (IoT) sensors, and industrial processes. Currently, in Zaidi et al. [39], a novel methodology is introduced for monitoring CoDa. This method integrates a multilayer feed-forward neural network with a T^2 control chart to enhance the detection of patterns within control charts. Demonstrated to be effective even amidst out-of-control conditions, its proficiency is corroborated through detailed simulations and analytical assessments. This advancement represents a significant stride in the development of ML techniques, challenging and potentially superseding the conventional assumptions embedded in traditional SPC. The efficiency and effectiveness offered by this novel approach underscore its potential in refining the monitoring of intricate data structures inherent in contemporary industrial processes. The direct future studies from this study can be the integration of MLFFNN with MEWMA, MCUSUM, and other advanced control charts (as mentioned in previous sections) for monitoring CoDa/Ratio.

For a comprehensive understanding of ML techniques in CCPR, one can refer to the detailed work of Tran et al. [54]. In their study, the authors provide an in-depth summary and analysis of the field, offering valuable insights into the application and effectiveness of various ML approaches in enhancing CCPR.

4.3 Interpreting out-of-control signal for CoDa control charts

In this subsection, we will give our perspective exclusively on interpreting out-of-control signals in control charts for monitoring CoDa. This focused approach stems from the fact that the ratio of two normal variables is effectively managed using univariate control charts and has a well-established framework for signal interpretation. Given this established understanding, there is no pressing need to delve into the interpretation of out-of-control signals for this type of chart. Therefore, we will only concentrate on the more complex and nuanced realm of CoDa, where the interpretation of signals is required due to its multidimensional nature and the constraints it presents.

Interpreting out-of-control signals when monitoring CoDa presents a particular challenge, especially when considering the special structure of CoDa and the presence of autocorrelation in data in general. Traditional multivariate control charts, such as Hotelling's T^2, MEWMA, and MCUSUM, for monitoring such data often only signal the general mean shifts without specifying the responsible variable(s). In general, some ML methods are being integrated into the process with promising results. NNs, SVM, decision trees, MLP, and DL have been explored to isolate and identify the specific causes of out-of-control conditions, see Niaki and Abbasi [55]; Cheng and Cheng [56]; Guh and Shiue [57], and Diren et al. [49] among many others for more information. These ML techniques not only offer a more precise identification of variables responsible for shifts but also improve upon the traditional charts' sensitivity to process anomalies. As these methods continue to evolve, they represent a significant potential in advancing the interpretability of control charts. These methods can be integrated into control charts for monitoring the ratio of two normal variables, however, due to the special structure of CoDa, these ML techniques can not be applied directly. Recently, there has been an attempt to interpret the out-of-control signals in CoDa monitoring by employing a NN technique in the work of[39]. In this study, a MLPNN with back-propagation learning has been introduced to interpret signals within Hotelling's T^2 control chart specifically for CoDa monitoring processes. This approach helps in pinpointing the specific variables causing deviations, rather than merely indicating shifts in the overall mean. Such focused detection is crucial for refining process control strategies, ultimately improving the efficiency of industrial operations. The model's effectiveness was validated in scenarios involving different components of CoDa, $p = 3$ and $p = 5$. This innovation represents a significant step forward in applying ML to ensure the precision and reliability of process monitoring in industrial settings. Future research directions could consider the integration of MLPNN with MCUSUM or MEWMA CoDa control charts to increase adaptability. Additionally, merging MLPNN with clustering algorithms or econometric models may offer new opportunities for innovation in the analysis of CoDa. Besides, the scarcity of robust methods for interpreting out-of-control signals in the monitoring of CoDa underscores a critical need for concentrated research efforts to develop new techniques in this field.

5 Conclusion

This chapter delivers an in-depth review and analysis of the methodologies for monitoring the ratio of two normal variables and CoDa within the Industry 5.0 framework. We began with a short introduction, setting the stage for the importance and challenges of CoDa/Ratio monitoring in industrial processes. Subsequently, we delved into an extensive examination of published research, analyzing 68 papers focused on monitoring the ratio of two normal variables and 19 dedicated to CoDa. In this exploration, we categorized the existing methodologies into relevant groups, offering a comprehensive analysis of these findings. This includes examining trends, assessing the distribution of publications across various journals, and highlighting the contributions of key authors in the field. Throughout each segment, we interjected our perspectives and insights, enriching the reader's understanding with deeper context and expert commentary on the results.

In our perspectives section, we offered insights on applying existing ML methodologies to CoDa and the ratio of two normal random variables. We stressed the importance of integrating ML techniques at various stages, from design to pattern recognition and interpretation of control charts. This perspective is essential for navigating the intersection of statistical control processes and advanced technological paradigms in Industry 5.0.

The paper concludes by emphasizing the need for ongoing research and development in this area. The integration of emerging technologies like AI and big data analytics with traditional monitoring methods promises further advancements in the field, potentially revolutionizing the efficacy and reliability of monitoring systems in smart manufacturing and Industry 5.0.

References

[1] G. Celano, P. Castagliola, A. Faraz, and S. Fichera. Statistical Performance of a Control Chart for Individual Observations Monitoring the Ratio of two Normal Variables. *Quality and Reliability Engineering International*, 30(8):1361–1377, 2014.

[2] J. Aitchison. *The Statistical Analysis of Compositional Data (Monographs on Statistics and Applied Probability).* Chapman & Hall Ltd., London, (Reprinted in 2003 with additional material by The Blackburn Press), 1986.

[3] D.C. Montgomery. *Statistical Quality Control: a Modern Introduction.* Wiley, New York, 2009.

[4] G. Celano and P. Castagliola. Design of a Phase II Control Chart for Monitoring the Ratio of two Normal Variables. *Quality and Reliability Engineering International*, 32, 2014.

[5] A.W. Spisak. A Control Chart for Ratios. *Journal of Quality Technology*, 22(1):34–37, 1990.

[6] D. Oksoy, E. Boulos, and L.D. Pye. Statistical Process Control by the Quotient of two Correlated Normal Variables. *Quality Engineering*, 6(2): 179–194, 1994.

[7] G. Celano and P. Castagliola. A Synthetic Control Chart for Monitoring the Ratio of Two Normal Variables. *Quality and Reliability Engineering International*, 2015.

[8] K.P. Tran, P. Castagliola, and G. Celano. Monitoring the ratio of two normal variables using Run Rules type control charts. *International Journal of Production Research*, ahead-of-print:1–19, 2015.

[9] Kim Phuc Tran, Philippe Castagliola, and G. Celano. Monitoring the ratio of two normal variables using ewma type control charts. *Quality and Reliability Engineering*, 10 2015.

[10] Kim Phuc Tran and Sven Knoth. Steady-state arl analysis of arl-unbiased ewma-rz control chart monitoring the ratio of two normal variables. *Quality and Reliability Engineering*, 34, 11 2017.

[11] Kim Phuc Tran, Philippe Castagliola, and G. Celano. Monitoring the ratio of population means of a bivariate normal distribution using cusum type control charts. *Statistical Papers*, 59, 03 2018.

[12] Huu Du Nguyen, Kim Phuc Tran, and Thong Ngee Goh. Variable sampling interval control charts for monitoring the ratio of two normal variables. *Journal of Testing and Evaluation*, 48, 12 2019.

[13] Huu Du Nguyen, Kim Phuc Tran, and Cédric Heuchenne. Monitoring the ratio of two normal variables using variable sampling interval expo-

nentially weighted moving average control charts. *Quality and Reliability Engineering International*, 35(1):439–460, 2019.

[14] Huu Du Nguyen, Kim Phuc Tran, and Henri L. Heuchenne. Cusum control charts with variable sampling interval for monitoring the ratio of two normal variables. *Quality and Reliability Engineering International*, 36(2):474–497, 2020.

[15] Kim Phuc Tran, Philippe Castagliola, and G. Celano. The performance of the shewhart-rz control chart in the presence of measurement error. *International Journal of Production Research*, 54, 05 2016.

[16] Huu-Du Nguyen and Kim Phuc Tran. Effect of the measurement errors on two one-sided shewhart control charts for monitoring the ratio of two normal variables. *Quality and Reliability Engineering International*, 36, 05 2020.

[17] H.D. Nguyen, K.P. Tran, and K.D. Tran. The effect of measurement errors on the performance of the Exponentially Weighted Moving Average control charts for the Ratio of Two Normally Distributed Variables. *European Journal of Operational Research*, 293(1):203–218, 2021.

[18] Sani Salihu Abubakar, Michael B. C. Khoo, Sajal Saha, Adamu Abubakar Umar, and Wai Chung Yeong. Run sum ratio chart when measurement errors are present. *Quality and Reliability Engineering International*, 38 (8):4049–4072, 2022.

[19] V. Pawlowsky-Glahn, J. J. Egozcue, and R. Tolosana-Delgado. *Modeling and Analysis of Compositional Data*. John Wiley & Sons, 2015.

[20] J. J. Egozcue, V. Pawlowsky-Glahn, G. Mateu-Figueras, and C. Barceló-Vidal. Isometric logratio transformations for compositional data analysis. *Mathematical Geology*, 35(3):279–300, 2003.

[21] J.J. Egozcue and V. Pawlowsky-Glahn. Groups of Parts and Their Balances in Compositional Data analysis. *Mathematical Geology*, 37(7):795–828, 2005.

[22] R. A. Boyles. Using the chi-square statistic to monitor compositional process data. *Journal of Applied Statistics*, 24(5):589–602, 1997.

[23] R. Guevara, J. Vargas, and D. Linero Segrera. Profile monitoring for compositional data. *Revista Colombiana de Estadística*, 37:159, 07 2014.

[24] M. Vives-Mestres, J. Daunis-I-Estadella, and J.A. Martin-Fernandez. Out-of-Control Signals in Three-Part Compositional T^2 Control Chart. *Quality and Reliability Engineering International*, 30(3):337–346, 2014.

[25] M. Vives-Mestres, J. Daunis-I-Estadella, and J.A. Martin-Fernandez. Individual T^2 Control Chart for Compositional Data. *Journal of Quality Technology*, 46(2):127–139, 2014.

[26] K. P. Tran, P. Castagliola, G. Celano, and Michael B.C. Khoo. Monitoring compositional data using multivariate exponentially weighted moving average scheme. *Quality and Reliability Engineering International*, 34(3): 391–402, 2018.

[27] M. Imran, J. Sun, F.S. Zaidi, Z. Abbas, and H.Z. Nazir. Multivariate cumulative sum control chart for compositional data with known and

estimated process parameters. *Quality and reliability engineering international*, 38(5), 2022-07. ISSN 0748-8017.

[28] T.T.V. Nguyen, C. Heuchenne, and K.P. Tran. Anomaly detection for compositional data using vsi mewma control chart. *IFAC-PapersOnLine*, 55(10):1533–1538, 2022. ISSN 2405-8963. 10th IFAC Conference on Manufacturing Modelling, Management and Control MIM 2022.

[29] M. Imran, J. Sun, F.S. Zaidi, Z. Abbas, and H.Z. Nazir. On designing efficient multivariate exponentially weighted moving average control chart for compositional data using variable sample size. *Journal of Statistical Computation and Simulation*, 93(10):1622–1643, 2023.

[30] C.A. Lowry, W.H. Woodall, C.W. Champ, and S.E. Rigdon. A Multivariate Exponentially Weighted Moving Average Control Chart. *Technometrics*, 34(1):pp. 46–53, 1992. ISSN 00401706.

[31] F.S. Zaidi, P. Castagliola, K.P. Tran, and M.B.C. Khoo. Performance of the hotelling t2 control chart for compositional data in the presence of measurement errors. *Journal of Applied Statistics*, 46(14):2583–2602, 2019.

[32] F.S. Zaidi, PY. Castagliola, K.P. Tran, and M.B.C. Khoo. Performance of the mewma-coda control chart in the presence of measurement errors. *Quality and Reliability Engineering International*, 36(7):2411–2440, 2020.

[33] M. Imran, J. Sun, F.S. Zaidi, Z. Abbas, and H.Z. Nazir. Effect of measurement error on the multivariate cusum control chart for compositional data. *CMES-Computer Modeling in Engineering & Sciences*, 136(2), 2023.

[34] M. Imran, J. Sun, F.S Zaidi, Z. Abbas, and H.Z. Nazir. Evaluating the performance of variable sampling interval hotelling t2 charting scheme for compositional data in the presence of measurement error. *Quality and Reliability Engineering International*, 2023.

[35] F.S. Zaidi, H.L. Dai, M. Imran, and K.P. Tran. Monitoring autocorrelated compositional data vectors using an enhanced residuals hotelling t2 control chart. *Computers & Industrial Engineering*, 181:109280, 2023.

[36] M. Vives-Mestres, J. Daunis-i Estadella, and J. A. Martin-Fernandez. Signal interpretation in hotelling's t 2 control chart for compositional data. *IIE Transactions*, 48(7):661–672, 2016.

[37] M. Imran, J. Sun, X. Hu, F.S. Zaidi, and A. Tang. Investigating zero-state and steady-state performance of mewma-coda control chart using variable sampling interval. *Journal of Applied Statistics*, pages 1–22, 2023.

[38] C.H. Weiß. Ordinal compositional data and time series. *Statistical Modelling*, page 1471082X231190971, 2023.

[39] F.S. Zaidi, H.L. Dai, M. Imran, and K.P. Tran. Analyzing abnormal pattern of hotelling t2 control chart for compositional data using artificial neural networks. *Computers & Industrial Engineering*, 180:109254, 2023.

[40] M. Imran, H.L. Dai, F.S. Zaidi, X. Hu, K.P. Tran, and J. Sun. Analyzing out-of-control signals of t2 control chart for compositional data using

artificial neural networks. *Expert Systems with Applications*, page 122165, 2023.

[41] J. Schmidhuber, S. Hochreiter, et al. Long short-term memory. *Neural Comput*, 9(8):1735–1780, 1997.

[42] C.K.I. Williams and C.E. Rasmussen. *Gaussian processes for machine learning*, volume 2. MIT press Cambridge, MA, 2006.

[43] A. Vaswani, N. Shazeer, N. Parmar, J. Uszkoreit, L. Jones, A.N. Gomez, L. Kaiser, and I. Polosukhin. Attention is all you need. *Advances in neural information processing systems*, 30, 2017.

[44] J. Wang and L. Liu. A new multivariate control chart based on the isolation forest algorithm. *Quality Engineering*, pages 1–17, 2023.

[45] H.D. Nguyen, A.A. Nadi, K.D. Tran, P. Castagliola, G. Celano, and K.P. Tran. The shewhart-type rz control chart for monitoring the ratio of autocorrelated variables. *International Journal of Production Research*, 61(20):6746–6771, 2023.

[46] G.A. Pugh. Synthetic neural networks for process control. *Computers & industrial engineering*, 17(1-4):24–26, 1989.

[47] R.S. Guh. Real-time recognition of control chart patterns in autocorrelated processes using a learning vector quantization network-based approach. *International Journal of Production Research*, 46(14):3959–3991, 2008.

[48] T. Li, S. Hu, Z.Y. Wei, Z.Q. Liao, et al. A framework for diagnosing the out-of-control signals in multivariate process using optimized support vector machines. *Mathematical Problems in Engineering*, 2013, 2013.

[49] D.D. Diren, S. Boran, and I. Cil. Integration of machine learning techniques and control charts in multivariate processes. 2020.

[50] V. Vapnik, I. Guyon, and T. Hastie. Support vector machines. *Machine Learning*, 20(3):273–297, 1995.

[51] S.Y. Lin, R.S. Guh, and Y.R. Shiue. Effective recognition of control chart patterns in autocorrelated data using a support vector machine based approach. *Computers & Industrial Engineering*, 61(4):1123–1134, 2011.

[52] C.S. Cheng. A neural network approach for the analysis of control chart patterns. *International Journal of Production Research*, 35(3):667–697, 1997.

[53] A. Addeh, A. Khormali, and N.A. Golilarz. Control chart pattern recognition using rbf neural network with new training algorithm and practical features. *ISA transactions*, 79:202–216, 2018.

[54] P.H. Tran, A. Ahmadi Nadi, T.H. Nguyen, K.D. Tran, and K.P. Tran. Application of machine learning in statistical process control charts: A survey and perspective. In *Control charts and machine learning for anomaly detection in manufacturing*, pages 7–42. Springer, 2022.

[55] S.T.A. Niaki and B. Abbasi. Fault diagnosis in multivariate control charts using artificial neural networks. *Quality and reliability engineering international*, 21(8):825–840, 2005.

[56] C.S. Cheng and H.P. Cheng. Identifying the source of variance shifts in the multivariate process using neural networks and support vector machines. *Expert Systems with Applications*, 35(1-2):198–206, 2008.

[57] R.S. Guh and Y.R. Shiue. An effective application of decision tree learning for on-line detection of mean shifts in multivariate control charts. *Computers & Industrial Engineering*, 55(2):475–493, 2008.

9. Energy Efficiency Scheduling of Flexible Flow Shop Using Group Technology

Zhiqiang Zeng, Le Gao,* Zhiyong Hong and Wenyi Liang

1 Introduction

Processing energy consumption and setup energy consumption are those generated during the processing and machine setup processes, accounting for the majority of the total energy consumption. Energy-aware scheduling (energy efficiency scheduling) is one of the effective ways to reduce these two-parts of energy consumption, which has been widely studied [1–5]. The multi-variety and small batch manufacturing mode widely exists in the real world, which poses a huge challenge to production scheduling. In the multi-variety and small batch manufacturing mode, the main challenge is the significant production switching that occurs during the manufacturing process, leading to an increase in manufacturing costs and a decrease in production efficiency. The use of group technology can effectively reduce production switching, thereby reducing costs and improving production efficiency. There are two main grouping strategies in group technology [6]. The first one is to combine all jobs of the same process into a group for processing. This strategy can minimize production switching, but may increase the makespan and leads to the imbalance of machine load. The second is to allow multiple setups in jobs of the same process, that is, jobs of the same process can be divided into multiple groups. This strategy will reduce the makespan and balance the machine load, but may increase the production switching.

Faculty of Intelligent Manufacturing, Wuyi University, Jiangmen 529020, China.
Email: zhiqiang.zeng@outlook.com
* Corresponding author: le.gao@nscc-gz.cn

Different group strategies have different sub problems to solve. For the first group strategy, the group scheduling problem needs to solve two subproblems: i) the traditional scheduling problem; ii) the jobs sorting problem within a group. For the second grouping strategy, group scheduling needs to solve three subproblems. In addition to the above two problems, it also needs to solve the jobs grouping problem of the same process. Flexible flow shop is a type of flow shop, which is a manufacturing system widely present in the real world. In the study of grouping scheduling in a flow shop. The first grouping strategy is the most studied [7–17]. There are few studies on the second grouping strategy. Shahvari and Logendran [18] optimized the completion time and tardiness simultaneously in the group scheduling of a flexible flow shop. They abandoned the assumption that each group has single setup, allowing multiple setups for each job type. Feng et al. [19] allowed multiple setups in the group scheduling of the flexible flow shop The makespan and preventive maintenance were considered simultaneously in their study.Jain et al. studied the group scheduling problem of flow shop with sequence dependent set-up time (SDST), which considered SDST among groups and among part types within a group [20].

At present, most of the researches on group scheduling in a flow shop consider that the process differences of jobs are consistent in all stages [21–24]. However, there are many manufacturing systems in reality, and the differences of job process are not consistent [6,25]. There are few studies on group scheduling of flow shops considering the process difference inconsistency of jobs. Andrés et al. [26] studied the group scheduling problem of tile industry, which is a three-stage flexible flow shop. They defined similarity coefficient for each job, and grouped jobs based on similarity coefficients to reduce setup time. Among the existing studies on group scheduling, most of them use group technology to reduce the setup time, makespan, earliness and tardiness and so on [27–37].However, setup time minimization is not the same as setup energy consumption minimization. Up to now, the research on group scheduling to reduce energy consumption is rare.He et al. studied the group scheduling problem of a flow shop,which optimized makespan,total flow time and total energy consumption at the same time[38].

Inspired by the production scheduling of tissue paper enterprises, this paper studies the problem of energy-aware scheduling of a two-stage flexible flow shop considering inconsistency of the process differences of jobs, which intends to reduce the energy consumption by using group technology. The studied manufacturing system consists of two processing stages. The jobs with the same processing technology in the second stage must have the same processing technology in the first stage. On the contrary, jobs with the same processing technology in the first stage may not have the same processing technology in the second stage. Considering that the problem being studied is a multi-objective optimization problem, an improved multi-objective scatter search algorithm is proposed. By combining group technology with scheduling, this study not only improves production efficiency by optimizing the makespan, but also optimizes processing energy consumption and setup energy consumption to improve energy efficiency.

2 Modeling

In the two-stage flexible flow shop, it is assumed that there are n jobs $J = \{J_1, J_2, ..., J_n\}$ and m machines $M = \{M_1, M_2, ..., M_m\}$.There are $m1$ machines $M^p = \{M_1^p, M_2^p, ..., M_{m1}^p\}$ in the first stage and $m2$ machines $M^c = \{M_1^c, M_2^c, ..., M_{m2}^c\}$ in the second stage. According to the process characteristics of jobs in the first stage, jobs can be divided into p types, according to these p types of jobs, all jobs can be divided into p' groups $G' = \{G'_1, G'_2, ..., G'_{p'}\}$. The Group G'_i contains p''_i jobs, each group contains jobs of the same type, different groups may have the same type of jobs. $\pi_{G'_i}$ represents the sequence of jobs in group G'_i , $\pi_{G'_i}(j)$ represents the j-th job in group G'_i. According to the process characteristics of jobs in the second stage, jobs can be divided into c types, according to these c types of jobs, all jobs can be divided into c' groups $G'' = \{G''_1, G''_2, ..., G''_{c'}\}$. The Group G''_i contains c''_i jobs, each group contains jobs of the same type, different groups may have the same type of jobs. Each G'_j is consist of multiple G''_i . $\pi_{G'_i}$ represents the sequence of jobs in group G''_i, $\pi_{G'_i}(j)$ represents the j-th job in group G''_i . Because jobs of the same process can be divided into multiple groups, so $p \leq p'$ and $c \leq c'$. Because of the jobs with the same process in the second stage, they must have the same process in the first stage. On the contrary, jobs with the same process in the first stage, they may not have the same process in the second stage. Therefore, $p \leq c$, but p' may not be less than c'. Combined with group technology, the energy-aware scheduling model of two-stage flexible flow shop can be represented by the following formula:

$$\begin{cases} min(C), \\ min(E). \end{cases} \tag{1}$$

$$E{=}E^p + E^s \tag{2}$$

$$C = \max\left(C_{i,h}\right) \forall i \in J; \forall h \in M \tag{3}$$

$$E^p = \sum_{i \in G'} \sum_{j \in M^p} \left(O_{i,j} U_{i,j}^p \sum_{k \in G'_i} T_{k,j}^p \right) + \sum_{i \in G''} \sum_{j \in M^c} \left(O_{i,j} U_{i,j}^p \sum_{k \in G''_i} T_{k,j}^p \right) \tag{4}$$

$$E^s = \sum_{i \in G'} \sum_{j \in G'} \sum_{h \in M^p} Q_{i,j,h} P_{i,j,h}^s T_{i,j,h}^s + \sum_{i \in G''} \sum_{j \in G''} \sum_{h \in M^c} Q_{i,j,h} P_{i,j,h}^s T_{i,j,h}^s \tag{5}$$

$$Q_{i,j,h}\left(B_{j,h}-C_{i,h}-T^{s}_{i,j,h}\right)=0\forall i,j\in G';\forall h\in M^{p} \quad (6)$$

$$\begin{cases}\sum_{h\in M^{c}} O_{j,h}B_{j,h}=\sum_{h\in M^{p}} O_{\pi_{G''_j}(1),k}C_{\pi_{G''_j}(1),k} \text{ if j is the first job group} \\ \quad \text{if j is the first job group on the machine } \forall j\in G'', \\ Q_{i,j,h}(B_{j,h}-max(C_{i,j},\sum_{h\in M^{p}} O_{\pi_{G''_j}(1),k}C_{\pi_{G''_j}(1),k})-T^{s}_{i,j,h})=0 \\ \quad \text{else } \forall i,j\in G'';\forall h\in M^{c}.\end{cases} \quad (7)$$

$$B_{\pi_{G'_i}(k+1),h}-C_{\pi_{G'_i}(k),h}=0\forall i\in 1,2,...,p';\forall k\in 1,2,...,p''_i-1;\forall h\in M^{p} \quad (8)$$

$$B_{\pi_{G''_i}(k+1),h}-max(C_{\pi_{G'_i}(k),h},\sum_{j\in M^{p}} O_{\pi_{G''_i}(k+1),j}C_{\pi_{G''_i}(k+1),j})=0$$
$$\forall i\in 1,2,...,c';\forall k\in 1,2,...,c''_i-1;\forall h\in M^{c} \quad (9)$$

$$T^{p}_{i,j}=\sum_{k\in i} T^{p}_{k,j} \quad \forall i\in G' \quad (10)$$

$$T^{p}_{i,j}=\sum_{k\in i} T^{p}_{k,j} \quad \forall i\in G'' \quad (11)$$

$$C_{i,h}-B_{i,h}=T^{p}_{i,h} \quad \forall i\in G'\cup J;\forall h\in M^{p} \quad (12)$$

$$C_{i,h}-B_{i,h}=T^{p}_{i,h} \quad \forall i\in G''\cup J;\forall h\in M^{c} \quad (13)$$

$$\sum_{j\in M^{p}} O_{i,j}=1 \quad \forall i\in J \quad (14)$$

$$\sum_{j\in M^{c}} O_{i,j}=1 \quad \forall i\in J \quad (15)$$

where C is the makespan, E is the total energy consumption, $C_{i,h}$ represents the complete time of group i or the complete time of job i on machine h. E^{p} represents the processing energy consumption, E^{s} represents the setup energy consumption. $O_{i,j}=\{0,1\}$. If it equals 1 it means that group i or job i is processed on machine j, it equals zero means that group i or job i is not processed on machine j. Because the jobs of the same group have the same process, the processing power of the jobs of the same group on the same machine is the same, $U^{p}_{i,j}$is the processing power of group i on machine j. $T^p_{k,j}$ represents the processing time of group k or job k on machine j. $O_{i,j,h} = \{0, 1\}$, when group i and group j are two consecutive processing job groups of machine h, $Q_{i,j,h} = 1$, otherwise, $Q_{i,j,h} = 0$. $P^s_{i,j,h}$ represents the power of group i switching to group j on machine h, it equals the power of last job of group i switching to the first job of group j on machine h. $T^s_{i,j,h}$ represents the setup time of group i switching to group j on machine h, it equals the setup time of the last job of group i switching to the first job of group j on machine h.

Formula 6 and 7 ensure that the machine can only process one group at the same time, where $B_{j,h}$ represents the start time of the group j or job j on machine h. In stage one, the start time of the next group is the end time of the previous group plus the setup time. In stage two, the start time of the next group is equal to the maximum value of the completion time of the first operation of first job of the group and the completion time of the previous group plus the setup time. In addition, formulas 8 and 9 indicate the jobs in each group must not be preempted, that means the start of the next job must wait until the end of the previous job. Moreover, in stage two, the start of the job must wait until the completion of the previous operation of the job. Equations 10 and 11 show that the processing time of each group is equal to the total processing time of all jobs in the group. Formulas 12 and 13 indicate that each job cannot be interrupted once it is started. Formulas 14 and 15 indicate that each job must and can only be processed once in each stage.

3 The improved multi-objective scatter search algorithm

The scatter search algorithm was first proposed by Glover [39] to solve the integer programming problem. The scatter search algorithm defines a flexible framework, which contains multiple modules, including the diversity generation module, solution improvement module, reference set update module, subset generation module, solution combination module, and more. The scatter search algorithm forms high-quality and diverse solutions in the reference set. Then it generates generates new solutions through the subset generation and combination methods, and improves the new solutions by the solution improvement method. Then it renews the reference set, and obtains the global optimal solution or satisfactory solution by continuously cycling the above steps. Figure 2 is the flow chart of the scatter search algorithm. According to the characteristics of the problem, combined with the fast non-dominated sorting algorithm, this paper develops an improved multi-objective scatter search algorithm (IMSS). First, a three-layer chromosome coding scheme is proposed, and a dynamic split strategy is used to split the job group dynamically in the process of chromosome decoding to ensure the load balance between machines. In order to ensure the diversity of the population, a method of population diversity generation is developed and applied to the reference set updating method. In addition, according to the characteristics of encoding, a combination and local improvement method are designed. Each step of the proposed algorithm is shown in Fig1.

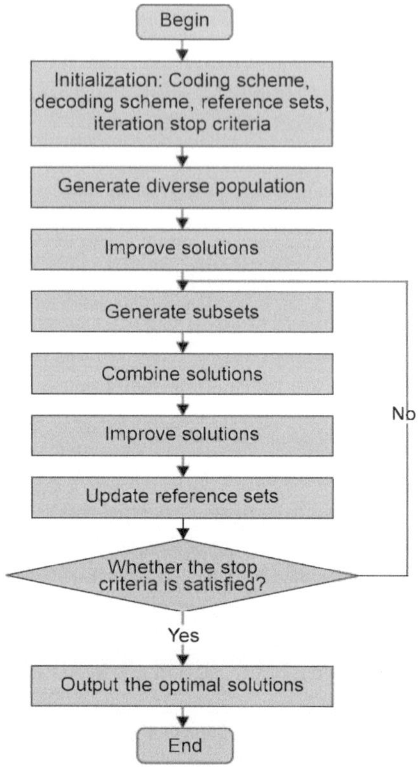

Figure 1: The flow chart of scatter search algorithm.

3.1 Encoding and decoding

According to the research problem, in the first stage, the problems of processing sequence of job groups $G' = \{G'_1, G'_2, ..., G'_{P'}\}$ and machine assignment need to be solved. Because the jobs in the same group are of the same type, the order of jobs in it does not affect the setup energy consumption and completion time. In addition, this paper uses the rule that the machine with the earliest available time is assigned first. When the order of job groups is determined, the problem of machine assignment is solved. Therefore, the processing sequence of job groups $G' = \{G'_1, G'_2, ..., G'_{P'}\}$ can be represented by a vector. In order to reduce setup energy consumption, we combine jobs with the same process in the first stage into a group. For example, when $p' = 3$, it can use [3,1,2] to represent the processing sequence of groups in stage one. The first chromosome determines the processing order of groups ,$G' = \{G'_1, G'_2, ..., G'_{p'}\}$ but it cannot determine the processing order of groups .$G'' = \{G''_1, G''_2, ..., G''_{c'}\}$ As the job sequence of each group G'_iwill affect the completion time and energy consumption of the second stage. Therefore, we use a chromosome to represent the jobs' sequence of each group .G'_i In order to reduce switching, the minimum unit of sequencing for each group G'_i is set to a G''_i . Suppose G'_1 consists of G''_1, G''_2 and G''_3 , G'_2 consists of G''_4 , G''_5 and G''_6, G'_3 consists of G''_7 and .G''_8 Then the second chromosome can be represented by a vector [8,7,1,2,3,4,6,5]. In addition, the sequence of jobs in each group G''_idoes not affect energy consumption, but affects the completion time. Since the jobs of each group G''_i are processed on the same machine in the first and second stages, in order to reduce the makespan, according to the NEH rule, the jobs of each group $G''_i$are sorted according to the processing time from small to large. So, the third layer chromosome represents the jobs' sequence of each group G''_i, and it is fixed. Fig. 2 shows a three-layer chromosome representation of jobs $J1 \sim J20$. Here, it is assumed that the processing time of jobs $J1 \sim J20$ increases from small to large.

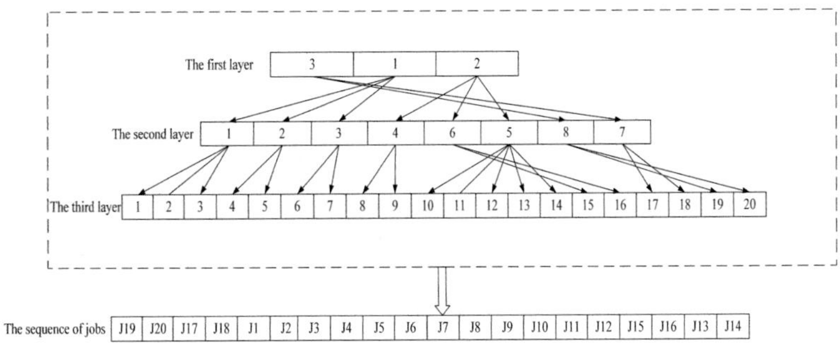

Figure 2: Three-layer chromosome scheme.

In the iteration process, chromosomes need to be decoded to form a feasible production schedule. The first layer of chromosomes represents the sequence of G'. Because the jobs of the same group G'_i have the same process in the first stage, in order to reduce the setup energy consumption, the jobs of the same group G'_i should be processed on the same machine in the first stage. Assume there are three machines $M^p = \{M^p_1, M^p_2, M^p_3\}$ in the first stage, M^p_1 has the earliest available time, then M^p_2 , and finally M^p_3 . Assume there are three machines $M^c = \{M^c_1, M^c_2, M^c_3\}$ in the second stage, M^c_1 has the earliest available time, then M^c_2, and finally M^c_3. According to the above chromosome coding, for the convenience of demonstration, it is assumed that the processing time of all jobs is the same. The first layer chromosome is [3,1,2], which means that the jobs of G'_3 is first arranged to be processed in M^p_1. According to the second layer chromosome and the third layer chromosome, the job sequence of G'_3 is [19,20,17,18], so the job processing sequence on M^p_1 is [19,20,17,18]. Similarly, the processing sequence on M^p_2 is [1,2,3,4,5,6,7], and the processing sequence on M_3^p is [8,9,15,16,10,11,12,13,14]. Fig. 3 is the job processing sequence diagram of the three machines. From the figure, it can be seen that the load of the machine is unbalanced, the load of M_1^p is much less than that of M_3^p . For this reason, a dynamic split strategy (DSS) is proposed, the J_{10} and J_{11} are moved to the M_1^p through DSS and thus balance the load of M_1^p and M_3^p. The detail of the DSS is described in Section 3.2. Although G_2' is split and the setup energy consumption is increased, but the load of the machine is balanced, the result is shown in Fig. 4.

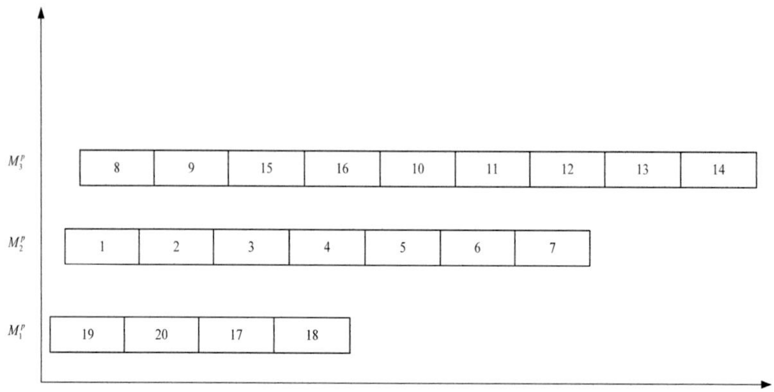

Figure 3: The jobs processing sequence in stage one.

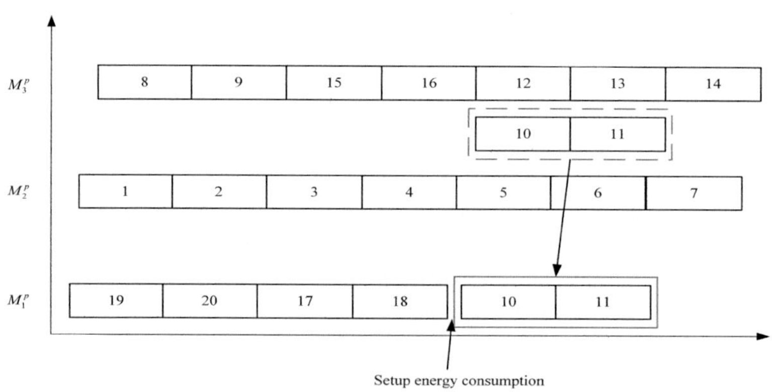

Figure 4: The adjusted job processing sequence of stage one.

According to the processing sequence of group $G'' = \{G_1'', G_2'', ..., G_{c'}''\}$ in the first stage, calculate the starting processing time of each group G_i'', sort $G'' = \{G_1'', G_2'', ..., G_{c'}''\}$ from small to large according to the start processing time, then the processing sequence of $G'' = \{G_1'', G_2'', ..., G_{c'}''\}$ in the second stage can be obtained. For example, according to the processing sequence of the first stage, the order of $G'' = \{G_1'', G_2'', ..., G_{c'}''\}$ is [8,1,4,7,6,2,5_1,5_2,3], 5_1 consists of $J10$ and $J11$, 5_2 consists of $J12$, $J13$ and $J14$, 5_1 and 5_2 are part of G_5'' and has the same process. According to the third layer chromosome and [8,1,4,7,6,2,5_1,5_2,3], it can obtain the jobs processing sequence of M_1^c is [19,20,17,18,13,14], and the jobs processing sequence of M_2^c is [1,2,3,4,5,6], and the jobs processing sequence of M_3^c is [10,11,12,15,16,7,8,9]. Fig. 5 shows the job processing sequence of the machines in the second stage. It can be seen from the Fig. 5, although the setup energy consumption is increased when move $J14$ from M_3^c to M_1^c via dynamic split strategy, but the load of the machines is more balanced, the result is shown as Fig. 6. In this paper, dynamic splitting strategy is used to achieve load balancing of the machines with the minimum increasement of setup energy consumption. It should be noted that when adjusting the job in stage two, the completion time of the job in stage one should be considered to ensure that the processing of the job follows the constraints of processing sequence.

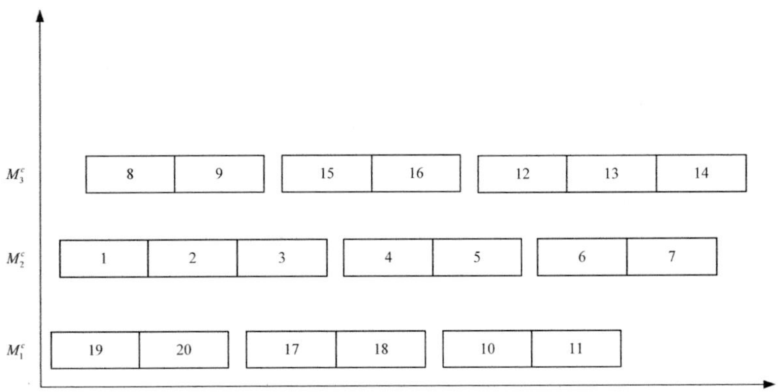

Figure 5: The jobs processing sequence in stage two.

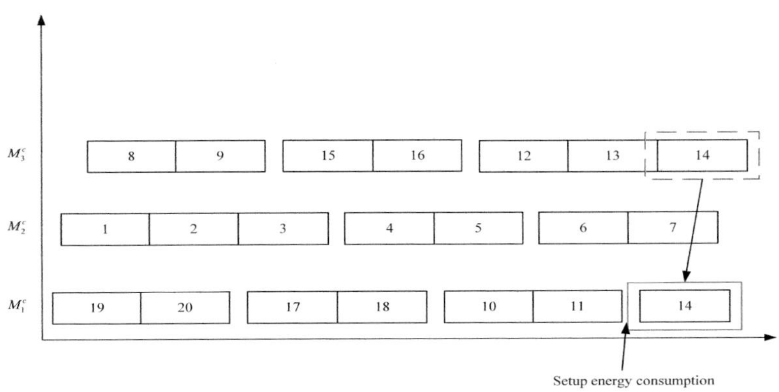

Figure 6: The adjusted job processing sequence in stage two.

3.2 Dynamic splitting strategy

It is easy to cause the load imbalance of the machines by scheduling jobs in groups as described above. The imbalance referred to here can be understood as that there are some jobs that can be processed earlier when they are moved to other machines for processing, such as J14 in Fig. 5. Therefore, the machine load balance here refers to that no job can be processed earlier by moving jobs to other machines. In this paper, a dynamic splitting strategy is developed to balance the load of the machines. The specific steps are as follows:

Algorithm 1 Dynamic splitting strategy

01. Inpu A feasible scheduling.
02. Outpu The adjusted scheduling.
03. While true do:
04. Find the job with the largest start time in the last job of all machines (*Js*) record the start time of *Js* (*Max_start*) and the machine that process *Js* (*Max_start_m*). Find the job with the smallest complete time in the last job of all machines (Je)record the complete time of *Je* (*Min_end*) and the machine that process *Je* (*Min_end_m*).
05. If *Max_start* > *Min_end*:
06 Traverse the jobs on machine *Max_start* m to find the first job with start time greater than *Min_end*(*Ja*). Move *Ja* to the machine *Min_end_m*, judge whether *Ja* can be processed in advance, if yes, then update the start time and completion time of the jobs after *Ja* on machine *Max_start_m*, and update the start time and complete time of the last job on machine *Min_end_m* . Otherwise, cancel the move, and find the next job with start time greater than *Min_end* on machine *Max_start_m*. Then, adjust the processing machine of the finding job according to the above method. Repeat the above steps until there is no job can be moved to other machines.
07. Else:
08. Terminate while loop.
09. End If
10. End While

3.3 Method of diversity population generation

Generating a diverse population is one of the conditions to ensure the global search ability of the scatter search algorithm. A diverse population means that each individual has a large distance. There are many indicators to measure individual distance, suppose $D(x_1, x_2)$ is the distance between individuals x_1 and x_2, in this paper, $D(x_1, x_2)$ is set as the sum of different edges of two individuals. For example, given two individuals $x_1 = [1, 3, 5, 4, 2]$ and $x_2 = [5, 4, 1, 2, 3]$. The edges of x_1 are (1,3), (3,5), (5,4), (4,2), the edges of x_2 are (5,4)(4,1)(1,2)(2,3), so $D(x_1, x_2) = 3$. Because the third layer chromosome is fixed, only the first and second layers of chromosomes need to be generated. The proposed population generation method is as follows.

Algorithm 2 The proposed population generation method

01. Input: population size N, p′ and
02. Output: diversity populati
03. For i=1 to
04. If i=
05. The first layer chromosome and the second layer chromosome of the new individual are randomly generated the new individual is added to the existing population.
06. Else
07. Randomly generate 20 chromosomes, and the distance between each generated chromosome and the first layer chromosome of the existing individuals is calculated, and the chromosome with the largest distance is selected as the first layer chromosome of the new individual.
08. For j =1 to p′ do:
09. Since each G′j corresponds to a segment of the second layer chromosome, 20 segments are randomly generated for each G′j , then the distance between each generated segment and the existing corresponding chromosome segment is calculated, and the segment with the largest distance is selected as the segment corresponding to G′j in the second layer chromosome of the new individual.
10. End For
11. Add the new individual to existing populations.
12. End If
13. End

3.4 The solution improvement method

Applying the improvement method to the solution can further improve the solution quality. The solution improvement method is generally designed based on the local search method. Although the complex method can improve the quality of the solution more effectively, it also increases the operation time. However, the simplified method has difficulties in improving the solution quality. In this paper, a solution improvement method is designed based on variable neighborhood search algorithm. The key of variable neighborhood search algorithm is neighborhood action. According to the characteristics of chromosome coding, three kinds of neighborhood actions are designed in this paper. The first neighborhood action is to use the exchange, inversion and insertion operators to disturb the first layer chromosome, so as to generate new individuals. The second neighborhood action is to use the exchange, inversion and insertion operators to disturb a segment of the second layer chromosome. This is a segment segment of the second layer chromosome that corresponds to the job group G_i' which is randomly selected from G' . The third neighborhood action is to combine the first two neighborhood actions. The following is the flow of the proposed solution improvement method.

Algorithm 3 The proposed solution improvement method

01. Input: a soluti
02. Output: the improved solut
03. F *i*=1:10 do:
04. While true do:
05. Use the first neighborhood acti to generate a new solution for the current solution, and judge whether the new solution is improved. If there is improvement, replace the current solution with the new solution and carry out the next cycle. Otherwise, judge whether there is no improvement 5 consecutive times. If so, exit the cycle. Otherwise, continue the next cycle.
06. E While
07. While true do:
08. Use the second neighborhood acti to generate a new solution for the current solution, and judge whether the new solution has been improved. If there is improvement, replace the current solution with the new solution and skip to Line 04. Otherwise, judge whether there is no improvement for 5 consecutive times. If so, exit the cycle, otherwise continue the next cycle.
09. E While
10. While true do:
11. Use the third neighborhood action to generate a new solution for the current solution, and judge whether the new solution is improved. If there is improvement, replace the current solution with the new solution and jump to Line 04. Otherwise, judge whether there is no improvement for 5 consecutive times. If so, skip to Line 03. Otherwise, continue the next cycle.
12. E While
13. End

3.5 Subset generation method

In the scatter search algorithm, the reference set Ref consists of two parts, Ref1 and Ref2, respectively. Here, Ref1 stores high-quality solutions, while Ref2 stores diverse solutions. In this paper, two subsets are generated, the first subset $Subset1 = \{p_1, p_2\}$ is composed of two solutions selected by the tournament selection mechanism from Ref1. The second subset $Subset2 =$

$\{p_3, p_4\}$ is also composed of two solutions. One is selected from Ref1 by the tournament selection mechanism, and the other is selected from Ref2 by the same mechanism.

3.6 The solution combination method

According to the characteristics of coding, this paper designs three solution combination methods. In the first method, the first layer chromosome of the two solutions is crossed and the second layer and the third layer chromosome remain unchanged. In the second method, the second layer chromosome of the two solutions is crossed, and the first layer and third layer chromosome remain unchanged. The third method is to cross the first layer and second layer chromosomes at the same time, and the third layer chromosome remain unchanged. The crossing method used in this paper is the linear order crossover (LOX) [40]. First, three methods are used to combine the solutions of subset one and subset two, and a total of six new solutions are obtained. Then the Pareto rank and crowding distance of each solution are calculated. Finally, according to the Pareto rank and crowding distance, the best one of the six new solutions is selected to inherit to the next generation.

3.7 Reference set updating method

Through the solution combination method and the solution improvement method, a new population *popn* will be obtained. Based on the old reference set and the new population, the new reference set can be obtained by using the reference set updating method. The developed reference set update method is as follows: 1) Merge the new population *popn* and Ref , remove the repeated solutions to get the remaining solutions *popu*, and calculate the Pareto rank and crowding distance of each of *popu* solution. 2) Assume that the size of Ref1 is Ref1_*num* , the size of Ref2 is Ref2_*num*, the size of *popu* is *popu_num* , if the *popu_num* is less than Ref1_*num*+Ref2_*num*, then Ref1_*num*+Ref2_*num*- *popu_num* solutions are generated by the diversity generation method and added to *popu* to obtain a new population *popun*, and the Pareto rank and crowding distance of each *popun* popun solution are calculated. Select Ref1_*num* best solutions as the new Ref1 and Ref2_*num* solutions with the largest distance as the new Ref2 from *popun*. Otherwise, select Ref1_*num* best solutions as the new Ref1 and Ref2_*num* solutions with the largest distance as the new Ref2 from *popu*.

4 Experiment

In order to prove the superiority of the proposed algorithm, this paper compares the proposed algorithm with NSGA-II, which is an outstanding multi-objective optimization algorithm. All algorithms are written in MATLAB

2015b and the operating environment: Windows 10, 8G RAM, 2.11GHz Intel Core i7-8650U.

4.1 Experiment data

Because of the particularity of the problem, there is no standard data set available. In this paper, based on the real data of tissue paper enterprise, Monte Carlo method is used to generate the required experimental data. A total of 25 scheduling problems are used to test the performance of the proposed algorithm. The number of jobs varies from 200 to 400. The number of process types in stage one varies from 20 to 40, and the number of process types in stage two varies from 100 to 200.The number of process types indicates the process selection range of the jobs of the scheduling problem in each stage. For example, the number of process types in the first stage is 20, indicating that the process in the first stage of each job of the scheduling problem is randomly selected from 20 processes. Similarly, the number of process types in the second stage is 100, indicating that the process in the second stage of each job of the scheduling problem is randomly selected from 100 processes. The number of machines for both the first and second stages is set to five. The size of each job is randomly generated from [120000,180000], and the processing speed of each machine is randomly generated from [900,1200], where the machines have the same speed when processing the job with identical processes. The processing energy consumption per unit time of machine identical generated from [1200,1500] and [120,150], respectively. The machines have the same processing energy consumption per unit time when processing the jobs with the same processes. The setup energy consumption per unit time of machine stage one and stage two is generated from [800,1000] and [80,100], respectively. The setup time in stage one and stage two is generated from [10,60]. The setup energy consumption per unit time and setup time relate to the processes of the two consecutive jobs and machine.

4.2 Evaluation methods

The multi-objective optimization algorithm obtains the non-dominated solution set. The quality of the non-dominated solution set can be measured from the uniformity, universality and the degree of approaching the optimal solution. At present, many evaluation methods only evaluate one of the indexes of the solution. The hypervolume (HV) index is one of the rare indexes that can evaluate the uniformity, universality and the degree of approaching the optimal solution simultaneously [41]. In this paper, hypervolume index is used to evaluate the quality of solution. The larger the hypervolume index, the better the quality of solutions is. In addition, the evolutionary algorithm is stochastic, and it is not reliable to compare the differences between the algorithms only through one run. In this paper, each algorithm is run 20 times, and the t-test is used to test whether there is significant difference between the two algorithms. The significance level is set to 95%, and the corresponding p value is 0.05, that is, when p value is less than 0.05, we can conclude that there is significant difference between the two algorithms.All algorithms have the same population size and number of iterations, that is, the population size is set to 100, and the maximum number of iterations is set to 100. In addition, the crossover and mutation rates of NSGA-II are set to 0.8 and 0.2, respectively.

4.3 Result and discussion

In order to prove the effectiveness of group technology in reducing energy consumption, 25 scheduling problems are solved by NSGA-II without group technology, and then 25 scheduling problems are solved by NSGA-II with group technology (NSGA-II-GT). Table 1 shows the average energy consumption and the average makespan obtained by NSGA-II and NSGA-II-GT. It can be seen from Table 3 that not only the energy consumption of NSGA-II-GT is lower than that of NSGA-II, but also the makespan of NSGA-II-GT is smaller than that of NSGA-II. This is because after grouping the same type of jobs, setup time is significantly r educed, t hus t he m akespan i s r educed. T able 1 shows the energy consumption reduction rate can reach 20.25%, while the makespan reduction rate can reach 13.68%. The average energy consumption and mae span reduction rates are 19.23% and 13.06% respectively. It can be seen from Table 1 that group technology has significant e ffect on reducing energy consumption and makespan.

Table 1: The average energy consumption and the average makespan obtained by NSGA-II and NSGA-II-GT.

Problems	Energy consumption			Makespan		
	NSGA-II	NSGA-II-GT	Metric	NSGA-II	NSGA-II-GT	Metric
J1	834450	**666911**	+	7707	**6728**	+
J2	830510	**671116**	+	7702	**6698**	+
J3	841519	**672746**	+	7845	**6828**	+
J4	826057	**658766**	+	7737	**6721**	+
J5	837281	**668368**	+	7698	**6701**	+
J6	1042808	**847716**	+	9619	**8350**	+
J7	1049222	**845061**	+	9678	**8459**	+
J8	1057701	**857602**	+	9626	**8359**	+
J9	1064636	**863027**	+	9653	**8332**	+
J10	1054478	**855426**	+	9671	**8440**	+
J11	1251866	**1011555**	+	11488	**9996**	+
J12	1274342	**1022181**	+	11611	**10092**	+
J13	1262525	**1025688**	+	11546	**9996**	+
J14	1270066	**1022274**	+	11577	**10039**	+
J15	1263347	**1019270**	+	11588	**10093**	+
J16	1489374	**1209501**	+	13524	**11779**	+
J17	1484925	**1212374**	+	13513	**11726**	+
J18	1495354	**1210407**	+	13632	**11868**	+
J19	1479235	**1195657**	+	13487	**11709**	+
J20	1478787	**1198872**	+	13527	**11798**	+
J21	1700021	**1383907**	+	15349	**13331**	+
J22	1700846	**1374400**	+	15411	**13364**	+
J23	1688963	**1362222**	+	15411	**13482**	+
J24	1721091	**1401151**	+	15507	**13462**	+
J25	1704244	**1371262**	+	15420	**13421**	+

"+" means NSGA-II-GT is better than NSGA-II, "-" means NSGA-II-GT is worse than NSGA-II, "=" means there is no significant difference between NSGA-II-GT and NSGA-II.Bold indicates the winner.

Group technology merges all kinds of similar jobs together, which may reduce the freedom of scheduling and leads to the imbalance in machine loads. For this reason, a dynamic splitting strategy is designed to reduce the machine imbalance caused by group technology. To analyze the performance of the dynamic splitting strategy, for each problem, 1000 individuals are randomly generated. Then decoded using the decoding scheme with dynamic splitting strategy (DS-DSS) and the decoding scheme with no dynamic splitting strategy (DS-NDSS) respectively. Figures 7 and 8 show the average makespan and energy consumption of DS-DSS and DS-NDSS. Figure 7 shows that the makespan of DS-DSS is smaller than that of DS-NDSS, but Fig. 8 shows that the energy consumption of DS-DSS is higher than that of DS-NDSS. Figure

9 shows the degree of unbalance of DS-DSS and DS-NDSS, it can be seen the degree of unbalance of DS-DSS is less than 4%, while the degree of unbalance of DS-DSS is greater than 10% and the highest is about 20%. From Fig. 9, it can be concluded that the dynamic splitting strategy can effectively balance the load of the machines. Although the dynamic splitting strategy leads to the increase of energy consumption, the makespan is reduced, and the machine load is more balanced.

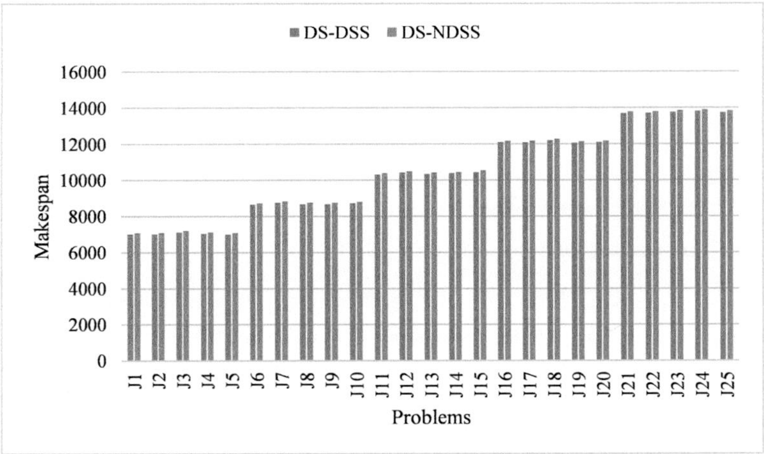

Figure 7: The average makespan of DS-DSS and DS-NDSS.

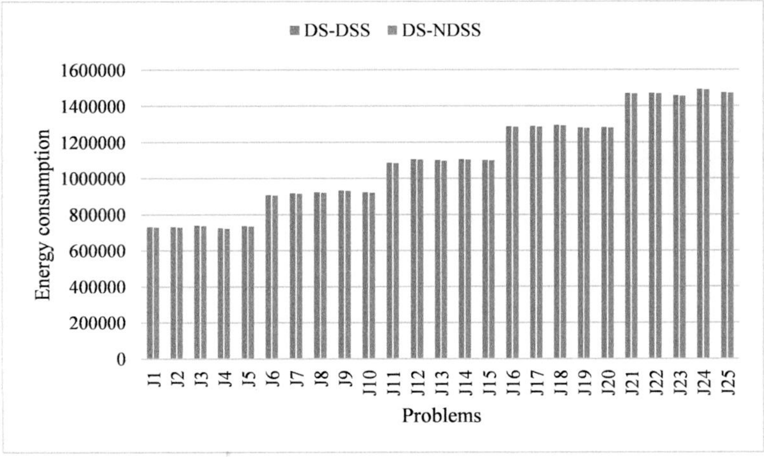

Figure 8: The average energy consumption of DS-DSS and DS-NDSS.

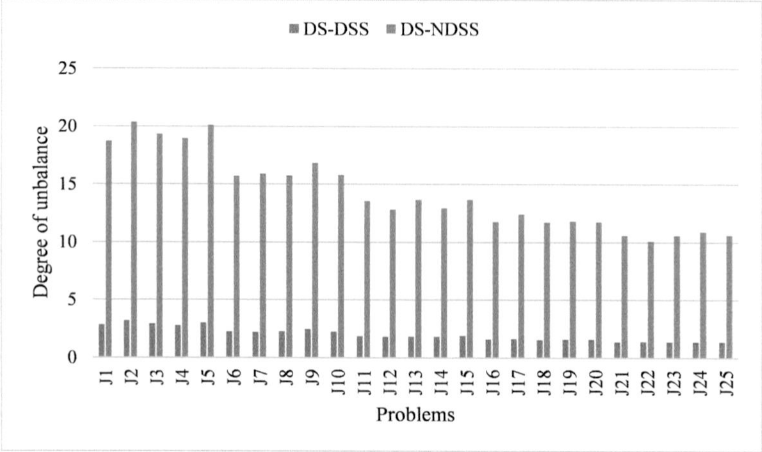

Figure 9: The degree of unbalance of DS-DSS and DS-NDSS.

In order to make a fair comparison between NSGA-II and IMSS, IMSS and NSGA-II adopt group technology and dynamic splitting strategy simultaneously (IMSS-GTDSS, NSGA-II-GTDSS). Because the solutions obtained by NSGA-II-GTDSS and IMSS-GTDSS are non-dominated solution sets, the HV index is used to measure the quality of the solutions of the two algorithms, and analyze the difference between the two algorithms with a t-test. Table 2 lists the HV metrics of the solutions obtained by the two algorithms and the result of the t-test. From Table 2, it can be seen that the HV index of the solutions obtained by IMSS-GTDSS algorithm is larger than that of NSGA-II-GTDSS, indicating that the solutions obtained by IMSS-GTDSS have a better quality. The p value of all t-tests is less than 0.05. This indicates that there is a significant difference between IMSS-GTDSS and NSGA-II-GTDSS. Table 3 shows the average energy consumption and the average makespan of NSGA-II-GTDSS and IMSS-GTDSS. It can be seen from the table that the average makespan obtained by IMSS-GTDSS are less than those obtained by NSGA-II-GTDSS. The highest average energy consumption by IMSS-GTDSS is less than those consumed by NSGA-II-GTDSS. There is no energy consumption and makespan obtained by NSGA-II-GTDSS that is less than that obtained by IMSS-GTDSS simultaneously. It can be seen from Table 3 that the energy consumption can be reduced by 20.18% and the makespan can be reduced by 15.64%. The average energy consumption and makespan reduction rates are 19.24% and 15.07% respectively. In comparison with with NSGA-II-GT, the energy consumption reduction rate is similar, but the makespan reduction rate is higher by about 2%. Additionally, the dynamic splitting strategy has a significant effect on the load balance of the machine. Thus, the dynamic splitting strategy can be considered for decoding.

Table 2: The HV metrics obtained by NSGA-II-GTDSS and IMSS-GTDSS and the result of the t-test.

Problems	NSGA-II- GTDSS	IMSS-GTDSS	Metric
J1	2.19E+07	**2.87E+07**	+
J2	2.25E+07	**3.11E+07**	+
J3	2.24E+07	**3.03E+07**	+
J4	1.36E+07	**2.36E+07**	+
J5	1.67E+07	**2.52E+07**	+
J6	2.99E+07	**3.96E+07**	+
J7	3.01E+07	**4.42E+07**	+
J8	2.87E+07	**3.86E+07**	+
J9	1.83E+07	**2.74E+07**	+
J10	3.60E+07	**4.90E+07**	+
J11	2.08E+07	**2.77E+07**	+
J12	2.35E+07	**3.05E+07**	+
J13	3.60E+07	**5.15E+07**	+
J14	2.25E+07	**2.94E+07**	+
J15	3.41E+07	**4.80E+07**	+
J16	4.70E+07	**6.07E+07**	+
J17	5.06E+07	**7.25E+07**	+
J18	3.38E+07	**5.94E+07**	+
J19	4.37E+07	**5.72E+07**	+
J20	4.10E+07	**6.04E+07**	+
J21	3.24E+07	**5.70E+07**	+
J22	5.74E+07	**8.51E+07**	+
J23	4.65E+07	**7.01E+07**	+
J24	5.57E+07	**8.10E+07**	+
J25	5.58E+07	**8.43E+07**	+

"+" means IMSS-GTDSS is better than NSGA-II-GTDSS, "-" means IMSS-GTDSS is worse than NSGA-II-GTDSS, "=" means there is no significant difference between IMSS-GTDSS and NSGA-II-GTDSS.Bold indicates the winner.

Table 3: The average energy consumption and the average makespan ob-tained by NSGA-II-GTDSS and IMSS-GTDSS.

Problems	Makespan			Energy consumption		
	NSGA-II GTDS	IMSS-GTDSS	Metric	NSGA-II GTDS	GTDSS	Metric
J1	6714	**6576**	+	672404	**670176**	=
J2	6672	**6511**	+	672429	**667819**	+
J3	6817	**6640**	+	676225	**671685**	+
J4	6715	**6557**	+	667581	**660282**	+
J5	6692	**6523**	+	**671641**	672427	+
J6	8335	**8145**	+	852751	**848866**	+
J7	8419	**8233**	+	845157	**841643**	+
J8	8348	**8150**	+	857827	**853628**	+
J9	8328	**8143**	+	**865993**	866018	=
J10	8403	**8217**	+	857387	**851939**	+
J11	9949	**9774**	+	**1003730**	1012080	-
J12	10080	**9855**	+	1031200	**1029610**	=
J13	9984	**9789**	+	1027110	**1026130**	=
J14	10019	**9826**	+	**1027650**	1030270	=
J15	10106	**9862**	+	1022650	**1018730**	=
J16	11746	**11529**	+	1201700	**1199850**	=
J17	11734	**11461**	+	1205840	**1204110**	=
J18	11853	**11612**	+	1211460	**1201950**	+
J19	11707	**11447**	+	1198710	**1196670**	=
J20	11737	**11525**	+	**1193380**	1197710	=
J21	13357	**13092**	+	1386680	**1383250**	=
J22	13343	**13055**	+	**1380890**	1382580	=
J23	13392	**13136**	+	1366540	**1365500**	=
J24	13446	**13206**	+	1404230	**1400350**	=
J25	13424	**13116**	+	1375460	**1370130**	=

"+" means IMSS-GTDSS is better than NSGA-II-GTDSS, "-" means IMSS-GTDSS is worse than NSGA-II-GTDSS, "=" means there is no significant difference b etween I MSS-GTDSS a nd N SGA-II-GTDSS.Bold indicates the winner.

5 Conclusion

This paper has studied the problem of energy-aware scheduling of a two-stage flexible fl ow sh op co nsidering pr ocess in consistency di fferences of tasks. Group technology is used to reduce setup energy consumption. Dynamic split strategy is used to solve the problem of unbalanced machine load caused by group technology. According to the characteristics of the research problem, an improved multi-objective scatter search algorithm is proposed to solve the problem. The experimental results show that group technology can not only reduce the energy consumption, but also reduces the makespan, and the dynamic split strategy can effectively balance the machine load. Finally, a comparison with existing algorithms that are excellent shows that the improved multi-objective scatter search algorithm performs better.

The personalized needs of consumers are constantly improving. Traditional manufacturing systems are bound to seek transformation and upgrading. Enterprise manufacturing systems must develop more advanced technology, which can effectively improve production efficiency and reduce costs on the premise of meeting the needs of consumers. This study can provide a benchmark for enterprises to reduce energy consumption cost and improve production efficiency at the operation levels in a the multi variety and small batch manufacturing mode.

Job delivery time is one of the important factors affecting production scheduling, and future work will consider the group strategy with job delivery time. The energy consumption cost is closely related to the electricity price. The optimization of energy consumption cost under time-of-use electricity tariffs is also one of the future works. In addition, in practical applications, the speed of the algorithm is not a small problem, and how to speed up the algorithm in solving large-scale problems is a challenge.

References

[1] Liu Y.Y., Liao X.Y., Zhang R.: An enhanced MOPSO algorithm for energy-efficient single-machine production scheduling. Sustainability, 11, 5381 (2019).

[2] Wu X.L., Shen X.L., Li C.B.: The flexible job-shop scheduling problem considering deterioration effect and energy consumption simultaneously. Comput. Ind. Eng. 135, 1004-1024 (2019).

[3] Zheng X., Zhou S., Xu R., Chen H.: Energy-efficient scheduling for multi-objective two-stage flow shop using a hybrid ant colony optimisation algorithm. Int. J. Prod. Res. 12, 1-18 (2019).

[4] Oukil A., El-Bouri A., Emrouznejad A.: Energy-aware job scheduling in a multi-objective production environment–An integrated DEA-OWA model. Comput. Ind. Eng. 168, 108065 (2022).

[5] Hong Z., Zeng Z., Gao, L.: Energy-efficiency scheduling of multi-cell manufacturing system considering total handling distance and eligibility constraints. Comput. Ind. Eng. 151, 106998 (2021).

[6] Wilson A.D., King R.E., Hodgson T.J.: Scheduling non-similar groups on a flow line: multiple group setups. Robot. Cim-Int. Manuf. 20, 505-515 (2004).

[7] Alireza K., Fariborz J., & Morteza B.: Integrating sequence-dependent group scheduling problem and preventive maintenance in flexible flow shops. Int. J. Adv. Manuf. Tech. 77, 173-185 (2015).

[8] Ebrahimi M., Ghomi S.M.T.F., Karimi B.: Hybrid flow shop scheduling with sequence dependent family setup time and uncertain due dates. Appl. Math. Model. 38, 2490–2504 (2014).

[9] Feng H.X., Xi L., Xiao L., Xia T., Pan E.: Imperfect preventive maintenance optimization for flexible flowshop manufacturing cells considering sequence-dependent group scheduling. Reliab. Eng. Syst. Safe. 176, 218-229 (2018).

[10] Keshavarz T., Salmasi N.: Makespan minimisation in flexible flowshop sequence-dependent group scheduling problem. Int. J. Prod. Res. 51, 6182-6193 (2013).

[11] Keshavarz T., Salmasi N., Varmazyar M.: Minimizing total completion time in the flexible flowshop sequence-dependent group scheduling problem. Ann. Oper. Res. 226, 351-377 (2015).

[12] Shahvari O., Salmasi N., Logendran R., Abbasi B.: An efficient tabu search algorithm for flexible flow shop sequence-dependent group scheduling problems. Int. J. Prod. Res. 50, 4237-4254 (2012).

[13] Costa A., Cappadonna F.V., Fichera S.: Minimizing makespan in a Flow Shop Sequence Dependent Group Scheduling problem with blocking constraint. Eng. Appl. Artif. Intel. 89, 103413 (2020).

[14] Yuan S., Li T., Wang B.: A co-evolutionary genetic algorithm for the two-machine flow shop group scheduling problem with job-related blocking and transportation times. Expert. Syst. Appl. 152, 113360 (2020).

[15] Dorfeshan Y., Tavakkoli-Moghaddam R., Mousavi S.M., Vahedi-Nouri B.: A new weighted distance-based approximation methodology for flow shop scheduling group decisions under the interval-valued fuzzy processing time. Appl. Soft Comput. 91, 106248 (2020).

[16] Cheng C.Y., Pourhejazy P., Ying K.C., Liao Y.H.: New benchmark algorithms for No-wait Flowshop Group Scheduling Problem with Sequence-Dependent Setup Times. Appl. Soft Comput. 111, 107705 (2021).

[17] Yuan S., Li T., Wang B.: A discrete differential evolution algorithm for flow shop group scheduling problem with sequence-dependent setup and transportation times. J. Intell. Manuf. 32(2), 427-439 (2021).

[18] Shahvari O., Logendran R.: Hybrid flow shop batching and scheduling with a bi-criteria objective. Int. J. Prod. Econ. 179, 239-258 (2016).

[19] Feng H., Tan C., Xia T., Pan E., Xi L.: Joint optimization of preventive maintenance and flexible flowshop sequence-dependent group scheduling considering multiple setups. Eng. Optimiz. 1-18 (2018).

[20] Jain, A., & Jain, A.: An approach for optimisation of flexible flow shop group scheduling with sequence dependent set-up time and preventive maintenance. Int. J. Comput. Aided Eng. Technol. 16(1), 40-66 (2022).

[21] Adressi A., Hassanpour S. T., Azizi V.: Solving group scheduling problem in no-wait flexible flowshop with random machine breakdown. Decis. Sci. Lett. 5, 157-168 (2016).

[22] Bozorgirad M.A., Logendran R.: Bi-criteria group scheduling in hybrid flowshops. Int. J. Prod. Econ. 145, 599-612 (2013).

[23] Karimi N., Zandieh M., Karamooz H.R.: Bi-objective group scheduling in hybrid flexible flowshop: a multi-phase approach. Expert Syst. Appl. 37, 4024-4032 (2010).

[24] Zandieh M., Karimi N.: An adaptive multi-population genetic algorithm to solve the multi-objective group scheduling problem in hybrid flexible flowshop with sequence-dependent setup times. J. Intell. Manuf. 22, 979-989 (2010).

[25] Nekoiemehr N., Selvarajah E., Zhang G.: Scheduling of jobs with cross families in two stage manufacturing systems. Int. J. Prod. Econ. 167, 88-96 (2015).

[26] Andrés C., Albarracín J.M., Tormo G., Vicens E., García-Sabater J.P.: Group technology in a hybrid flowshop environment: a case study. Eur. J. Oper. Res. 167, 272-281 (2015).

[27] Costa A., Cappadonna F. A., Fichera S.: A hybrid genetic algorithm for minimizing makespan in a flow-shop sequence-dependent group scheduling problem. J. Intell. Manuf. 28, 1269-1283 (2015).

[28] Huang X.: Bicriterion scheduling with group technology and deterioration effect. J. Appl. Math. Comput. 60, 455-464 (2018).

[29] Li, W.X., Zhao, C.L.: Single machine scheduling problem with multiple due windows assignment in a group technology. J. Appl. Mat.h Comput. 48, 477-494 (2015).

[30] Liou C.D., Hsieh Y.C.: A hybrid algorithm for the multi-stage flow shop group scheduling with sequence-dependent setup and transportation times. Int. J. Prod. Econ. 170, 258-267 (2015).

[31] Wang J.J., Liu Y.J.: Single-machine bicriterion group scheduling with deteriorating setup times and job processing times. Appl. Math. Comput. 242, 309-314 (2014).

[32] Yu C., Ji Y., Qi G., Gu X., Tao L.: Group-based production scheduling for make-to-order production. J. Intell. Manuf. 26, 585-600 (2013).

[33] Goli A., Keshavarz T.: Just-in-time scheduling in identical parallel machine sequence-dependent group scheduling problem. J. Ind. Manag. Optim. (2021).

[34] Liao B., Song Q., Pei J., Yang S., Pardalos P.M.: Parallel-machine group scheduling with inclusive processing set restrictions, outsourcing option and serial-batching under the effect of step-deterioration. J. Global. Optim. 78(4), 717-742 (202).

[35] Liao B., Wang H., Zhu X., Yang S., Pardalos P.M. Improved DE search for competing groups scheduling with deterioration effects. Optim. Lett. 15(2), 469-494 (2021).

[36] Wang J.B., Jia X., Yan J.X., Wang S.H., Qian J.: Single machine group scheduling problem with makespan objective and a proportional linear shortening. RAIRO-Oper. Res. 56(3), 1523-1532 (2022).

[37] Wang W., Wang J.J., Wang J. B.: Solution algorithms for single-machine group scheduling with learning effect and convex resource allocation. Complexity, 2021, 6615824 (2021).

[38] He X., Pan Q.K., Gao L., Wang L., Suganthan, P.N.: A greedy cooperative co-evolutionary algorithm with problem-specific knowledge for multi-objective flowshop group scheduling problems. IEEE T. Evolut. Comput. (2021).

[39] Glover F.: A template for scatter search and path relinking. In: European Conference on Artificial Evolution. Springer, Berlin, Heidelberg (1997).

[40] Wu X.L., Sun Y.L.: A green scheduling algorithm for flexible job shop with energy-saving measures. J. Clean. Prod. 172, 3249-3264 (2018).

[41] While L., Hingston P., Barone L., Huband S.: A faster algorithm for calculating hypervolume. IEEE Trans. Evol. Comput. 10, 29–38 (2006).

10. Optimal Operation Wind-solar-thermal Synergy Considering Carbon Trading and Energy Storage Systems

Chang Liu,[1] Zhenglei He,[1,*] Yi Man,[1,2] Jigeng Li[1] and Mengna Hong[1]

1 Introduction

In September 2020, China officially proposed the dual carbon target at the United Nations General Assembly[1]. Other countries have not tried to achieve the speed of net zero emission in 40 years. Therefore, the path of carbon neutrality in China will be arduous and fast in the foreseeable future[2]. However, China's power generation is dominated by coalthermal[3] that results in a high proportion of carbon emissions, approximately 40% of the national carbon emissions, are accounted for by the power sector[4]. The promotion and optimization of the energy structure of power generation is of immediate importance, to implement its carbon reduction strategies[5;6].

Due to the significant difference of carbon emission intensity among different generation technologies, it is theoretically possible to achieve generation-side decarbonization targets using a combination of different generation technologies[7;8]. In terms of current optimistic estimates of the future cost of such technologies, renewable plus storage can provide an effective solution to climate change mitigation in the future[9]. Development of renewable energy and improvement of energy utilization efficiency are the keys to low-carbon power generation processes[10]. Whereas, generation of wind power, solar power and

[1] State Key Laboratory of Pulp and Paper Engineering, South China University of Technology, Guangzhou,
[2] Guangdong Artificial intelligence and Digital Economy, Laboratory (Guangzhou) (Pazhou Lab), Guangzhou.
* Corresponding author: hezhenglei@scut.edu.cn

other similar renewable power are deterministically influenced by environmental and climatic factors, leading to the strong instability in their power generation process[11;12]. This uncertainty can easily cause a shock to the re-liability of the normal operation of the power system[13].

The peaking capacity of thermal power generation can provide a compromise for counteracting the instability caused by the renewable energy generation[14]. Energy storage technologies are another critical option for improving the low-carbon level of power systems because they can reduce renewable curtailment and carbon emissions of electricity systems[15]. Literature points out that the construction of a dispatching model on wind-solar-thermal synergy power generation exploiting the thermal power peaking capacity, can assist in connecting the generated wind and solar power to the grid on a large scale. This will promote the consumption level of wind and solar power[16]. Therefore, it is imperative to study the specific characteristics of different power sources, and disclose their complementary potential on the basis of their characteristics. The complementary potential of different renewable power has been investigated with regard to wind power and solar power complementary[17;18], wind power-solar power and hydropower complementary[19;20], wind power-solar pow-er-hydropower- thermal power and energy storage complementary[21;22], these studies reported that the complementary application of different power sources can promote the consumption of clean energy generation, making them extremely meaningful and promising[23].

To examine the impacts of environmental costs on power systems, it is necessary to study the interaction between coordinated energy systems operation and carbon emissions with multiple factors such as integrated energy and carbon prices[24;25], thus evaluating the impact of carbon prices on emissions from the energy systems' operations. This approach focuses on the direct impact of carbon pricing on the cost of power generation. However, due to the unclear mechanisms of carbon pricing, it is impossible to determine the potential impact of carbon pricing on power generation costs. Comparatively, analyzing the costs associated with the implementation of carbon pricing mechanisms can provide information on the potential impact of carbon pricing mechanisms on the cost of direct power generation[26;27]. It is also possible to determine whether carbon trading is effective in adjusting the energy consumption structure by analyzing the impact of carbon trading on energy consumption[28]. While both can determine the impact of carbon trading mechanisms on the cost of power generation and energy mix, neither has identified the mechanisms by which carbon trading mechanisms and carbon pricing mechanisms affect the economic and environmental benefits of entities emitting carbon dioxide.

Further study on the economic benefits of the main body of carbon emission shows that the carbon trading mechanisms promote carbon emission reduction by guiding the energy consumption of enterprises rather than by changing the industrial structure[29], which will have a significant effect on the economic benefits of the main body of carbon emission reduction[30;31]. In general, the introduction of carbon trading mechanisms can promote changes in the operating status of the power generation side. This can lead to the consumption of more clean energy generation on the power generation side, resulting in an overall reduction of carbon emissions on the power generation side. Nevertheless, since carbon trading is essentially designed to increase the cost of carbon emissions in an effort to encourage power generation entities to save fossil fuels. This may have an adverse effect on the overall economic benefits of power generation.

As a result, considering that thermal power is still the main form of power generation for a certain period[32], and the power generation technology for wind power and solar power is an essential support to propel energy reforms[33]. So, multi-energy synergy is of critical significance for power generation. However, due to the high carbon emissions of thermal power and the instability of wind and solar power, the multi-energy synergy may have an adverse impact on the environment and society. Hence, considering the impact of carbon trading and clean energy on the power generation system, this paper proposes a multi-energy synergy dispatching model considering econo-my-low carbon, which can reduce the economic loss to the main body of carbon reduction while improving the level of clean energy consumption.

2 Model Research Program

Based on machine learning technology and the kernel density estimation method, this paper establishes a multi-energy synergy dispatching model considering carbon trading and energy storage systems. Since it can be very computationally intensive for multi-scale energy dispatch optimi-zation[34], there is repeatability and consistency in industrial power consumption. Therefore, in order to reduce the amount of calculations, machine learning techniques are used to process electricity load data throughout the year to obtain representative days. In view of the fact that clean energy generation is a continuous process, discrete data for clean energy generation cannot be used to measure the process of generation. Thus, the kernel density estimation method is used for continuous discrete data analysis.

The proposed research program is shown in Figure 1, which is divided into four steps: (1) load data processing based on machine learning, (2) solving the mathematical expectation of wind and solar power generation based on kernel density estimation, (3) establishing a multi-energy synergy dispatching model, and (4) analyzing the model operation results. In the first step, a representative day acquisition model is established based on annual power load data. This model is used to obtain representative days to reduce the amount of dispatching model calculations; In the second step, the annual wind and solar power generation data are used to obtain the kernel density estimation function of wind and solar power generation by season and time scale, so as to solve the corresponding mathematical expectation of wind and solar power generation; In the third step, establish a dispatching model and obtain the output power of each generation unit, the input power of the grid power, as well as charging and discharging activities of energy storage systems during the operation of the dispatching model; In the fourth step, analyze the impact of carbon trading and energy storage systems on the operation of multi-energy synergy power generation.

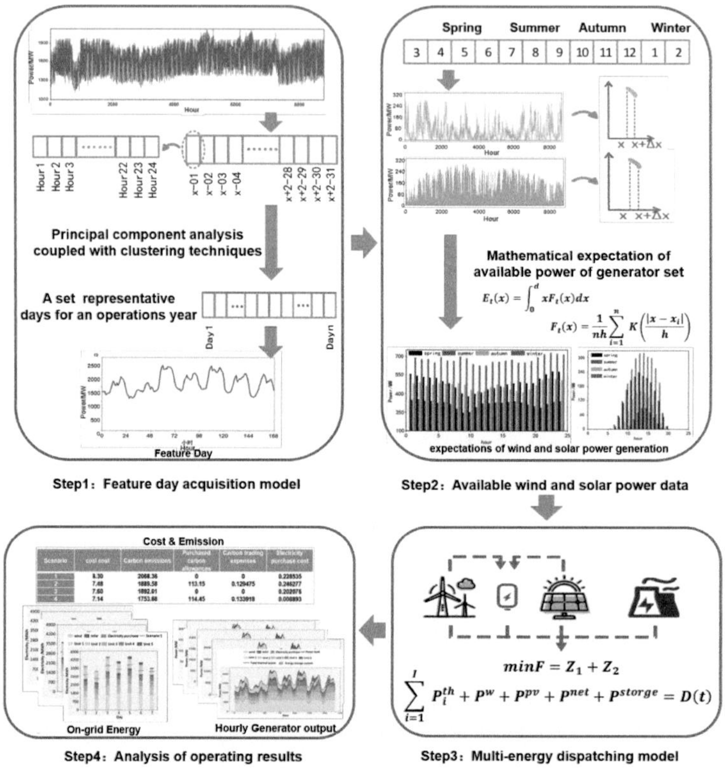

Figure 1: Research Program

3 Models and Methods

3.1 Representative Day Acquisition Model

To reduce the amount of calculations, the representative day method can be employed to represent the yearly electricity load on the basis of a group of typical days[35;36]. In this paper, principal component analysis (PCA) is combined

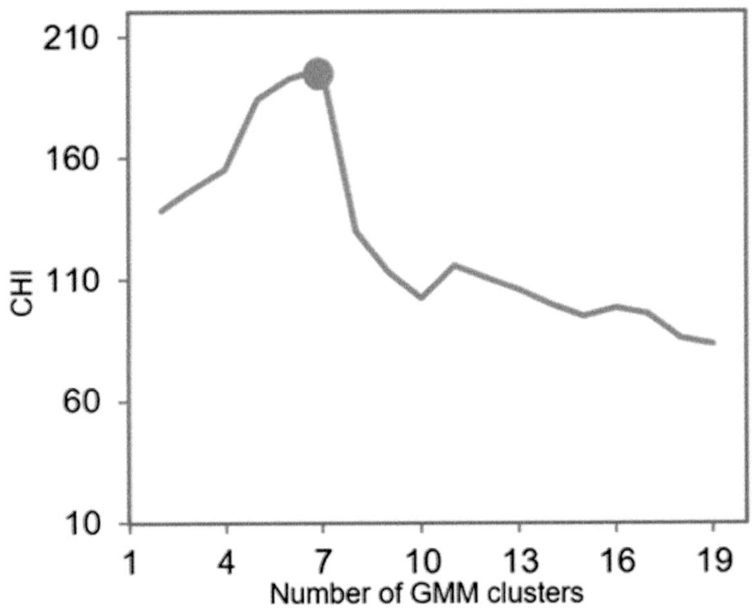

Figure 2: GMM Clustering Performance

with a gaussian mixture model (GMM) clustering The calinski harabasz index (CHI) is used to measure clustering performance. As the CHI value increases, the intra-cluster variance value decreases and the inter-cluster variance value increases

In Figure 2, a curve shows the values of CHI using GMM with dif-ferent numbers of clusters. Additionally, the markers on the curve indicate optimal clustering results. Based on the optimal clustering number, GMM clustering was carried out with PCA, and its performance was shown in Figure 3. The marks on the curve correspond to the optimal number of principal components.

From the performance evaluation results shown in Table 1, using PCA coupled with GMM clustering techniques could provide more effective results compared with using GMM clustering techniques without PCA. Additionally, a significant feature of the performance metrics is that better data grouping results are achieved when the number of principal components is less. The reason for this is that there is an extremely strong linear relationship between the load data for each hour, as shown in Figure 4. On the other hand PCA can compress the data with a strong linear relationship[37], resulting in a data

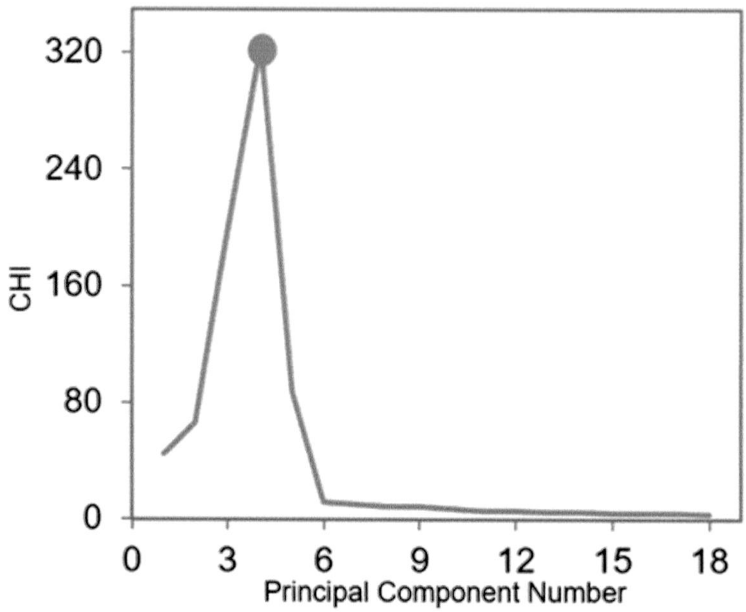

Figure 3: PCA Combined with GMM Clustering Performance

projection of only a few principal components to ef-fectively reflect the differences between the original data points. Hence, the clustering effect can be greatly enhanced by lowering the dimensionality of load data characteristics. On the other hand, combined with Figure 5, it can be seen that the selection of a higher number of principal component scores, although reflecting more information of the original data, may cause the features to be easily biased in projection due to the over-correlation of the features. Therefore, reducing the feature dimensionality of the load data is needed to retain a certain amount of data information in order to effectively improve the clustering effect.

Table 1: Clustering Results

GMM (CHI)	GMM+PCA (CHI)	Clustering Number	Principal Component Number	Feature Contribution Component
96.8	325.8	7	4	88.92

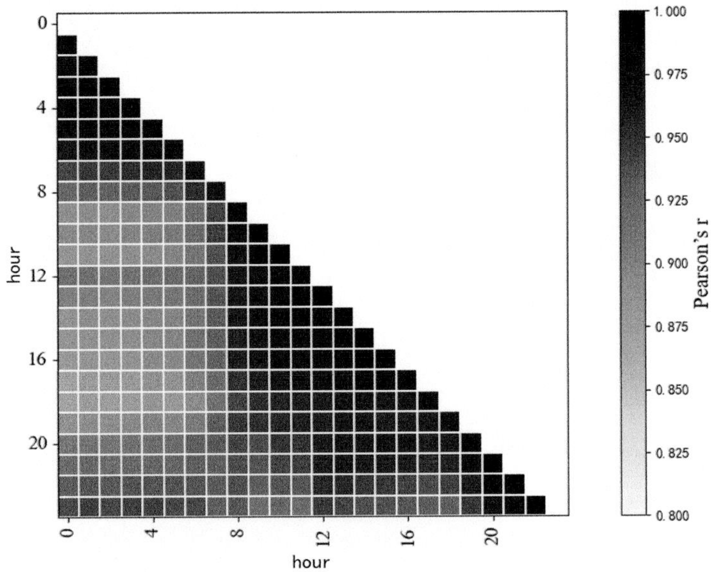

Figure 4: Power Load Correlation

3.2 Representative Day Acquisition Model

In the current combined wind and solar power generation model, it is difficult to predict exactly how wind and solar power will be generated due to their inherent unpredictability. In view of the uncertainty caused by the wind and solar power prediction error, stochastic optimization is often used to deal with it[38]. However, the purpose of this paper is to optimize a multi-energy dispatch strategy to consume re-newable energy power generation, which is why this research is focused on optimizing the dispatching of thermal power units with different carbon emission indexes. Hence, in order to portray the output of wind and solar power generation, this paper employs the kernel density estimation method to determine the mathematical expectation of power generation for each season and hour based on the wind and solar power generation data of the entire year.

As a result of technological limitations, wind and solar power generation data sampling intervals are 1 hour, which results in discrete power generation data that cannot accurately reflect continuous power generation scenarios. To reduce the impact of discrete data in each hour of the four seasons of the sampling year, the kernel density estimation method is employed in this paper.

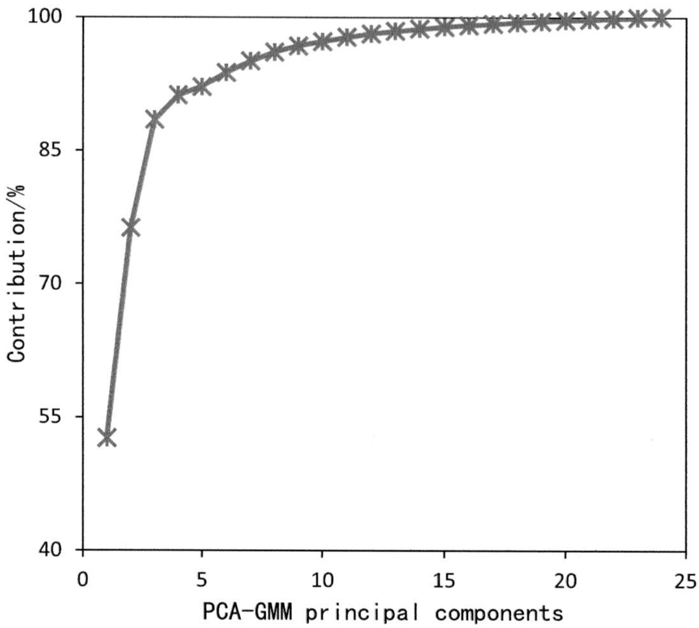

Figure 5: PCA-GMM Principal Component Contribution

The kernel density estimation function of wind and solar power generation per hour is calculated as shown in Eq.1:

$$F_t(x) = \frac{1}{nh}\sum_{i=1}^{n} K\left(\frac{|x - x_i|}{h}\right) \tag{1}$$

Where $F_t(x)$ is the kernel density estimate of wind or solar power generation x at moment t; x_i is one of the n sample points $(x_i, \ldots, x_i, \ldots x_n)$; h is the bandwidth, obtained empirically[39], as shown in Eq.2:

$$h = \left(\frac{4\hat{\sigma}^5}{3K}\right)^{\frac{1}{5}} \approx 1.06\hat{\sigma}n^{-\frac{1}{5}} \tag{2}$$

where σ is the standard deviation corresponding to n sample points at moment t;K is the Gaussi-an kernel located in (1) and (2) used to calculate the probability distribution as shown in Eq.3:

$$K(x) = \frac{1}{\sqrt{2\pi h^2}}e^{-\frac{(x-\mu)^2}{2h^2}} \tag{3}$$

where μ is the mean of the corresponding n sample points at moment t.

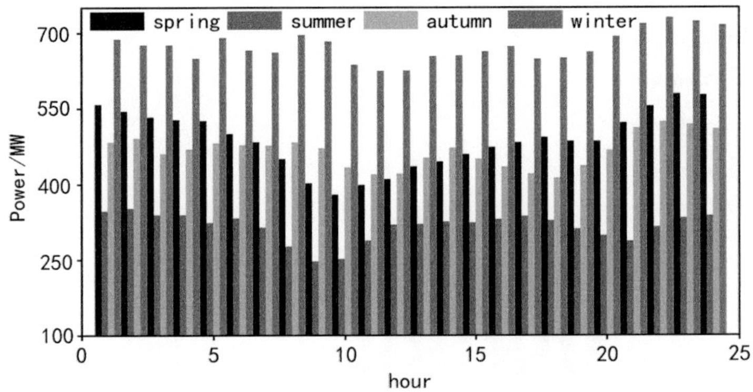

Figure 6: Mathematical Expectations of Wind Power Generation

Finally, the mathematical expectation of the generated power per hour is obtained as shown in Eq.4:

$$\mathrm{E}_t(x) = \int_0^d xF_t(x)dx \tag{4}$$

A comparison of the mathematical expectations of wind and solar power generation during the four seasons is shown in Figures 6 and 7.

3.3 Wind-solar-thermal-storage Combined Dispatching Model

In order to ensure the economic benefits of society and enterprises, power production units need to optimize their operation strategies in accordance with the requirements of the power grid, while maximizing the level of clean energy consumption and minimizing the cost of power generation[40;41]. As part of its efforts to promote the decarbonization of the power system, this paper investigates the carbon trading mechanisms along with wind power, solar power, thermal power, and energy storage systems as its research objects and sets up an optimal dispatching modal.

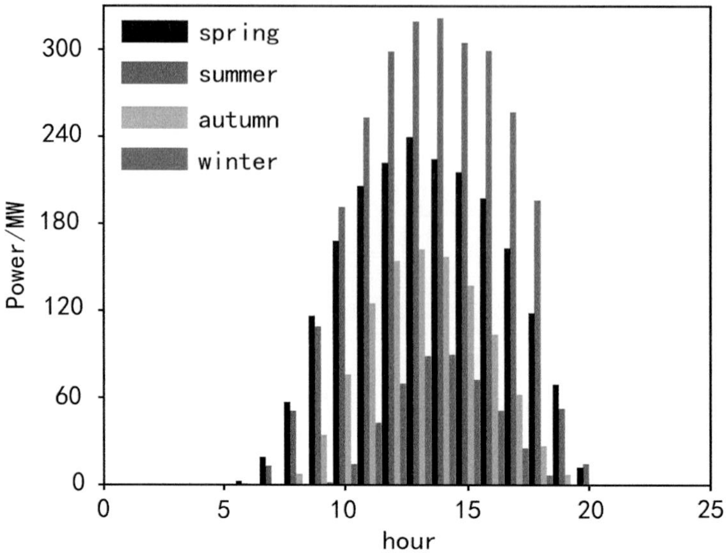

Figure 7: Mathematical Expectations of Solar Power Generation

3.3.1 Objective Function

In this paper, thermal power units and energy storage systems are used as tools to consume wind and solar power generation. Furthermore, this paper considers the fact that the largest amount of carbon emissions result from the combustion of thermal coal in power systems[42], and that reducing the consumption of fossil fuels would be the most effective method of improving efficiency when it comes to maximizing the benefits of society and enterprises[43]. Thus, this paper aims at integrating the carbon trading mechanism into the power generation system so that economy and low-carbon can be combined, starting with the cost of power generating and the carbon emissions. The cost of power generation and coal consumption are shown in Eq.5 and Eq.6:

$$Z_1 = \gamma^c \sum_{t=1}^{T} \sum_{i=1}^{I} \left[S_i(t) \left[1 - S_i(t)\right] SU_i + S_i(t) G_i \left(P_i^{th}(t)\right)\right] + \gamma^{\text{net}} \, P^{\text{net}} \, (t) \Delta t \tag{5}$$

$$G_i(P_i^{th}(t)) = a_i^c (P_i^{th}(t))^2 + b_i^c P_i^{th}(t) + c_i^c \tag{6}$$

Here T is the number of dispatching cycle periods; $P_i^{th}(t)$ is the output of thermal power unit i in time period t; $S_i(t)$ is the operating state of thermal power unit i in time period t. $S_i(t)$ takes the value of 1 when the unit is in the start-up state and 0 when the unit is in the shutdown state; SU_i is the coal consumption for unit i starting up; γ^c is the converted standard coal price of 115 USD/t[44]; $G_i(P_i^{th}(t))$ is the coal consumption function of unit i in the context of generation output; a,b and c are the relevant parameters; γ^{net} is the purchased electricity price; which is taken as 110 USD/MWh; $P^{net}(t)$ is the purchased power, and Δt is the dispatch time, which is taken as $1h$ in this paper.

Non-fossil energy power generation does not directly generate carbon emissions, so the cost of carbon emissions is only for thermal power in power generation systems. Carbon emissions can be obtained from Eq.7:

$$E\left(P_i(t)\right) = a_i^{co_2}\left(P_i(t)\right)^2 + b_i^{co_2}P_i(t) + c_i^{co_2} \tag{7}$$

where, $E\left(P_i(t)\right)$ is the carbon emission of output $P_i(t)$ of unit i at time slot t; $a_i^{co_2}$, $b_i^{co_2}$ and $c_i^{co_2}$ are relevant parameters.

We also consider that carbon allowances are crucial to the operation of the carbon market. Thus, in this paper, a certain amount of car-bon emission quotas are allocated without charge according to the generation capacity and type of units used[45], which can be obtained from Eq.8:

$$E_{fr} = \beta_i \sum_{t=1}^{T}\sum_{i=1}^{I} P_i(t)\Delta \mathrm{t} \tag{8}$$

where E_{fr} is the free carbon allowance obtained by unit i in time period t; β_i is the carbon allowance obtained per unit of electricity generated by unit i.

Therefore, the additional carbon allowances that need to be bought and the resulting carbon trading costs are given by Eq.9 and Eq.10:

$$E_{sp} = E(P_i(t)) - E_{fr} \tag{9}$$

$$Z_2 = \phi E_{sp} \tag{10}$$

where, ϕ is the carbon trading price of 11.5 USD/t[46].

The final objective function of the multi-energy synergy dispatching model considering economy and low-carbon is Eq.11:

$$minF = Z_1 + Z_2 \tag{11}$$

where, Z_1 is the cost of energy use; Z_2 is the cost of carbon trading.

3.3.2 Binding Conditions

(1) The load balance constraint is represented by Eq.12:

$$\sum_{i=1}^{I} P_i^{th}(t) + P^w(t) + P^{pv}(t) + P^{net}(t) + P^{bdch}(t) - P^{bch}(t) = D(t) \tag{12}$$

where, $P^{w}(t)$ and $P^{pv}(t)$ are the power outputs of the wind turbine and PV unit in time period t; $P^{bdch}(t)$ and $P^{bch}(t)$ are the battery discharges and charging power; $D(t)$ is the power load of the corresponding time period. (2) Thermal power unit climbing capacity constraints are represented by Eq.13:

$$|P_i^{th}(t+1) - P_i^{th}(t)| \leq P_{(i,lim)} \tag{13}$$

Where $P^{w}(t)$ and $P^{pv}(t)$ are the output power of wind turbine and PV unit in time period t; $P^{bdch}(t)$ and $P^{bch}(t)$ are the battery discharge and charging power; $D(t)$ is the power load of the corresponding time period. (3) The output constraints of each generator set are shown in Eq.14 to Eq.16:

$$P_{min}^{th} \leq P_i^{th}(t) \leq P_{max}^{th} \tag{14}$$

$$P_{min}^{w} \leq P^{w}(t) \leq P_{max}^{w} \tag{15}$$

$$P_{min}^{pv} \leq P^{pv}(t) \leq P_{max}^{pv} \tag{16}$$

where, P_{min}^{th} and P_{max}^{th} are the minimum and maximum outputs in thermal power units i; P_{min}^{w} and P_{max}^{w} are the minimum and maximum outputs of the wind turbine;P_{min}^{pv} and P_{man}^{pv} are the minimum and maximum outputs of photovoltaic units. (4) The storage capacity constraints of the energy storage system are shown in Eq.17 to Eq.18:

$$P^{bsc}(t-1) - P^{bdch}(t)\Delta t + P^{bch}(t)\Delta t = P^{bsc}(t) \tag{17}$$

$$P^{bsc}(t) \leq P^{bsc,max} \tag{18}$$

where, $P^{bsc}(t)$ is the available energy stored in the energy storage system at time period t; $P^{bch}(t$ and $P^{bdch}(t)$ are the charging power and discharging power of the energy storage system at time period t. (5) The charge and discharge constraints of the energy storage system are shown in Eq.19 to Eq.22:

$$P^{bch}(t)\Delta t + P^{bsc}(t-1) \leq P^{bsc,max} \tag{19}$$

$$P^{bdch}(t)\Delta t - P^{bsc}(t-1) \leq 0 \tag{20}$$

$$P^{bch}(t) \leq P^{bch,max} \tag{21}$$

$$P^{bdch}(t) \leq P^{bdch,max} \tag{22}$$

Where $P^{bdch,max}$ and $P^{bch,max}$ are the maximum discharge and charge powers respectively, of the energy storage system.

4 Results and Analysis

4.1 Numerical Example

In this paper, the simulation is based on the annual electricity load and wind and solar power generation data of a location. The specific equipment parameters are shown in the literature[47;48]. Seven feature days were obtained on the basis of Chapter 3, which are spring: repre-sentative day 2 and feature day 3; summer: representative day 4 and representative day 5; autumn: representative day 6; and winter: repre-sentative day 1 and representative day 7. In addition, this paper considers four typical scenarios of dispatch, namely Scenario 1: no energy storage systems and carbon trading mechanisms are considered; Scenario 2: only carbon trading mechanisms is considered; Scenario 3: only energy storage systems are considered; Scenario 4: energy storage systems and carbon trading mechanisms are considered.

4.2 Dispatching Cost Analysis

As shown in Table 2, the dispatch results of Scenario 1 and Scenario 2 illustrate that Scenario 1 does not consider energy storage mechanisms and carbon trading mechanisms, which leads to the power generation system disregarding the effects of wind and solar power on carbon emissions. In Scenario 1, there is a great deal of pressure on the power supply of the thermal power units during the peak period of energy consumption. This pressure is necessary to maintain high load operations for a long time, and that results in high overall operation costs and system carbon emissions. Compared to Scenario 1, the coal cost and system carbon emissions in Scenario 2 decreased by 8.83% and 8.64% re-spectively, and it has been shown that carbon trading mechanisms are not only effective at reducing power generation costs, but also at reducing carbon emissions.

Comparing scenario 3 with scenario 1, the curtailment of wind and solar power generation in scenario 3 is smaller than in scenario 1. This is mainly due to the ability of the energy storage system to act as a flexible load that can participate in the peak-shaving and valley-filling dispatch process. As compared to Scenario 1, the coal cost in Scenario 3 is reduced by 8.47%, and the system emissions are reduced by 8.53%. This demonstrates the effectiveness of the energy storage system for the wind and solar power consumption levels and saves coal costs to a certain extent. In contrast to scenario 3, scenario 2 has a lowe coal cost 3, but the need to purchase carbon allowances externally as well as the higher purchase of electricity in scenario 2 makes it 0.7% more costly than scenario 3 when comparing them.

In Scenario 4, the use of energy storage technology has led to a further rise in the consumption level of wind and solar power. Compared to Scenario 2, Scenario 4 expenses for coal consumption are reduced by 6.84%, while system carbon emissions are reduced by 7.19%. Furthermore, the final scenario 4 has a 7.79% reduction in total costs and a 7.31% re-duction in system carbon emissions in comparison to scenario 3, and that shows the carbon trading mechanism is sufficient to promote the system to consume wind and solar power generation.

Table 2: Dispatching Results in Different Scenarios

Scenario	Coal Cost/Millon USD	Carbon Emissions/Million Tons	Purchased Carbon Allowances /Million Tons	Carbon Trading Expenses /Million USD	Electricity Purchase Cost /Mil-lion USD
1	8.30	2068.36	0	0	0.228535
2	7.48	1889.58	113.15	0.129475	0.246277
3	7.60	1892.01	0	0	0.202076
4	7.14	1753.68	114.45	0.133918	0.008893

4.3 On-grid Energy

In order to analyze the positive effect of the multi-energy synergy dispatching model on energy conservation and emission reduction, this paper takes the on-grid power of thermal power in scenario 1 as the reference value and compares it with the dispatching results in other scenarios.

Based on Figure 8 10, it can be concluded that thermal power units with a low-carbon emission index prioritize output and have a higher on-grid power than those with a high carbon emission index. Taking scenario 2 as an example, the capacity of thermal power unit 1 is 50% higher than that of unit 2 and 70% higher than that of unit 3. Nevertheless, thermal power unit 1 has 79.96% more on-grid energy than unit 2 and 135.12% more than unit 3. It demonstrates that the model is capable of reducing carbon emissions by optimizing the output levels of each thermal unit while maintaining the same overall output of the thermal power units.

From Figures 8 and 9, Scenario 2 shows a 0.1% decrease in on-grid energy of thermal power and a 0.3% increase in the utilization share of wind and solar generation compared to Scenario 3. However, according to Table 2, the cost of purchased electricity in Scenario 2 is 43604USD higher than that in Scenario 3. It is due to the fact that the energy storage system is charged during low load times or periods when wind and solar power generation are high, and discharged during peak loads, that results in reduced power purchase demand. In general, the introduction of carbon trading mechanism reduces the output of thermal power units. When the load peaks, the system cannot meet the power demand, so power purchase increases.

Comparing Figure 10 to the other three scenarios, it is clear that integrating energy storage systems and carbon trading mechanisms can significantly increase the on-grid wind energy and solar power and reduce the electricity. Since the energy storage system ensures peak shaving capacity, the carbon trading mechanism encourages the system to reduce the output of thermal power units as much as possible. As a result, the combination of energy storage systems and carbon trading mechanisms has a positive impact on the

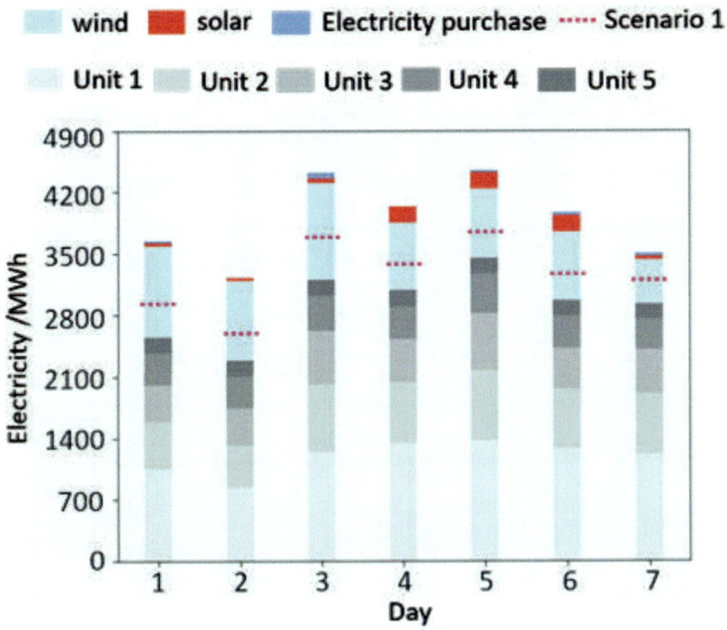

Figure 8: On-grid Energy for Scenario 2

consumption of renewable energy and emphasizes the quantity and quality of wind and solar energy power generation.

In general, the curtailment of wind and solar power can be reduced by energy storage systems and carbon trading mechanisms, and a dispatching model that considers a combination of the two can maximize on-grid wind and solar power energy.

4.4 Dispatching Results in 4 Typical Scenarios

By comparing the operation results of Scenario 1 and scenario 2, it is evident that peak regulation of thermal power units occurs more frequently in scenario 2 than in scenario 1, as shown in Figure 11 and Figure 12. The reason for this is that scenario 2 considers a carbon trading mechanism, which increases the cost of thermal units, causing thermal units to prioritize dispatching units with lower carbon emission factors, and a lack of energy storage systems results in thermal power units taking on all peaking tasks, requiring frequent adjustment of output to consume wind and solar power. Thus, thermal power units cannot be dispatched in time to meet power loads due to peaking capacity constraints, and that forces the system to purchase more electricity.

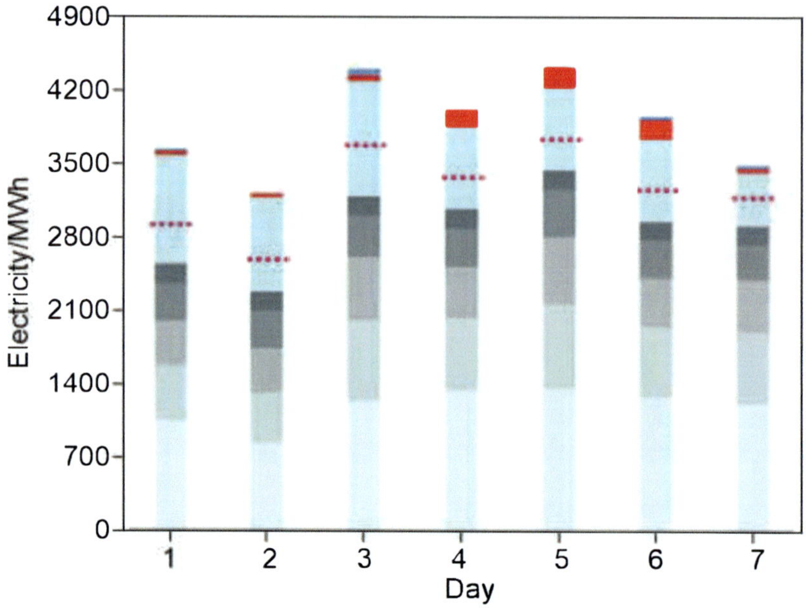

Figure 9: On-grid Energy for Scenario 3

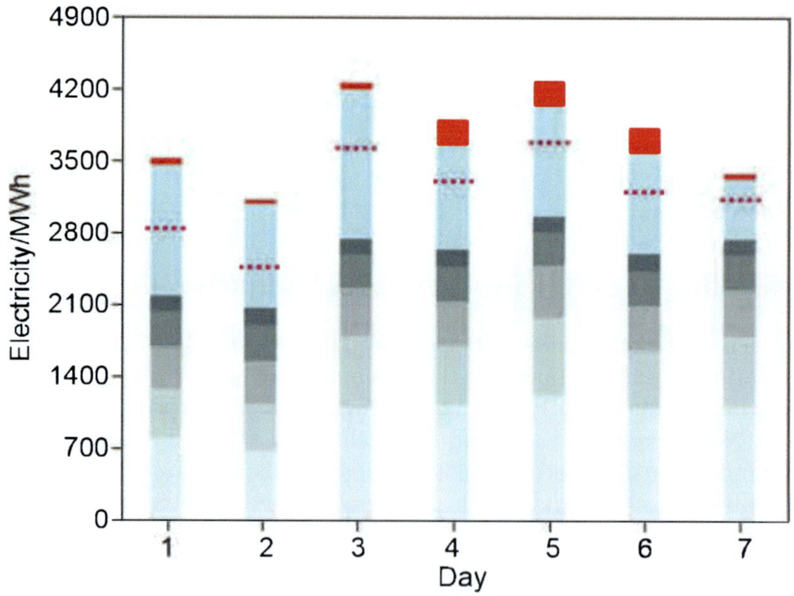

Figure 10: On-grid Energy for Scenario 4

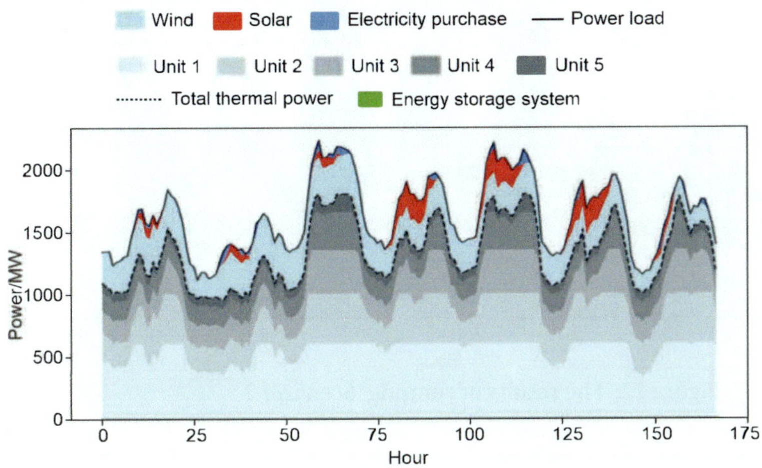

Figure 11: The results of running Scenario 1

From Figure 13, the system regulation capacity is improved after the introduction of the energy storage system, and the output of thermal power units is significantly reduced compared with Scenario 1, while the output of wind and solar power generation increases in proportion. Further analysis shows that energy storage systems are often charged at night, which is because wind power generation power is affected by wind speed, which is relatively high at night, electricity loads are low, so it is very convenient for energy storage equipment to charge and store energy. Furthermore, the energy storage sys-tem discharges its energy primarily at noon, when the load reaches its peak and the thermal power unit is required to significantly increase its output. As the gap between valley and peak periods of electricity load is large, in order to consume wind and solar power generation, the output of thermal power units is too low during the valley load period, which is affected by the climbing rate of units and cannot meet the load needs during the peak period. It can be concluded that the rapid response capability and flexible bidirec-tional regulation of the energy storage system prompt the system to reduce the output of thermal power units. So, the power stored in the energy storage system replaces purchased electricity to a certain extent, thereby improving the stability of the power generation system itself.

As shown in Figure 14, considering both the energy storage system and the carbon trading mechanism can maximize the use of wind and solar power

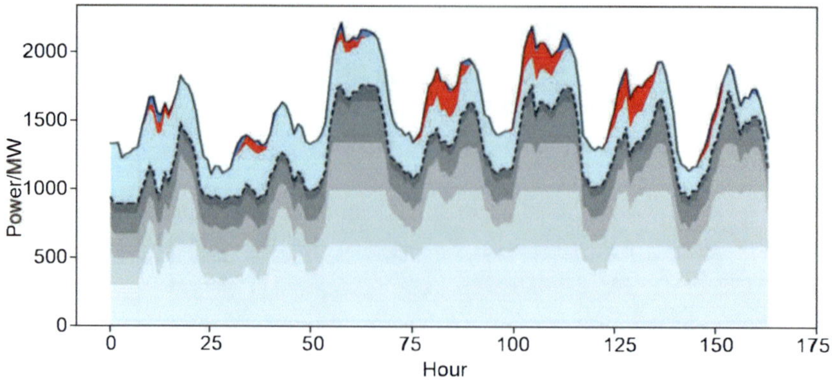

Figure 12: The results of running Scenario 2

generation. As in scenario 3, the power stored in the energy storage system in scenario 4 can replace a portion of the purchased elec-tricity. By combining the output timing of thermal power units, it can be seen that when the power load is consistently lower than 1500MW, each thermal power unit maintains the minimum output to ensure that wind and solar power replace thermal power, and it shows that the combination of the carbon trading mechanism and energy storage system can minimize the output of thermal power to consume wind and solar power generation.

In summary, both energy storage systems and carbon trading mechanisms can be used to promote the wind consumption level and solar power as well as enhance the low-carbon power generation characteristics of systems. However, compared with the carbon trading mechanism, the energy storage system can flexibly c arry o ut p eak l oad s hifting d ue t o i ts fl exible lo ad characteristics, which can improve the stability of the system. In addition, the combination of energy storage systems and carbon trading mechanisms can be used to optimize the internal output plan of the power generation system based on improving its peaking capacity, thus maximizing the wind and solar power consumption levels and minimizing the cost of power generation.

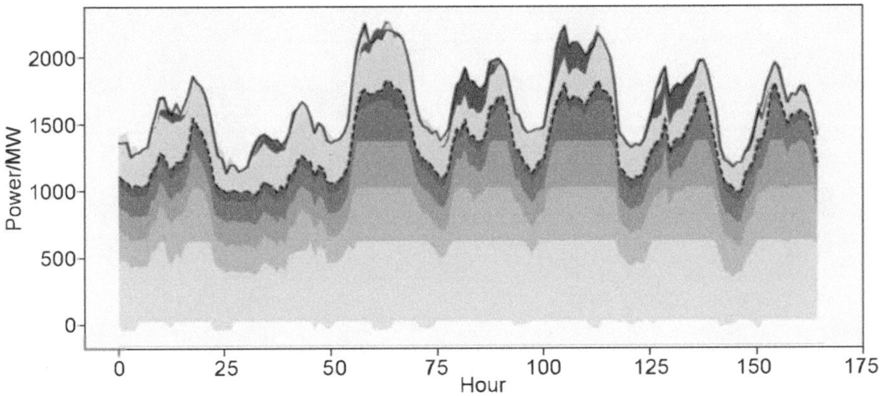

Figure 13: The results of running Scenario 3

5 Conclusion

Based on the proposed representative day acquisition model, and kernel density estimation of wind and solar power generation, this paper establishes a multi-energy synergy dispatching model considering the combination of economy and low-carbon, and compares the model operation benefits in four typical scenarios. The conclusion is as follows:

(1) By using the multi-energy synergy dispatching model presented in this paper, carbon emissions and generation costs can be effectively reduced while meeting the power load, and the power generation system can be made more efficient and low-carbon.

(2) In promoting the wind and solar power consumption levels, both energy storage systems and carbon trading mechanisms can effectively improve the consumption capacity. As shown in the example presented in this paper, the carbon trading mechanism and energy storage system increase consumption by 38.3% and 37.9% and reduce power generation costs by 8.4% and 8.6%, respectively, compared to the case without these mechanisms. By combining the two, consumption can be improved by 79.6% and the electricity generatin costs can be reduced by 16.7%.

(3) The representative day acquisition model established by PCA combined with GMM can be used to represent the yearly electricity load on the

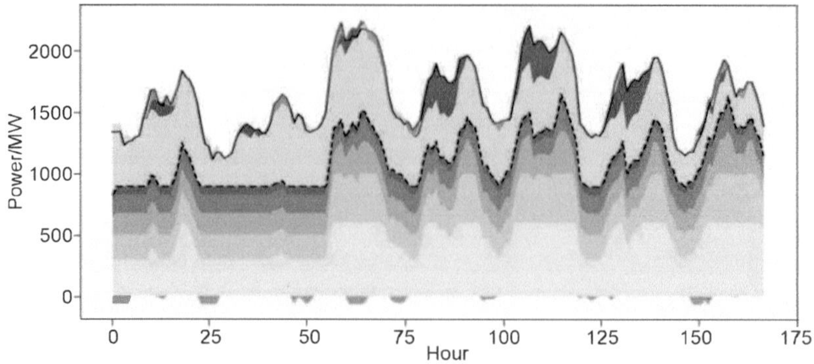

Figure 14: The results of running Scenario 4

basis of a group of typical days that effectively simplifies the model computation.

References

[1] S. Zhang, X. Bai, C. Zhao, Q. Tan, G. Luo, L. Wu, H. Xi, C. Li, F. Chen, C. Ran, et al., China's carbon budget inventory from 1997 to 2017 and its challenges to achieving carbon neutral strategies, Journal of Cleaner Production 347 (2022) 130966.

[2] S. Mallapaty, et al., How china could be carbon neutral by mid-century, Nature 586 (2020) 482–483.

[3] J. B. Horton, Parametric insurance as an alternative to liability for compensating climate harms, Carbon & Climate Law Review 12 (2018) 285–296.

[4] J. Gao, Q. Meng, F. Guo, Z. Wang, Optimal dispatch of integrated energy system with adjustable thermoelectric ratio considering wind-photovoltaic uncertainty, two-stage power-to-gas and reward-penalty ladder-type carbon trading, Two-Stage Power-to-Gas and Reward-Penalty Ladder-Type Carbon Trading (????).

[5] Y. Yan, X. Zhang, J. Zhang, K. Li, Emissions trading system (ets) implementation and its collaborative governance effects on air pollution: The china story, Energy Policy 138 (2020) 111282.

[6] Y. Tang, X. Wu, C. Zhou, G. Zhu, J. Song, G. Liu, Electricity-carbon graph modeling and analysis, in: 2022 7th Asia Conference on Power and Electrical Engineering (ACPEE), IEEE, pp. 1256–1260.

[7] M. H. Zrelli, Renewable energy, non-renewable energy, carbon dioxide emissions and economic growth in selected mediterranean countries, Environmental Economics and Policy Studies 19 (2017) 691–709.

[8] S. Asumadu-Sarkodie, P. A. Owusu, The causal effect of carbon dioxide emissions, electricity consumption, economic growth, and industrialization in sierra leone, Energy Sources, Part B: Economics, Planning, and Policy 12 (2017) 32–39.

[9] S. Sgouridis, M. Carbajales-Dale, D. Csala, M. Chiesa, U. Bardi, Comparative net energy analysis of renewable electricity and carbon capture and storage, Nature Energy 4 (2019) 456–465.

[10] T. Qi, X. Zhang, V. J. Karplus, The energy and co2 emissions impact of renewable energy development in china, Energy Policy 68 (2014) 60–69.

[11] J. Gao, Q. Meng, F. Guo, Z. Wang, Optimal dispatch of integrated energy system with adjustable thermoelectric ratio considering wind-photovoltaic uncertainty, two-stage power-to-gas and reward-penalty ladder-type carbon trading, Two-Stage Power-to-Gas and Reward-Penalty Ladder-Type Carbon Trading (????).

[12] Y. Yan, X. Zhang, J. Zhang, K. Li, Emissions trading system (ets) implementation and its collaborative governance effects on air pollution: The china story, Energy Policy 138 (2020) 111282.

[13] J.-N. Kang, Y.-M. Wei, L.-C. Liu, R. Han, B.-Y. Yu, J.-W. Wang, Energy systems for climate change mitigation: A systematic review, Applied Energy 263 (2020) 114602.

[14] Y. Gu, J. Xu, D. Chen, Z. Wang, Q. Li, Overall review of peak shaving for coal-fired power units in china, Renewable and Sustainable Energy Reviews 54 (2016) 723–731.

[15] M. Arbabzadeh, R. Sioshansi, J. X. Johnson, G. A. Keoleian, The role of energy storage in deep decarbonization of electricity production, Nature communications 10 (2019) 3413.

[16] S. S. Reddy, P. Bijwe, Real time economic dispatch considering renewable energy resources, Renewable Energy 83 (2015) 1215–1226.

[17] W. Hou, X. Zhang, M. Wu, Y. Feng, L. Yang, Integrating stability and complementarity to assess the accommodable generation potential of multiscale solar and wind resources: A case study in a resource-based area in china, Energy 261 (2022) 125312.

[18] A. Gangopadhyay, A. Seshadri, N. Sparks, R. Toumi, The role of wind-solar hybrid plants in mitigating renewable energy-droughts, Renewable Energy 194 (2022) 926–937.

[19] B. Liu, J. R. Lund, S. Liao, X. Jin, L. Liu, C. Cheng, Optimal power peak shaving using hydropower to complement wind and solar power uncertainty, Energy Conversion and Management 209 (2020) 112628.

[20] K. Huang, P. Liu, B. Ming, J.-S. Kim, Y. Gong, Economic operation of a wind-solar-hydro complementary system considering risks of output shortage, power curtailment and spilled water, Applied Energy 290 (2021) 116805.

[21] S. S. Reddy, Optimal scheduling of thermal-wind-solar power system with storage, Renewable energy 101 (2017) 1357–1368.

[22] A. Panda, K. B. Aviso, U. Mishra, I. Nanda, Impact of optimal power generation scheduling for operating cleaner hybrid power systems with energy storage, International Journal of Energy Research 45 (2021) 14493–14517.

[23] A. Shokri, M. S. Fard, Water-energy nexus: Cutting edge water desalination technologies and hybridized renewable-assisted systems; challenges and future roadmaps, Sustainable Energy Technologies and Assessments 57 (2023) 103173.

[24] Y. Cheng, N. Zhang, B. Zhang, C. Kang, W. Xi, M. Feng, Low-carbon operation of multiple energy systems based on energy-carbon integrated prices, IEEE Transactions on Smart Grid 11 (2019) 1307–1318.

[25] Y. Fu, G. Huang, Y. Xie, R. Liao, J. Yin, Planning electric power system under carbon-price mechanism considering multiple uncertainties–a case study of tianjin, Journal of Environmental Management 269 (2020) 110721.

[26] L. Liu, Z. Wang, Y. Wang, J. Wang, R. Chang, G. He, W. Tang, Z. Gao, J. Li, C. Liu, et al., Optimizing wind/solar combinations at finer scales to mitigate renewable energy variability in china, Renewable and Sustainable Energy Reviews 132 (2020) 110151.

[27] A. Trinks, M. Mulder, B. Scholtens, External carbon costs and internal carbon pricing, Renewable and Sustainable Energy Reviews 168 (2022) 112780.

[28] D. Yang, S. He, Q. Chen, D. Li, H. Pandžić, Bidding strategy of a virtual power plant considering carbon-electricity trading, CSEE Journal of Power and Energy Systems 5 (2019) 306–314.

[29] D. Linghu, X. Wu, K.-H. Lai, F. Ye, A. Kumar, K. H. Tan, Implementation strategy and emission reduction effectiveness of carbon cap-and-trade in heterogeneous enterprises, International Journal of Production Economics 248 (2022) 108501.

[30] Z. Ding, W.-J. Lee, A stochastic microgrid operation scheme to balance between system reliability and greenhouse gas emission, in: 2015 IEEE/IAS 51st Industrial & Commercial Power Systems Technical Conference (I&CPS), IEEE, pp. 1–9.

[31] B. Lin, C. Huang, Analysis of emission reduction effects of carbon trading: Market mechanism or government intervention?, Sustainable Production and Consumption 33 (2022) 28–37.

[32] S. Zeng, B. Su, M. Zhang, Y. Gao, J. Liu, S. Luo, Q. Tao, Analysis and forecast of china's energy consumption structure, Energy Policy 159 (2021) 112630.
[33] L. Hongtao, L. Wenjia, The analysis of effects of clean energy power generation, Energy Procedia 152 (2018) 947–952.
[34] N. Zhao, F. You, Toward carbon-neutral electric power systems in the new york state: a novel multi-scale bottom-up optimization framework coupled with machine learning for capacity planning at hourly resolution, ACS Sustainable Chemistry & Engineering 10 (2021) 1805–1821.
[35] M. G. Prina, G. Manzolini, D. Moser, R. Vaccaro, W. Sparber, Multi-objective optimization model eplanopt for energy transition analysis and comparison with climate-change scenarios, Energies 13 (2020) 3255.
[36] H. Teichgraeber, A. R. Brandt, Clustering methods to find representative periods for the optimization of energy systems: An initial framework and comparison, Applied energy 239 (2019) 1283–1293.
[37] J. Ma, Y. Yuan, Dimension reduction of image deep feature using pca, Journal of Visual Communication and Image Representation 63 (2019) 102578.
[38] P. Pravin, X. Wang, Stochastic optimization approaches in solving energy scheduling problems under uncertainty, IFAC-PapersOnLine 55 (2022) 815–820.
[39] P. Calvo-Bascones, M. A. Sanz-Bobi, T. M. Welte, Anomaly detection method based on the deep knowledge behind behavior patterns in industrial components. application to a hydropower plant, Computers in Industry 125 (2021) 103376.
[40] W. Chu, W. He, Q. Jiang, S. Zhang, Z. Hu, G. Xu, Y. Yuan, J. Fan, Optimization of operation strategy for a grid interactive regional energy system, Energy and Buildings 250 (2021) 111294.
[41] D. Wu, Z. Han, Z. Liu, P. Li, F. Ma, H. Zhang, Y. Yin, X. Yang, Comparative study of optimization method and optimal operation strategy for multi-scenario integrated energy system, Energy 217 (2021) 119311.
[42] B. Li, et al., Timing decision of low-carbon technology investment adoption by high energy consuming enterprises under carbon trading and subsidies, Journal of Environmental and Public Health 2022 (2022).
[43] J. Hao, F. Gao, X. Fang, X. Nong, Y. Zhang, F. Hong, Multi-factor decomposition and multi-scenario prediction decoupling analysis of china's carbon emission under dual carbon goal, Science of The Total Environment 841 (2022) 156788.
[44] Y. Meng, Y. Cao, J. Li, C. Liu, J. Li, Q. Wang, G. Cai, Q. Zhao, Y. Liu, X. Meng, et al., The real cost of deep peak shaving for renewable energy accommodation in coal-fired power plants: Calculation framework and case study in china, Journal of Cleaner Production 367 (2022) 132913.
[45] Z. Shuang, L. Hai-Yan, Comparative study on power sector allowance allocation among china's emissions trading scheme pilots and its implica-

tions for national carbon market, Advances in Climate Change Research 16 (2020) 748.

[46] J. Xiao, G. Li, L. Xie, S. Wang, L. Yu, Decarbonizing china's power sector by 2030 with consideration of technological progress and cross-regional power transmission, Energy Policy 150 (2021) 112150.

[47] I. Pies, S. Hielscher, S. Everding, Do hybrids impede sustainability? how semantic reorientations and governance reforms can produce and preserve sustainability in sharing business models, Journal of Business Research 115 (2020) 174–185.

[48] Y. Ma, C. Li, J. Zhou, Y. Zhang, Comprehensive stochastic optimal scheduling in residential micro energy grid considering pumped-storage unit and demand response, Journal of Energy Storage 32 (2020) 101968.

11. Adaptive Dempster-Shafer Theory for Evidence-based Trust Models in Multiagent Systems

Xiangyao Liang,[1] Ningkui Wang,[2] Mengna Hong,[1]
Yi Man[1] and Zhenglei He[1,*]

1 Introduction

Trust and reputation estimation is classic daily social behavior. Before purchasing particular items, the product's quality has to be determined by rational customers who accumulate relevant information. The decision-maker depends heavily on the comprehensive trustworthiness of the product provider. Similarly, the same mechanism fits MASs, peculiarly those with deceptive agents in a distributed environment.

Plenty of protocols are used to emphasize the trustworthiness issue in MASs. Normally, four models related to trusted systems are currently applied. The logical models depict how one agent trusts another according to mathematical logic. The social-cognitive models decide agents' trustworthiness inspired by human psychology. Organizational models maintain trust through personal relationships in the system, whereas numerical models understand trust from a mathematical probability perspective[1;2]. Each of the four models has advantages for trust estimation.

An agent's trustworthiness is represented by a real number in the numerical models, signifying the possibility that the interaction with it is a success. The service provider with a high degree of trust is more probable to be selected. In order to obtain the honesty of the agent, the trust model needs to

[1] State Key Laboratory of Pulp and Paper Engineering, South, China University of Technology, Guangzhou, China.
[2] School of Mathematics and Statistics, Hubei Minzu University, Enshi, China.
*Corresponding author: hezhenglei@scut.edu.cn

gather some information about the possible cooperation. Generally speaking, the trust estimation mainly comes from the following types of information: direct trust, which is based on its experience of direct interactions[3;4]; indirect reputation, which is based on the testimonials provided by third-party agents[2;5]. In addition, social cognitive information and organizational knowledge are other important aspects[6;7;8]. However, the collected information from the third-party agents must contain significant uncertainty; therefore, the collected information must have reasonable reliability to be serviceable.

In terms of distinguishing reliable information collected from third-party agents, some argue that the majority of agents are reliable[2]. Some identify dishonest agents through personal direct experience[9] or filter out the unreliable advice and devalue their evidence by assigning adaptive weights[9;10;11;12]. Though some of them have excellent security features, most assume that service providers have a list of static performance. That is to say, service providers are trustworthy and always maintain their honest behavior. What is more, the testimony given by the third party is always reliable. However, there is no guarantee that service providers and witnesses would always present static behavior. They may decide to go from sincere to malicious or the other way around. Thus, the general assumption might not be sufficient due to the dynamic behavior of agents in MASs. In addition, most of the systems we reviewed only consider reliability, certainty, or consistency as separate models in MASs. These models do not utilize an integrated model that covers all aspects of trust for trustworthiness estimation.

In summary, agents, whether resource providers or third-party agents, can change their performances in MASs. A large amount of information is collected in the numerical model to assess direct trust and indirect reputation. However, many uncertainties arise from incompleteness, randomness, fading, and so forth in direct trust, whereas the collected information from third-party agents must have reasonable confidence to be serviceable. This paper proposes an adaptive trust model in MASs where agents' behaviors might change over time. Agents in this model can be service providers or witnesses who provide testimony for trust estimation. The Dempster–Shafer evidence theory is adopted to obtain direct trust. Based on the quality of information collected, credibility and certainty of witnesses, dynamic adaptive weights are used to describe confidence in modifying indirect reputations. The ultimate trust is a combination of direct trust and indirect reputation. Our model provides a metric for the customer agents to decide which agent to interact with. The rest of the paper is organized as follows. Some review of related researches are introduced in Section 2. The primary tool, the Dempster–Shafer evidence theory is discussed in Section 3. Section 4 explains the proposed trust model in detail. The last section gives the comparisons and conclusions.

2 Review of Related Researches

Trust is a subjective judgment, and in most of our reviewed articles, it refers to confidence in one aspect or another of a relationship [4]. It was first introduced to computer science in 1994 as a measure of agency[13]. Trust can be classified into three types[14;15]. The first kind of trust refers to truth and belief, that is, trusting the behavior or quality of the other object. Next, trust means that a declaration is true. Finally, "Trust as commerce" indicates the buyer's ability or intention to buy in the future. In this paper, we adopt the first approach, which is that trust can be explained by the behavior and quality of the believing identity.

The literature is plentiful with various approaches to trust estimation in MASs[16;2;15;17;18;19]. In general, trust should be based on evidence consisting of positive and negative experiences with it[20;21;15]. Jøsang, A and Ismail, R Obtain trust in MASs based on the beta probability density functions that depend on the number of negative and positive interactions[2]. Use probabilistic expectation to represent trust value. The literature[22] has systematically studied when the combination of direct trust and indirect reputation is beneficial. TRAVOS[9] models trustworthiness probabilistically through a beta distribution computed from all interaction outcomes observed by trustees, and it uses the probability of successful interaction between two agents to obtain the value of trustworthiness. Fortino, Giancarlo, et al. studied trust and reputation in the Internet of Things[23]. In addition, some references assess trustworthiness by taking advantage of recent performances. In the literature[24;12], they were trusted by evaluating recent performances and historical assessments. However, trust assessment may encounter uncertainty deviation from cognitive impairment, inefficient evidence, randomness and incomplete information[25;26]. Wang and Singh use evidence theory to present uncertainty for trust estimation[27;28;29]. In Zuo and Liu's model[30], they adopt a three-value opinion (belief, disbelief, uncertainty) to estimate trust in the opinion-based model. However, these models are doubtful for trust estimation in a MAS with deceptive agents involved. At the same time, most of them define uncertainty through the medium evaluations. However, as explained, uncertainty stems from cognitive impairment, inefficient evidence, randomness, and incomplete information[25;26]. Therefore, a more detailed exploration that exploits the uncertainty needs to be carried out.

In the case of insufficient direct experience, it is natural and reasonable to query third-party information. However, it is risky to rely entirely and directly on the information gathered in a distributed MAS to make a final decision, especially when deceptive agents are involved. Thus, we need to make sure that the accumulated information is reliable before using it. Therefore, the concept of confidence is used to capture the reliability of the collected information[28;9;20;31;15;24;12]. We define confidence as a probability that the trust applies, supported by evidence, consistency, credibility, and certainty.

However, most of these models consider only certain aspects as a separate trust model in MASs.

First, some trust models assume that most third -party agents are reliable[2]. Thus, those opinions that are in agreement with the majority can be considered reliable[2;32]. Others identify unreliable agents by comparing them with direct experience[28;9]. In this article, we define this property as consistency. Consistency indicates the differences or similarities with other evidence. This is helpful, especially for those agents whose trust and certainty are unknown. However, most of our reviewed articles did not focus on the consistency for trust estimation in MASs.

Credibility is widely used in distributed MASs to gain confidence. In most cases, credibility can be interpreted as a weight that indicates the reliability of the agent[12]. It is updated based on the provided information and the interactive feedback[28;29;24;12]. However, in the literature[12;28], the credibility of the third party agents is gradually declines that is to say, the truster will no longer trust the agents with sufficient interactions. In the literature[29], the Dempster–Shafer evidence theory is used to achieve trust. However, the credibility of the third-party agents increases if the feedback is successful, and the other way around. In dynamic MASs, false information might also be reliable, and it can be caused by changing performances. Thus, improving the credibility of the agent through current interactions can be problematic and insufficient.

Certainty is "a measure of confidence that an agent may have in the trust information"[15]. Measuring certainty can effectively filter out reliable information[33]. As mentioned earlier, trust always contains uncertainty caused by the lack of certainty, a lack of correct and complete information, or randomness[34]. Therefore, the information that comes from a different degree of certainty may affect the agent's confidence. For example, agent A always trusts agent B, but agent B is uncertain about its information towards some event. Thus, agent A would be apprehensive about the collected information from agent B. In the literature[31], the trust model uses fuzzy sets and assigns certainty to each evaluation. In the literature[15], the model analyzed certainty to gain confidence. Most trust models assume that certainty and trust are independent. However, we believe that trust is the rate of successful interactions. Thus, uncertainty arises from the conflict between the interactions of success and failure, and certainty should depend on evidence.

To sum up, the influential factors and their explanations for trust estimation in MASs are shown in Table 1. As we can see, from our perspective, we value uncertainty, evidence consistency, witness' credibility, and certainty and the motivation for a witness as influential factors for trust estimation in MASs. In the following sections, we will present some of the recent trust estimation models related to confidence estimation.

FIRE is an integrated trust and reputation model for open MASs according to interaction trust (IT), role-based trust (RT), witness trust (WT), and certified trust (CT)[35]. It is direct trust calculated from direct experience; RT, WT, and CT are about witness experience. FIRE shows good performance. However, the model does not emphasize uncertainty and assumes that all agents provide accurate information. In addition, many static parameters must be used in the model, which limits the application of the model[36;24]. SecuredTrust is a dynamic trust computation model in MASs[32]. The model involves many factors, including differences and similarities. However, feedback credibility generated from similarity is used to capture the accuracy degree of the feedback information, which is the same as consistency. However, certainty and personal credibility are not well emphasized. At the same time, encouraging agents to be reliable should use an incentive mechanism. The actor–critic model selects reliable witnesses through reinforcement learning[37]. It dynamically selects credible witnesses and parameters related to direct and indirect trust evidence based on interactive feedback. Nevertheless, bootstrapping errors occur when only one known witness is available for indirect reputation[38]. More importantly, in dynamic systems, it is necessary to capture uncertainty in a perfect trust estimation model. By analyzing the uncertainty, a certain degree of certainty can be obtained and reliable agents can be filtered out effectively. The model combines direct trust and indirect reputation and also evaluates the credibility of witnesses. Moreover, it is reasonable for the model to consider the uncertainty when generating direct trust. However, witnesses' certainty is essential for a trust model, especially a dynamic trust model. Despite the advantages presented by the models noted above, these trust models are still imperfect and can be extended. This paper proposes a trust model based on evidence theory. It first improves the method of updating the credibility of witnesses which is not appropriate in previous articles. When capturing direct trust, the model manages uncertainty from the perspective of cognitive impairment, inefficient evidence, randomness, and incomplete information[25;26]. More importantly, consistency, credibility, and certainty are declared to achieve confidence. An incentive mechanism is used to motivate agents to provide reliable information.

3 Dempster–Shafer Theory of Evidence To Handle Uncertainty

The Dempster–Shafer evidence theory, proposed by Dempster and Shafer to handle uncertainty[39;40], has a wide range of applications in risk assessments, medical diagnosis, and target recognition[41;42;43]. This method has the adjustability of uncertainty modeling and is a typical uncertain information fusion method[44]. Some preliminaries are introduced below.

The frame of discernment and BPAs. Evidence theory is defined in the frame of discernment denoted by Ω, which consists of n mutually exclusive and collectively exhaustive elements; 2^{Ω} is the set of all subsets of Ω, which is represented by $2^\Omega = \{\emptyset, \{\theta_1\}, \{\theta_2\}, \ldots, \{\theta_n\}, \{\theta_1, \theta_2\}, \ldots, \{\theta_1, \theta_2, \ldots, \theta_n\}\}$. Mathematically, a basic probability assignment (BPA, also known as mass function) is $2^\Omega \to [0, 1]$, with the following conditions satisfied:

$$\begin{cases} \sum_{A \in 2^\Omega} m(A) = 1; \\ m(\emptyset) = 0, \end{cases} \tag{1}$$

where $\emptyset$ is the empty set, the subsets $A \in 2^\Omega$ such that $m(A) > 0$ is named the focus element, and the set of all focus elements is called core. The other two functions (belief function and plausibility function) are defined as follows.

Belief function. The belief function Bel of BPA m is defined as follows:

$$Bel_m(A) = \sum_{B \in 2^\Omega : B \subseteq A} m(B) \tag{2}$$

Plausibility function. The plausibility function Pl of BPA m is defined as follows:

$$Pl_m(A) = \sum_{B \in 2^\Omega : B \bigcap A \neq \emptyset} m(B) \tag{3}$$

In the Dempster–Shafer evidence theory Dempster's combination rule can be used to combine multiple evidence pieces to achieve the final evaluation. Given two BPAs m_1 and m_2, the Dempster's combination rule denoted by $m = m_1 \oplus m_2$ can fuse them as follows.

Dempster's combination rule.

$$m(A) = \begin{cases} \frac{1}{1-K} \sum_{B \cap C = A} m_1(B) m_2(C), & A \neq \emptyset; \\ 0, & A = \emptyset; \end{cases} \tag{4}$$

with $K = \sum_{B \cap C = \emptyset} m_1(B) m_2(C)$, A, B, and C are elements in 2^Ω. The normalization constant K shows the conflicting degree between defined BPAs. If $K = 1$, the two pieces of evidence are in total conflict, and Dempster's combination rule is not applicable. If $K = 0$, the two pieces of evidence are not inconsistent.

Evidence distance. With the full application of evidence theory, distance of evidence research has attracted increasing interest. The distance measure of evidence can represent the dissimilarity between two BPAs, and several definitions of distance in evidence theory have also been proposed, such as Jousselme's distance[45], Wen's cosine similarity[46], and Smets' transferable belief model (TBM) global distance measure[47]. Jousselmes distance[45] is one of the rational and popular definitions of distance, which is identified based on Cuzzolin's geometric interpretation of evidence theory[48]. The power set of the frame of discernment space and a distance and vectors are defined with the BPA as a particular case of vectors[45]. Jousselme's distance is defined as

$$d(m_i, m_j) = \sqrt{\frac{1}{2}(\vec{m_i} - \vec{m_j})^T D (\vec{m_i} - \vec{m_j})} \tag{5}$$

with m_i, m_j being two BPAs under the frame of discernment Ω matrix. The element in D is defined as $D(A,B) = |A \cap B|/|A \cup B|$, $A, B \in \Omega$, and $|A|$ represents cardinality. Jousselme's distance is an efficient tool to quantify the dissimilarity of two BPAs.

Entropy of Dempster–Shafer theory.

The concept of entropy comes from physics as a measure of uncertainty and disorder[49;50]. The Shannon entropy has been widely accepted for uncertainty measurement (or information measurement) in information theory[51]. However, the uncertainty estimation of BPAs of the Dempster–Shafer evidence theory still requires trial and error. Since the first definition of a BPA in the evidence theory was given by Höhle in 1982[52], many definitions of entropy of the Dempster–Shafer theory have raised, for instance, the definitions proposed by Smets[53], Yager[54], Nguyen[55], Dubois and Prade[56], Deng[57], and so on. However, these definitions are still imperfect. For example, Deng entropy has some limitations when the propositions are for the intersection. Therefore, several methodologies have been applied for Deng entropy modification[58;59]. Recently, Jiroušek and P. Shenoy proposed a new definition, in which five of the six listed properties of entropy are satisfied[60]. Whether there exists a definition that satisfies the six properties is still an open question, but the proposed entropy in[60] is acceptable and could be considered as near-perfect. As a result, we use the definition of entropy of the Dempster–Shafer theory in the literature[60] for uncertainty estimation in this paper. For the frame of discernment defined in Ω where $x \in \Omega$ and A is a focal element, the entropy of the BPA m is defined as follows:

$$\begin{aligned} H(m) = H_s(Pl\_P_m) + H_d(m) = \sum_{x\in\Omega} Pl\_P_m(x) log(\frac{1}{Pl\_P_m(x)}) \\ + \sum_{A\in 2^\Omega} m(A) log(|A|) \end{aligned} \quad (6)$$

where the first part is employed to measure conflict[60;61], and $Pl\_P_m(x)$ is defined as the plausibility transform[62;60], where

$$Pl\_P_m(x) = \frac{Pl_m(x)}{\sum_{x\in\Omega}(Pl_m(x))} \quad (7)$$

The second part in Equation (6) can be considered as non-specificity of the BPA m[56]. In other words, the uncertainty of the Dempster–Shafer evidence theory can be captured from two aspects, namely the conflicts between exclusive elements in set Ω and the non-specificity uncertainty. Both components are on the scale $[0, log(|\Omega|)]$, thus, the maximum entropy of a BPA is $2log(|\Omega|)$.

An example of the Dempster–Shafer theory of evidence is given as follows. Suppose an animal stands far away from us, perhaps a cat or a dog. Then the frame of discernment is presented as $\Omega = \{cat, dog\}$, and three passers-by show their opinions. Their opinions are shown as

1. $m_1(Cat) = 0.6$, $m_1(Dog) = 0.3$, $m_1(Cat, Dog) = 0.1$
2. $m_2(Cat) = 0.4$, $m_2(Dog) = 0.4$, $m_2(Cat, Dog) = 0.2$
3. $m_3(Cat) = 0.1$, $m_3(Dog) = 0.4$, $m_3(Cat, Dog) = 0.5$

In terms of evidence distance, we have the following calculation process:

1. $\overrightarrow{m_1} = \{Cat, Dog, (Cat, Dog)\} = (0.6, 0.3, 0.1)$. Similarly, $\overrightarrow{m_2} = (0.4, 0.4, 0.2)$, $\overrightarrow{m_3} = (0.1, 0.4, 0.5)$;
2. D =cccc Cat Dog (Cat, Dog)
   c[ccc] Cat 1 0 0.5
   Dog 0 1 0.5
   (Cat, Dog) 0.5 0.5 1
3. $(\overrightarrow{m_1} - \overrightarrow{m_2}) = (0.2, -0.1, -0.1)$;
4. $d(m_1, m_2) = \sqrt{\frac{1}{2}(\overrightarrow{m_1} - \overrightarrow{m_2})^T D(\overrightarrow{m_1} - \overrightarrow{m_2})} = 0.158$. Similarly, $d(m_1, m_3) = 0.361$ and $d(m_2, m_3) = 0.212$.

The three passers-by are all not one hundred percent sure what the animal is. Nevertheless, we may study their degrees of uncertainty, respectively. Let us take the first witness as an example.

1. $Pl\_P_m(Cat) = \frac{Pl_m(Cat)}{\sum_{x \in \Omega}(Pl_m(x))} = \frac{0.6+0.1}{0.6+0.1+0.3+0.1} = \frac{7}{11}$. In the same manner, $Pl\_P_m(Dog) = \frac{4}{11}$;
2. Thus, H(m_1)= $\frac{7}{11}log_2(\frac{7}{11})$ $+\frac{4}{11}log_2(\frac{4}{11})$+ $0.6 * log2(|Cat|) + 0.3 * log2(|Dog|)$+ $0.1 * log2(|(Cat, Dog)|) = 1.046$. Similarly, H($m_2$) = 1.2, and H($m_3$) = 1.47. As we can see, the testimony given by the third passer-by is the most ambiguous and uncertain.

4 The Trust Model for Trust Estimation

This section introduces the trust model for trust estimation in MASs, where agents' behaviors might differ over time. The proposed method depends on direct trust learned from direct experience and indirect reputation by integrating all indirect experiences provided by third-party agents (witnesses). We study the confidence of witnesses and then combine direct trust and indirect reputation. In more detail, the direct satisfaction statement based on interactive feedback is first illustrated, focusing on the fact that recent interactions are more prominent by assigning higher weights. In this way, the proposed method is suitable for dynamic trust estimation. The uncertainty that arises from randomness, incomplete information, and inefficient interactions is then addressed, and the Dempster–Shafer evidence theory is adopted to express direct trust. Then the confidence in determining the personal evidence of indirect reputation was studied. Subsequently, direct trust and indirect reputation were combined to obtain the resource provider's trustworthiness. The agent with the highest trust value was selected to raise interactions. The credibility of witnesses is updated through the interactive feedback and the average evaluation after the interaction. Finally, credits are reassigned to witnesses, who can use them to purchase information for trustee evaluations. This way, information providers have an incentive to provide accurate information. For simplicity, the detailed processes are shown in Figure 1, and some related parameters are listed in Table 2.

4.1 Generation of Direct Trust

Direct trust is a performance prediction for the next interaction of the selected agent based on previous interaction feedback. The obtained interactive feedback could be binary, i.e., a success or a failure, or any real number in the range $[0, 1]$ indicating that the service received is satisfactory. This real number can also be interpreted as the trustworthiness of the information or service provider. For the sake of clarity, the service requester is represented by service client agent c_j, the service provider is modeled as service provider agent s_j, and those who provide information to help assess the performance are represented as witness agent w_k. As is shown in Figure 2, the client agent (evaluator) c_i estimates the trustworthiness of the resource providers s_j according to its direct experience and the indirect reputation provided by the witness w_k.

After interactions, each service client updates the corresponding satisfaction degree for each service provider based on its current degree of compensation and the feedback outcome. Of the previous literature, they average all the evaluations to get the final satisfactory degree. In[27], an evidence theory-based trust model, two parameters are used to divide evaluations into trust, uncertainty, and distrust, and the medium evaluations are designated as uncertainty. It could work, but we think recent interactions in dynamic MASs should significantly affect the final assessment, and many reasons may lead to uncertainty, including medium evaluation and conflict evaluations. Subsequently in the literature[2;9], it seems reasonable to assign higher weights to the latest interactions. From our point of view, uncertainty causes randomness, incomplete information, and the resulting environment is inevitable and should be taken into account in dynamic MASs. In addition, clients could not have sufficient interactions with one fixed agent over time. Therefore, we will

Table 1: Influential factors for trust estimation in MASs.

	Definition	**Why the Factor Is Influential**
Uncertainty (Direct trust)	Uncertainty caused by fading, randomness, incompleteness, and more.	To ensure trust accuracy in dynamic MASs
Consistency (Confidence)	The degree of similarity with third-party testimony	To present high-quality evidence; avoidance of sudden changes
Credibility (Confidence)	The credibility degree of third-party witnesses	To distinguish trusted witnesses and avoid group deception
Certainty (Confidence)	The certainty degree of third-party testimony	To capture the certainty of testimony
Motivation	Inspire witnesses to be honest	Ensure the system runs correctly

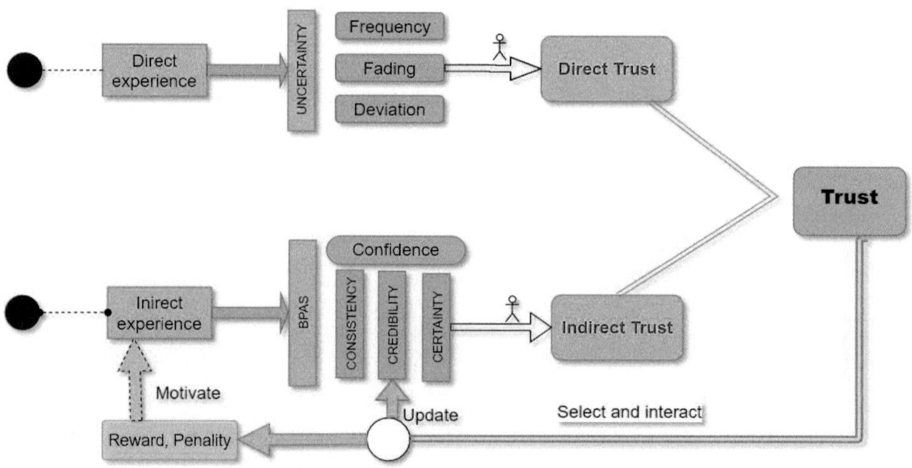

Figure 1: Trust estimation processes.

Table 2: Explanation of main variables.

c_i	The resource customer agent c_i
s_i	The resource supplier agent s_i
w_k	The information provider (witness) w_k
$Scur_{ij}(t)$	The evaluation of the tth interaction
$S_{ij}(t)$	The satisfaction degree before the tth interaction
$\mu_{ij}(t)$	The relationship between evaluation and current satisfaction
$fq_{ij}(R)$	The uncertainty factor
$\Omega = \{T, nT\}$	The frame of discernment
m_{ij}^{d}	The direct evidence
m_{ij}^{ind}	The indirect evidence
$Conf(c_i, w_k)$	The confidence degree of w_k from the perspective of c_i
$Dis(m_{kj}^{ind})$	The distance of m_{kj}^{ind} in the group of L witness
$Sim(m_{kj}^{ind})$	The similarity of m_{kj}^{ind} in the group of L witness
$Cred(c_i, w_k)$	The credibility value of w_k from the perspective of c_i
$Cer(m_{kj})$	The certainty of w_k

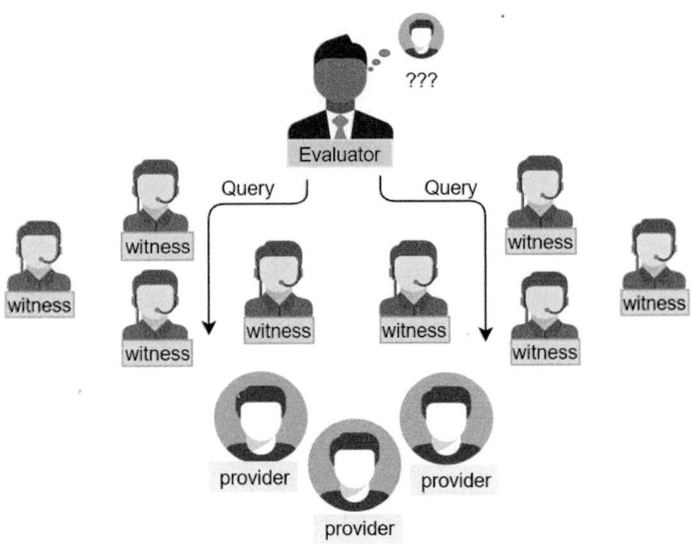

Figure 2: Trust evaluation in MASs.

use the uncertainty created by the number and frequency of interactions and the current deviation between trust and performance to generate direct trust.

Let T imply that the service client agent perceives the service provider agent to be trustworthy. Accordingly, nT indicates that the supplier agent is untrustworthy. Thus, the frame of discernment is represented by $\Omega = \{T, nT\}$. When c_i needs to appraise the performance of s_j, $S_{ij}(t)$ and $Scur_{ij}(t)$ are employed to represent the satisfaction before the tth interaction and of the tth interaction, separately. The satisfaction is updated as follows:

$$S_{ij}(t+1) = \mu_{ij}(t)S_{ij}(t) + (1 - \mu_{ij}(t))Scur_{ij}(t) \tag{8}$$

As noted above, $Scur_{ij}(t)$ indicates the current interactive feedback, which could be any number in the range of $[0, 1]$, namely,

$$Scur_{ij}(t) = \begin{cases} 1, & c_i \text{ is completely satisfied with } s_j; \\ 0, & c_i \text{ is completely unsatisfied with } s_j; \\ v \in (0,1); & \text{others.} \end{cases} \tag{9}$$

In Equation (8), it is challenging to decide an appropriate value of $\mu_{ij}(t)$. In the literature[2], a forgetting factor is used. However, the assignment is appropriate for the beta reputation system. In[24], they believe that weight is related to interactive times and frequency. Das et al. hold that $\mu_{ij}(t)$ should depend on the accumulated deviation[32]. From our perspective, we suppose that $\mu_{ij}(t)$ should be determined mainly based on the current performance deviation. That is to say, we pay much attention to its previous performance if agents are stable. If agents frequently demonstrate behaviors different from their actual attributes in order to hide their performance, the evaluation should play a more important role in the final estimation. Usually, $\mu_{ij}(t)$ is close to 0, for instance, $\mu_{ij}(t) = 0.1$, indicating that most agents consistently react, which is rational and intuitive in reality. In this paper, we define $\mu_{ij}(t)$ as follows:

$$\mu_{ij}(t) = \frac{exp(1)}{exp(|S_{ij}(t) - Scur_{ij}(t)| - 1) + exp(1)} \tag{10}$$

During the generation of evidence, there is great uncertainty due to incomplete information, the interactive frequency, and randomness. A frequency factor $fq_{ij}(t)$ is defined to show the uncertainty of c_i's direct experience in the last period, and is defined as

$$fq_{ij}(t) = 1 - [\frac{2 * \lambda}{\pi} * atan(\kappa_{(i,j)})]^2 \tag{11}$$

where $\kappa_{(i,j)}$ shows the quantum of interactions between c_i and s_j in the last considered period. The value of λ is influenced by the interaction type and the willingness to trust others; it is usually close to 1. As displayed in Equation (11), uncertainty caused by the amount of interaction, incompleteness, and randomness is described, and the uncertainty decreases to 0.2 when two agents have at least eight interactions. Therefore, the generated BPAs can be represented as follows:

$$\begin{cases} m_{ij}^{d}(T) = S_{ij}(t) * fq_{ij}(t) \\ m_{ij}^{d}(nT) = (1 - S_{ij}(t)) * fq_{ij}(t) \end{cases} \tag{12}$$

Then $m_{ij}^{d}(T, nT) = 1 - (m_{ij}^{d}(T) + m_{ij}^{d}(nT))$. The direct experience can also be rewritten as $bpa_{ij}^{d} = (m_{ij}^{d}(T), m_{ij}^{d}(nT), m_{ij}^{d}(T, nT))$.

4.2 Indirect Reputation Management

In the last subsection, we discussed how to measure direct trust based on interactive feedback. The unavoidable factors, which are considered to be uncertainty caused by low interactive frequency, randomness, or incomplete information, are addressed during trust estimation. In this subsection, we focus on indirect reputation management. In some studies, agents use direct trust only if they have direct experience; otherwise, the agent would claim indirect reputations. Some literature employs only direct trust if it is available; otherwise, it would then claim indirect reputations. From our point of view, we must make full use of both direct trust and indirect reputation for trust estimation, especially in dynamic and distributed MASs. However, it is challenging to rely on the received information directly for the following reasons:

- Some agents might be dishonest;
- A group of agents can conspire to be deceptive;
- Some agents might have incorrect or insufficient information;
- Agents may be uncertain about their provided information;
- Agents might be used for deception.

Accordingly, the trust evaluator needs to have confidence in every piece of collected information. Therefore, we determine the indirect reputation from the following aspects.

1. The estimated trust from the witness' experience: This aspect could be modeled in the same way as direct trust. The framework of the Dempster–Shafer evidence theory is adopted to capture direct trust from the perspective of the information provider (witness).
2. Evidence consistency: As mentioned earlier, both witnesses and service providers might present misleading information. For instance, the

client c_i trusts the witness w_k. However, w_k does not have sufficient testimony about the service provider s_j, or w_k presents false information deliberately because of a conflict of interest. Under these circumstances, the quality of received information is significant. The conflict is applied to identify the differences between evidences, and—as opposed to the differences–the similarities between received evidences could be used to judge the quality of the provided information.

3. The credibility of witnesses: This factor is measured by a real number in the range of $[0, 1]$. We initialize it to a fixed number that represents the degree to which the witness can be trusted. Subsequently, it is updated through the interaction of interactive feedback. The credibility of close to 1 shows that it is reliable. Otherwise, the witness is of low trustworthiness, and the interaction should be avoided. This way, a group of agents no longer have the opportunity to conspire to cheat.

4. Certainty of a witness: This factor plays an important role in interaction-based trust estimation among MASs. For example, agent A trusts agent B completely. However, it is questionable to trust agent B if B is uncertain of its evidence to agent C. In this paper, we argue that certainty depends on evidence, and we refine certainty from the received evidence from the entropy perspective. We assume that trusted, witness-provided, good-quality, high-certainty evidence makes a greater contribution to trust estimation.

We use $(m^d_{kj}(T), m^d_{kj}(nT), m^d_{kj}(T, nT))$ to present the direct trust of the service provider agent s_j from the perspective of the witness w_k. As is asserted in the previous part, agents should have reasonable confidence in the received information. In our view, evidence availability changes accompanying confidence. Low-confidence evidence has a smaller impact on the final decision. At the same time, high confidence evidence plays a more crucial role in the final decision-making. We assume that the confidence of the received information $(m^d_{kj}(T), m^d_{kj}(nT), m^d_{kj}(T, nT))$ is represented by $Conf(c_i, w_k)$, then the received information is revised by the degree of confidence as follows:

$$\begin{aligned} m^{ind}_{ij}(T) &= m^d_{kj}(T) * \mathbf{Conf}(c_i, w_k); \\ m^{ind}_{ij}(nT) &= m^d_{kj}(nT) * \mathbf{Conf}(c_i, w_k); \end{aligned} \tag{13}$$

where $m^{ind}_{ij}(T, nT) = 1 - m^{ind}_{ij}(T) - m^{ind}_{ij}(nT)$. Here, $m^{ind}_{ij}(T)$ shows the trust evaluation of the agent s_j from the service client agent c_i' perspective according to the information provided by the witness agent w_k. As defined, the discounted parts are interpreted as uncertainty. It is logical that if the provided information is not convincing, and more information is assigned as uncertainty. Thus, more information is needed to eliminate the uncertainty. After modification, Dempster's combination rule is adopted to merge all the evidence.

However, determining an appropriate confidence value is not easy, especially in a dynamic environment. In the literature[28], they give weight to witnesses who represent their credibility. Certainty is also used to gain confidence[15]. However, it is not enough to consider these aspects separately to decide confidence. From our point of view, we have to express the consistency of the obtained information because we think the agents may not have enough interactions. At the same time, even if an agent interacts with a target multiple times, its performance might diversify. Moreover, in the reviewed work, we rarely find that consistency, credibility, and certainty have been integrated to determine confidence. Therefore, in order to provide a near-perfect trust estimation model, this paper studies the consistency, credibility and certainty of confidence estimation.

4.2.1 Evidence Difference

Distance represents the distance between two objects in real life. The distance of BPAs indicates the conflicts or differences between two BPAs. The shorter distance means that the two BPAs are more likely to support the same target (service provider). For trust estimation models in MASs, we use average distance to capture the consistency of some evidence. Assume that the demander received BPAs from L different witnesses w_k, each of which is represented by m^d_{kj}. Therefore, the consistency (the difference) between these pieces of evidence is explored below. First, the average distance of m^d_{kj} is represented by $Dis(m^d_{kj})$, which is defined as

$$Dis(m^d_{kj}) = \frac{\sum_{p=1,p\neq k}^{L} d(m^d_{kj}, m^d_{pj})}{L-1} \tag{14}$$

where $d(m^d_{kj}, m^d_{pj})$ indicates the distance of m^d_{kj} and m^d_{pj}, and $Dis(m^d_{kj})$ represents the average distance to the evidence provided by the witness w_k, which also expresses its conflict or difference in the group of L evidences. Obviously, the distance between two pieces of evidence falls in the range $[0, 1]$ in this paper. At the same time, we assume that the opposite of difference is similarity. Consequently, we can use the following operations to represent similarity:

$$Sim(m^d_{kj}) = 1 - Dis(m^d_{kj}) \tag{15}$$

where $Sim(m^d_{kj})$ represents the consistency towards the service provider s_j from the perspective of the witness w_k in the group of L witnesses.

4.2.2 An Analysis of The Credibility of Individual Witnesses

In this subsection, the credibility of individual witnesses in MASs is analyzed in detail. For a group of witnesses $w_1, w_2, \ldots, w_k \ldots w_L$, testimony is provided to the resource client c_i if c_i urges them to evaluate the trustworthiness of the resource provider s_j. From the perspective of c_i, to rely on the witness or not is a thought-provoking question. Therefore, we use the credibility of the agent to represent the reliability of the witness. In this paper, we force the witness credibility to be in the range $[0, 1]$, and initialize this value to 0.5 if no cooperation has occurred. As interactions are undergone, the credibility is updated based on the interactive feedback.

It is necessary to explain how credibility is renewed. Of the previous Dempster–Shafer theory based trust estimations models, if the witness is found to be lying, the credibility value decreases with interactions all the time[63]. This model works under the assumption that all agents are acting in a constant manner. Moreover, never forgiving an agent, it seems unreasonable to give him a second chance. In the literature[29], the learning of credibility is mainly carried out from two aspects, namely, positive and negative feedback. That is to say, rewards are expressed to all agents when positive feedback is reached, and when negative feedback is received, all agents are punished. However, some witnesses may have negative attitudes when positive feedback is raised, and so it is the same when negative feedback is received. Thus, the credibility of these witnesses needs to be updated separately. What is more, updating values based solely on current interactive feedback is sometimes questionable, especially in dynamic systems. Individual reliability has a great influence on indirect reputation. We consider the credibility $Cred(c_i, w_k)$ of the witness w_k from the perspective of c_i when generating indirect reputation. More details about learning from the experience to update personal credibility are discussed in Section 4.4.

4.2.3 Evidence Certainty

The certainty of witnesses also counts for trust estimation in MASs. It is emphasized that uncertainty due to the lack of certainty and complete information is imperative when generating BPAs. Thus, certainty is independent of evidence (BPAs). From our perspective, certainty comes from two sources.

- First, certainty decreases as the degree of conflict increases in the evidence (we manage certainty by the ratio of positive and negative observations, namely, trust and distrust). Thus, evidence certainty decreases as the proportion of positive and negative interactions increases.

- Second, certainty decreases with the increase of uncertain information. That is to say, evidence certainty increases if $m(T, nT)$ decreases.

At the same time, we wish the value of certainty to fall in the $[0, 1]$ range. For the above reasons, the entropy of the Dempster–Shafer evidence theory is an ideal tool for uncertainty measurement.

A number of definitions of entropy have been proposed for the space of the Dempster–Shafer evidence theory, such as Deng entropy[57] and Dubois and

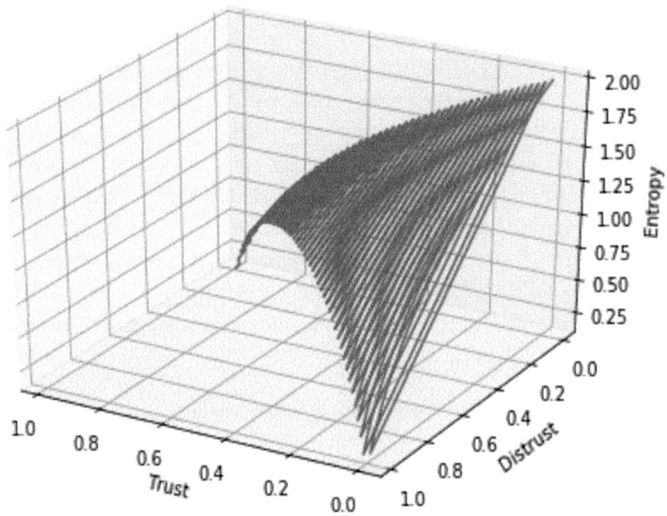

Figure 3: The entropy when frame of discernment is $\Omega = \{T, nT\}$. As is shown, entropy is in the range of $[0, 2]$.

Prade's definition[56]. In this paper, we adopt the entropy definition proposed by Radim and Prakash[60]. The main reasons are that five of the six listed properties are satisfied, which could be considered as near-perfect in some perspectives. Moreover, the entropy given by Radim and Prakash is designed from two components. The first component is used to measure conflicts in the BPAs, and the second is used to measure non-specificity in BPAs, which conforms to the main cause of uncertainty in this paper. As explained, we believe that uncertainty yields arise from the conflict of positive and negative interactions, and a lack of information could also lead to uncertainty. For the frame of discernment defined in this paper, i.e., $\Omega = \{T, nT\}$, we have $m(T) = x$, $m(nT) = y$ and $m(T, nT) = 1 - x - y$, which is under the conditions, of $0 \leq x \leq 1$, $0 \leq y \leq 1$, and $0 \leq 1 - x - y \leq 1$. The relationship between the entropies[60] is presented in Figure 3. As is shown, the uncertainty is in the range of $[0, 2 * log2|\Omega|]$. Thus, we define the certainty of the information m_{kj},

$$Cer(m_{kj}^{d}) = \frac{2 - H(m_{kj}^{d})}{2} \tag{16}$$

For example, a BPA is given as $(1, 0, 0)$ or $(0, 1, 0)$, then the witness is one hundred percent confident about the evidence. However, the witness is one hundred percent uncertain of its judgments if the evidence is (0, 0, 1).

4.2.4 Model The Ultimate Indirect Reputation

In this subsection, we explain how to learn the ultimate indirect reputation. As mentioned above, indirect reputation consists of two aspects, namely, the quality evaluation, which is based on performance, and the confidence of the witness. Three sub-factors determine the value of confidence, i.e., the consistency of the provided information (Equation (15)), the credibility of agents (Section 4.2.2), and the certainty (Equation (16)). These three sub-factors are used to correct the generated evidence and $\mathbf{Conf}(c_i, w_k)$ in Equation (13). Here, we use $\mathbf{Conf}(c_i, w_k)_t$, $Sim(m_{kj})_t$, $Cred(c_i, w_k)_t$, and $Cer(m_{kj})_t$ to represent the value of confidence, the consistency, the credibility, and the certainty for pretreatment of the tth interaction, respectively. A detailed definition is given as follows:

$$\mathbf{Conf}(c_i, w_k)_t = Sim(m_{kj})_t * Cred(c_i, w_k)_t * [(Cer(m_{kj})_t)^{|\triangle|}] \tag{17}$$

where $\triangle = min\{0, Sim(m_{kj}^{dir})_t * Cred(c_i, w_k)_t - \frac{1}{L}\sum(Sim(m_{kj}^{dir})_t * Cred(c_i, w_k)_t)\}$ is the smaller value between 0 and the difference of confidence obtained from consistency and credibility. In more detail, the confidence of a piece of evidence presented by the witness is directly proportional to the credibility of a witness, quality, and certainty of the evidence. In terms of certainty, as $\triangle$ defines, the confidence $\mathbf{Conf}(c_i, w_k, s_j)_t$ of the evidence provided by the witness w_k decreases with the certainty $Cer(m_{kj}^{dir})_t$ if its credibility $Cred(c_i, w_k)_t$ and quality $Sim(m_{kj}^{dir})_t$ are below the average level, which is calculated by $Sim(m_{kj}^{dir})_t * Cred(c_i, w_k)_t - \frac{1}{L}\sum(Sim(m_{kj}^{dir})_t * Cred(c_i, w_k)_t)$. Otherwise, the confidence depends only on the credibility and quality. In other words, the resource customer agent is more confident in the good-quality indirect experience provided by credibility agents, as shown in Figure 4. Those experiences that originate from unreliable agents must be discounted and are no longer reliable. After modification, agents' indirect reputations are fused using Dempster's combination rule, presented in Equation (4).

4.3 Model The Overall Value of Trust

In this section, we demonstrate how to model the overall value of trust. The overall evaluation of the service provider s_j from the perspective of the service client c_i derives from two parts, namely direct trust and indirect reputation. In general, a client trusts itself more than the information provided by third-party agents. At the very least, agents do not have self-conception. However, if a system is unstable, a client agent should better trust others. Thus, the pretreatment of direct trust and indirect reputation proposed in[64] is adopted; Equation (18) presents the detailed processes. In the equation, ψ indicates the weight of direct trust, and $1 - \psi$ corresponds to the indirect reputation. The value of ψ depends completely on the stability in the MASs;

a small ψ implies, agents' performances often change. Generally, $\psi > (1-\psi)$, we adopt m_{ij}^{pre} to represent the pretreated BPA, then we have,

$$\begin{array}{l} m_{ij}^{pre}(T) = \psi m_{ij}^{d}(T) + (1-\psi)m_{ij}^{ind}(T) \\ m_{ij}^{pre}(nT) = \psi m_{ij}^{d}(nT) + (1-\psi)m_{ij}^{ind}(nT) \\ m_{ij}^{pre}(T,nT) = 1 - m_{ij}^{pre}(T) - m_{ij}^{pre}(nT) \end{array} \tag{18}$$

After the pretreatment, Dempster's combination rule is applied to combine $(m_{ij}^{pre}(T), m_{ij}^{pre}(nT), m_{ij}^{pre}(T,nT))$ one time as[64] stated. Therefore, the client agent can accept the final evaluation by the service provider agent. Subsequently, the resource client selects one of the most reliable service providers to increase interactions. The interactive feedback outcome is obtained after the interaction, which could be a binary result (either success or failure) or an evaluation score within the range $[0,1]$. In the next section, we describe how personal credibility is updated.

4.4 Updating The Corresponding Credibility

Before updating the corresponding credibility, we first have to distinguish the viewpoint of the witness agent w_k on whether it stands with the service provider agent s_j . The evaluation of s_j from the perspective of w_k is represented by $(m_{kj}^{d}(T), m_{kj}^{d}(nT), m_{kj}^{d}(T,nT))$, where $m_{kj}^{d}(T)$ indicates that w_k believes that s_j is reliable and $m_{kj}^{d}(nT)$ represents that the resource provider is untrustworthy. The expression $m_{kj}^{d}(T,nT)$ is adopted to show the uncertainty caused by delay, interactive frequency, and incomplete information. It is necessary to emphasize that uncertainty does not mean knowing anything; it could be reassigned to trust or distrust according to Equation (7), which is represented as follows:

$$\begin{array}{l} Pl\_P_m(T) = \frac{Pl_m(T)}{Pl_m(T)+Pl_m(nT)} \\ Pl\_P_m(nT) = \frac{Pl_m(nT)}{Pl_m(T)+Pl_m(nT)} \end{array} \tag{19}$$

A value of $Pl\_P_m(T)$ greater than 0.5 symbolizes that the service provider is reliable from the perspective of the witness w_k. Otherwise, the witness w_k means that the service provider s_j is untrustworthy. Hence, an agent's credibility needs to be updated according to the difference between the provided information and the interactive feedback outcome. Generally, the interactive outcome is represented by a success $(1,0,0)$ or a failure $(0,1,0)$ if the binary evaluation system is employed; otherwise $(s, 1-s, 0)$. As is emphasized, agents in MASs might act erratically. Therefore, the obtained error information is probably caused by two conditions: the witness is dishonest, or the service provider deliberately provides a different service quality. Therefore, a client needs to investigate the consistency to update the credibility, and the following four scenarios are defined to conduct the operation.

- If the interactive feedback is positive:

  case 1: $Pl\_P_m(T) \geq 0.5,\ Cred(c_i, w_k)_{t+1} = Cred(c_i, w_k)_t * (1 + \upsilon D_k);$

  case 2: $Pl\_P_m(T) < 0.5,\ Cred(c_i, w_k)_{t+1} = Cred(c_i, w_k)_t * (1 - \upsilon D_k);$

- If the interactive feedback is negative:

  case 3: $Pl\_P_m(T) \geq 0.5,\ Cred(c_i, w_k)_{t+1} = Cred(c_i, w_k)_t * (1 - \upsilon D_k);$

  case 4: $Pl\_P_m(T) < 0.5,\ Cred(c_i, w_k)_{t+1} = Cred(c_i, w_k)_t * (1 + \upsilon D_k).$

The expression D_k is the feedback difference, which can be calculated from two aspects. First, its testimony and actual feedback outcome. For instance, the evaluation of service provider s_j given by witness w_k is represented by $(m_{kj}^d(T), m_{kj}^d(nT), m_{kj}^d(T, nT))$, and the first aspect is represented by

$$D_k^1 = d(m_{kj}^d, bpa(outcome)) \tag{20}$$

where $d(m_{kj}^d, bpa(outcome))$ indicates the differences between the provided testimony and the feedback. It could also be interpreted as a result of dependence on the received information.

The second aspect is the conflict of the evidence in the group of L evidences. From our point of view, it is important to highlight these differences, especially in distributed MASs with deceptive agents, which may deliberately react to hide their actual properties. Thus, the second aspect is familiar with the definition with Equation (15), which can be stated as follows:

$$D_k^2 = \frac{\sum_{p=1, p \neq k}^{L} d(m_{kj}^d, m_{pj}^d)}{L - 1}. \tag{21}$$

In summary, D_k is updated with the equation as follows, where ζ indicates the weight of the evaluation difference and the feedback. In general, $\zeta > 0.7$ if multiagents do not change their behaviors frequently.

$$D_k = \zeta * D_k^1 + (1 - \zeta) * D_k^2. \tag{22}$$

4.5 Incentives Mechanism

One of the important challenges in MASs involving deceptive agents is to stimulate agents to bestow accurate information. In this paper, we are inspired by the credit mechanism to do this operation. Each agent in MASs has a fixed number of credits, whether it acts as a resource client, a witness, or a resource provider, which are managed and displayed by a blackboard. However, credits only make sense if an agent acts as a resource client or a witness. A witness is only allowed to give its testimony if the resource client has credits. At the same time, the witness may gain or lose credit depending on the interactive feedback from the resource client. As a resource client, once the credit is exhausted, the agent can no longer solicit any testimony from others because it acts as a witness. Certainly, it has the opportunity to encourage information by providing convincing information and accumulating credits.

We assume that gaining or losing credits is influenced by interactive feedback. The exact amount is determined by its original credit and the confidence value is decided by the credibility value and consistency. To explain the credit mechanism in detail, we use CR_i to represent the credit of agent c_i. We force the value to be in the range $[0, 2]$, and is initialized to 1. It is capable of demanding information unless CR_j is bigger than the threshold, for example, $CR_j > 0.2$. After the interaction, the credits of all witnesses are updated as follows:

1. If the interactive feedback matches the testimony, then

$$CR_i = CR_i[1 + Cred(c_i, w_k) * Cer(c_i, w_k)]$$

2. If the interactive feedback does not match the testimony, then

$$CR_i = CR_i[1 - Cred(c_i, w_k) * Cer(c_i, w_k)]$$

It is worth explaining what "match" means here. Namely, if the feedback is successful, then the witness supports the resource provider and the other way around as well. We set two bounds because the witness is not allowed to accumulate or lose much credits in a dynamic MAS. An agent with high credit loses credits rapidly if it keeps lying. Another advantage of our trust estimation model is that it does not require another "second chance" mechanism. Once a dishonest agent reaches the minimum level of credit, it understands that it can no longer demand information for trust estimation. Therefore, in order to maximize its own social welfare, it must provide honest information to gain credit.

5 Simulation Experiments

In order to test the performance of the proposed model, we carried out simulation experiments from the following aspects. First, we had to see if the BPA generation method is sensitive to an unexpected performance change. Then, we tested wheter the entropy-based certainty approach runs correctly and rationally. Finally, the assigned credibility method in the proposed method is compared with previous work to test its efficiency for trust estimation.

5.1 The Efficiency of The Proposed Direct Trust Generation Approach

The proposed model is designed to maintain the trust of agents, in which agents can react differently over time. That is to say, both resource provider agents and information provider agents (witnesses) can decide to turn from truth to malicious or the other way around, as stated in Section 1. As a result, the proposed method enables agents to notice the agility difference and give sharp responses when the agent changes its performance deliberately. This part mainly investigates whether the resource client agent can react violently when facing an unstable resource provider agent, namely, to investigate the efficiency of the proposed direct trust generation approach.

In the experiment we fix one resource provider agent and one resource client agent. In each round, the resource client agent requests some resources from the resource provider, and, correspondingly, the resource provider agent contributes its resources cooperatively. In addition, for the resource provider agent, we also fix the probability of allocating satisfied resources to the decimal number of P_s ($P_s \in [0, 1]$), which is unknown to the resource client agent. After a fixed number of interactions, for instance, N_{i1}, the resource provider agent decides to go from genuine to malicious (or from malicious to genuine). Therefore, we fix the probability of distributing satisfied resources to $1 - P_s$ correspondingly.

In each round, the client agent uses the transaction feedback to estimate the trust of the resource provider, and if the randomly generated number is smaller than P_s, the transaction feedback could be a success. Otherwise, interaction feedback is a failure. As a comparison, we made a comparison to[65], as they map the evidence (r, s) to the trust represented by $(\frac{r}{r+s+1}, \frac{s}{r+s+1}, \frac{1}{r+s+1})$ in terms of the absolute error; here, r and s indicate the number of success interactions and failed interactions, separately. In the case of a performance change we pay much attention to the estimated value of trust in the proposed method. Figure 5 presents the accumulation of absolute errors (five rounds starting with N_{i1}) because of the resource provider changing its performance.

We can see that the accumulative absolute errors obtained by the proposed method are smaller than the $\frac{r}{r+s+1}$ in Figure 5, no matter whether 10 or 100 rounds are conducted. In other words, the proposed method is sensitive to the sudden change of the resource providers. This improvement enables agents to manage different performance in a dynamic MAS.

Nonetheless, there are some issues to be highlighted in this experiment. First, we fixed the number of resource provider agents and resource client agents to one, which would also be appropriate for any other number of agents. We simplify the number of agents because we wish to check the performance facing sudden changes. Second, the proposed approach usually works well when the probability of presenting high-quality services is high or low (0.7 and 0.9). But if the resource provider agent acts mediocre (for example, $P_s =$

0.5), the estimated trust by[65] might have a better performance because it doesn't seem to make a difference when the agent decides to go from malicious ($P_s = 0.5$) to genuine ($1 - P_s = 0.5$). As a comparison, we run the test 1000 times, and we set $P_s = 0.5$ and $N_{i1} = 10$. We assume it is a win if the accumulated five-round absolute error by the proposed method is smaller than that of $\frac{r}{r+s+1}$ starting with N_{i1}. We received an average of 0.47 for winning the comparison, and the method in[65] wins by a narrow margin. However, the proposed method wins with 0.96 and 0.999 if $P_s = 0.7$ and $P_s = 0.9$, respectively. Nevertheless, rational agents tend to cooperate with the resource provider agents who perform well, rather than the averagely performing agents in real life. Therefore, this is why we focus on 0.1, 0.3, 0.7, and 0.9 to present satisfying resources in the test. In the following part, we discuss evidence certainty from the point of view of entropy.

5.2 Understand Trust Certainty with Entropy

In Section 4.1, we explained that satisfaction is mainly determined by evidence consisting of positive and negative feedback (or the evaluation) and fluctuation. We use the satisfactory degree of S_{ij} to represent the performance, which could also be understood as the number of positive and negative feedbacks represented by r and s, respectively. The related evaluation is used to show trust with the Dempster–Shafer evidence theory (Equations (8) and (11)). Many uncertainties caused by randomness, incompleteness, uncertainty, and dynamic environments have been solved. This direct trust is generally shared with the third-party resource clients to aid in trust estimation. However, the certainty of the evidence received or presented must be established. In this paper, we use the entropy of the Dempster–Shafer evidence theory to understand certainty.

In previous work, some literature assumes that certainty is independent of evidence[15], and a third-party agent provides evidence and the relevant certainty towards the evaluation. In terms of the evidence (r, s),[65] map this evidence to a trust represented by $(\frac{r}{r+s+1}, \frac{s}{r+s+1}, \frac{1}{r+s+1})$, and $\frac{1}{r+s+1}$ is designated as uncertainty. Yu and Singh used two parameters in the range of 0 and 1 to model all evidence (evaluations) to positive, negative, or neutral by the Dempster–Shafer evidence theory [27], and neutral is regarded as uncertainty. However, it does not reflect the conflicting evidence (the amount of positive or negative interactions) that increases uncertainty. Furthermore, the number of interactions has no effect on certainty. Subsequently in[66;20], certainty is a statistical measure defined by a probability–certainty density function. However, this transformation from evidence to trust neglects implementing a thorough analysis of uncertainty caused by a fading factor, lacking information, incompleteness, and randomness. Moreover, the above work was carried out under the condition that trustees have stable performances. We transform the evidence to trust, providing a comprehensive evaluation appropriate for dynamic agents' trust estimation, and the certainty of the evidence is well-studied due to the definition of entropy. We find that the certainty estimated by entropy accords well with the general properties listed in Section 4.2.3; more details are expressed in the following subsections.

5.2.1 Certainty Rises with Increasing Experiences under Fixed Conflict

Assuming that agent a_i perceives evidence with the fixed satisfactory degree S_{ij} towards agent a_j, we can determine how evidence certainty changes with an increasing amount of interaction. Logically, the certainty of evidence increases with the interaction. For instance, compare 3 positive feedback instances out of 5 with 12 positive feedback instances out of 20, and we consider that the satisfactory degree S_{ij} is the same ($\frac{3}{5} = \frac{12}{20} = 0.6$) in both cases. Nevertheless, it makes intuitive sense that the second case is more certain than the first one. Definition 12 defines the generated BPAs, and Equation (16) is employed to calculate the certainty with the entropy of the Dempster–Shafer theory of evidence. We have a certainty of 0.60 from $(3, 2)$, and as a comparison, certainty of 0.69 for $(12, 8)$. In more detail, Figure 6 shows how certainty changes with an increasing amount of interaction (x-axis: from 0 to 20) and the fixed satisfaction $S_{ij} = 0.5$.

As displayed in Figure 6, certainty increases with the amount of interaction when the satisfaction degree is fixed to 0.5, and it reaches a fixed value of 0.7 with sufficient interaction. Next, we examine how certainty changes with increasing conflict under a fixed number of interactions.

5.2.2 Certainty Rises with Increasing Conflict under Fixed Experience

Under the condition of a fixed interaction amount $(r + s)$, another characteristic of certainty is that the certainty degree is different from the satisfaction, which is also imperative. For example, two agents have interacted ten times, and the feedback $(10, 0)$ is indeed more specific (certain or confident) than that of $(5, 5)$. Therefore, certainty increases when the satisfaction degree approaches 1 or approaches 0, and it decreases when the satisfactory degree is close to 0.5. Figure 7 shows the relevant result.

Four interactions were conducted between agent a_i and agent a_j, and the number of positive feedbacks increases from 0 to 4 (the amount of negative feedbacks decreases from 4 to 0). Table 3 displays the certainty calculated by Yu and Singh[27], Jøsangsang et al.[65], and Wang and Singh[66;20].

Table 3: Certainty computed by different approaches for different satisfaction degrees with a fixed amount of interactions at 4.

	(4, 0)	**(3, 1)**	**(2, 2)**	**(1, 3)**	**(0, 4)**
Yu and Singh[27]	0	0	0	0	0
Jøsangsang et al.[65]	0.8	0.8	0.8	0.8	0.8
Wang and Singh[66;20]	0.54	0.35	0.29	0.35	0.54
Proposed entropy-based approach	0.69	0.62	0.59	0.62	0.69

5.2.3 Comparison and Discussion

This paper discusses trust certainty from the perspective of evidence entropy. The previous literature studied certainty from the observed trust, namely, the amount of positive or negative feedback. For example, ref.[66;20] analyzed certainty from the point of view of probability–certainty density function[65], and the estimated certainty was used to generate evidence. That is to say, a piece of trust represented by belief, disbelief, and the unknown should come from the amount of positive and negative feedbacks, and the estimated uncertainty will be employed to gain the trust of the unknown. We offer another perspective on the use of certainty; we first analyzed a variety of conditions that lead to uncertainty, including lacking certainty, randomness, incompleteness, and varying performances. According to the Dempster–Shafer evidence theory BPAs could represent this observation. However, how much we could depend on (or trust) the generated evidence when making decisions is a great challenge, and an important aspect of this is the certainty of the evidence estimated by entropy.

Shannon entropy[61] is a well-known information-theoretic measure of uncertainty that is based on a discrete probability distribution over a finite set of alternatives. However, entropy is said to be inappropriate for certainty estimation of the observed evidence represented by the binary event (r, s)[66;20]. On the one hand, Shannon entropy's value ranges from 0 to ∞, which does not fit our intuition. On the other hand, the confidence placed in a probability estimation is required. Finally, the defined Shannon entropy cannot be used to capture the probability estimation's uncertainty represented by binary events (positive and negative feedback). In this paper, we used entropy to estimate the certainty of the observed evidence represented by BPAs, and this certainty is an important component of making full use of BPAs. First, entropy models bits of missing information, which can also be interpreted as the information provided by the evidence. Then more information evidence transmits, making it more valuable. Second, the estimated entropy value is in the range of 0 to 1, which can be used to represent the evidence's certainty. Last but most importantly, in our view, the uncertainty of trust arising from two events (r, s) consists of two main aspects: the conflict between each judgment (conflict between r and s) and the uncertain part of the judgment (assigned to uncertainty). Then the entropy definition (Definition 6)[60] is designed to capture the certainty of BPAs from these two aspects, which are extremely suitable for learning uncertainty. Furthermore, as shown in this section, the comparison results have proved the two fundamental properties of certainty. That is to say, certainty rises with increasing experiences beyond a fixed conflict (Section 5.2.1), and uncertainty rises with the increasing conflict under a fixed amount of interactions (Section 5.2.2). As a result, according to the Dempster–Shafer evidence theory the entropy is suitable for measuring the uncertainty of the generated evidence.

5.3 The Proposed Overall Trust Model

In this section, a simulation test is designed to estimate the proposed trust model chosen by the service provider in cloud manufacturing. The test triggers interactions between different cloud manufacturing agents, each of which would evaluate the different cloud manufacturing providers and filter with which agent to interact.

The simulation is assumed to have $Wnum$ cloud manufacturing client agents and $Pnum$ cloud manufacturing provider agents. After the interaction is initiated, each client first interacts with each of the providers and collects and records the feedback three times. This information will later be used as evidence and shared with other information demanders. Simulation returns show us how the system can adapt to fluctuations in the behavior of information providers using our model.

At the beginning of the simulation, assume that all cloud manufacturing clients are honest. That is, once the other cloud manufacturing client agents demand recommendations, one honest agent would contribute whatever it has to help the demander make the right decision. However, after several interactions $Cn1$ (for example, 30 interactions), some of the agents (for instance, the client agents a_1 and a_2) become malicious and entirely dishonest in sharing information. Namely, they would present excellent evaluations of agents performing badly and vice versa. Using the proposed method, malicious information providers are identified, and their average credibility decreases gradually, which implies that their contribution or recommendation will have a lower impact on the final decision. While these malicious agents are always offering recommendations, a rational decision maker would never trust or rely on these recommendations to make decisions. After a while (for instance, at $Cn2$ rounds and $Cn2 = 70$), the malicious agents decide to become normal and contribute honest information again. When they decides to change their performance, the average credibility increases and their contributions become more influential. Simulation results and figures are discussed in detail in the next paragraph. Table 4 explains the detailed parameters in the simulation. As we can see, there are 10 resource client agents and 25 service provider agents. All agents in the first 30 rounds ($Cn1 = 30$) act honestly. From round 30, some agents become dishonest, sharing evidence that was fake, and later they become honest again ($Cn2 = 70$). In each round, a client agent may decide to initiate interaction with a provider with a 50% probability. In each round one of the ten client agents is appointed as the evaluator, and seven agents ($Snum = 7$) are selected to share their information, and the evaluator may be included in the information providers; $Cnum$ best-evaluated service providers are selected to carry interactions. The evaluation process is conducted 200 times, and each of the evaluations is administered ten times. We get the same results, and Figure 8 is used to present the results of the clients' average credibility.

Figure 8 presents the average credibility of the cloud manufacturing service clients, and one of which, defined as a_1, coloured in blue, changes its performance from honest to malicious or the other way around, in rounds $Cn1$ and $Cn2$, respectively. The agent's credibility is updated at interaction 30 ($Cn1 = 30$), where it is redefined as a malicious information provider. As a result, we can recognize that its average credibility is reduced from the point of view of other customers, and therefore, rational information demanders decrease the confidence level in the provided information. After forty rounds, that malicious agent reverts to being honest and begins to tell the truth. As a result, its credibility is gradually rebuilt, and the other agents begin to trust its recommendations. However, we also find that the average credibility declines quickly (in approximately 40 rounds, it decreases from 0.75 to 0.15) and rises very slowly (it takes approximately 130 rounds to increase from 0.15 to 0.7), reflecting the fact that trust is easily lost and difficult to achieve.

More specifically, we find that only one cloud manufacturing agent out of ten information provider agents has changed its performance in round 30; that is, most agents also provide right information to facilitate making the right decisions. As mentioned earlier, some trust estimation models assume that most information providers are honest and provide right information[2]. Thus, in round 30, we force five honest cloud manufacturing agents to become malicious and start sharing incorrect information. Figure 9 presents the result. As is shown, the average credibility of each of the five agents (coloured in dark blue, pink, red, purple and light green) all decreased from round 30 and increased at round 70. That is, a rational evaluator can distinguish honest information providers who are no longer trustworthy if they consistently tell lies. However, if they provide right information, their information provider would become valuable again.

As a comparison, in the application of the Dempster–Shafer evidence theory to trust estimation in a system of multiagents, ref.[28;29] take personal credibility into consideration. However, in[28], we do not discuss the appropriateness of the BPAs' generation, especially in a dynamic and open MAS; we are concerned with updating the weights of third-party agents. As explained, the weight is updated by $w_i^{'} = \theta * w_i$, where w_i is the agent's weight, and $\theta < 1$. Therefore, the credibility of third-party agents decreases gradually.

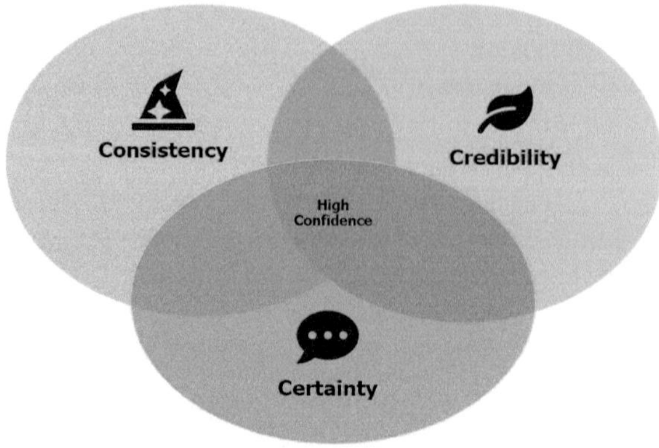

Figure 4: Intergrated confidence.

Table 4: Parameters in the simulation.

Parameters	Value	Explanation
Wnum	10	Number of cloud manufacturing client agents
Pnum	25	Number of cloud manufacturing provider agents
Snum	7	Number of selected agents to share information
Cnum	3	Number of selected provider agents to conduct interactions
*Cn*1	30	From when some agents become malicious
*Cn*2	70	From when the malicious return to being honest
Rounds	200	Total round

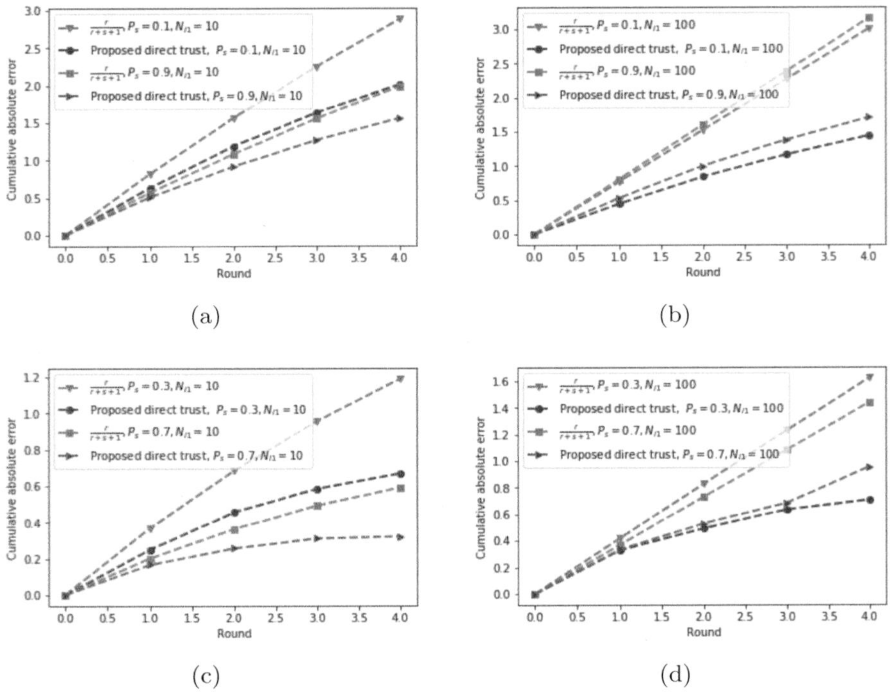

Figure 5: The recent five-round cumulative absolute error, obtained in different interactive rounds of changing performance, under different probabilities: (**a**): 10 rounds and the probability equals 0.1 or 0.9; (**b**): 100 rounds and the probability equals 0.1 or 0.9; (**c**): 10 rounds and the probability equals 0.3 or 0.7; (**d**): 100 rounds and the probability equals 0.3 or 0.7.

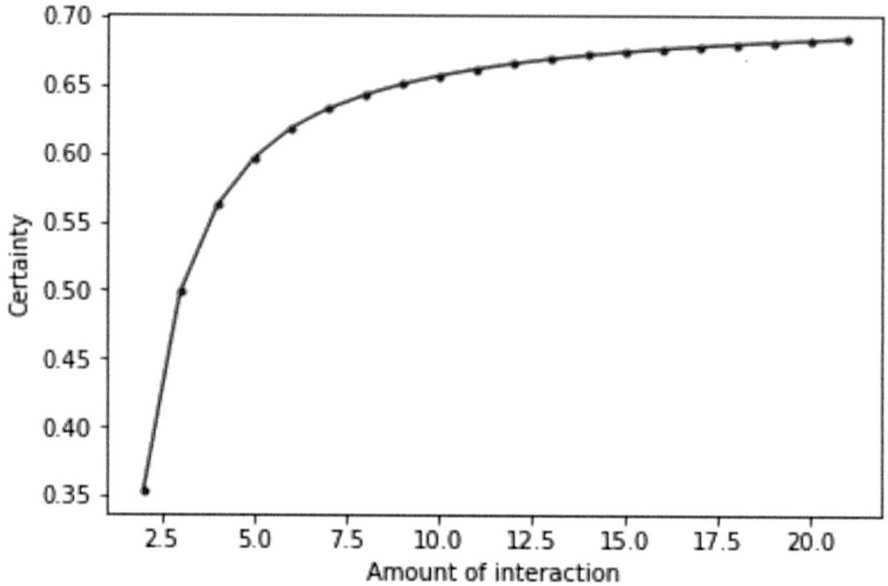

Figure 6: Evidence certainty increases with the amount of interaction $(r + s)$ when satisfaction 0.5 is fixed; x-axis: amount of interaction; y-axis: certainty degree.

As a result, in an open and dynamic system this seems inappropriate. In[29], the agents' weight is defined as increasing gradually to 1 when receiving positive feedback; in contrast, for negative feedback, the weight decreases linearly. Although evidence consistency and evidence certainty, are not considered in this literature, we also run simulations in the same test explained in Table 4. Figure 10 presents the average credibility of each information provider from the other perspective. As we can see, even when agent a_1(coloured in blue) chose to provide dishonest information from round 30 to round 70, its average credibility increased. In other words, no evaluator has recognized this cheater. The main reason for gaining increasing average credibility is that most honest agents share correct information that has helped the decision maker choose the well-evaluated service provider. As a comparison, Figure 11 presents the average credibility change when the five agents (coloured in dark blue, pink, red, purple and light green) are set to vary their performance in round 30 and round 70.

As Figure 11 shows, all agents' average credibility dropped from round 30 to round 70. In other words, during this period, it is difficult for evaluators to select service providers performing well and to identify dishonest information providers.

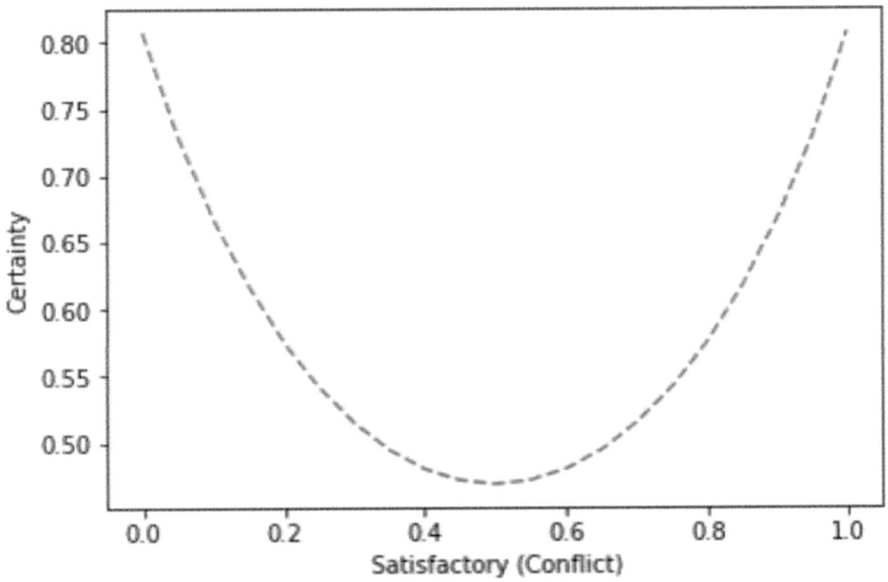

Figure 7: Evidence certainty varies with satisfaction when the amount of interaction 20 is fixed, and the satisfaction degree 0.5 leads to the lowest certainty.

6 Conclusions and Future Work

Introducing trust into the MASs for decision making eliminates many of the untrustworthy agents and significantly improve decision-making efficiency. In this paper, we proposed a trust estimation model in MASs, and its main contributions can be summarized as follows. First, direct trust resolves uncertainty caused by many factors, and it enables agents to react quickly to unexpected changes in performance. Next, we introduced rational confidence in the accepted testimony in three aspects: consistency, which reflects the quality of the evidence; credibility, indicating its reliability in the light of past experience; and certainty of the testimony estimated from the perspective of entropy. In addition, we used a credit mechanism to incentivize agents to share truthful information. These aspects are essential for trust estimation in dynamic MASs with deceptive and unstable agents. Simulation results show that: (1) the proposed direct trust has fast intuition regarding agents' different performances, which fits dynamic MASs correctly; (2) the entropy-based certainty evaluation is correct and reasonable; (3) the proposed method is efficient in dynamic trust management in MASs. Simulations in some industrial applications are demonstrated to show its efficiency. The proposed method also has limitations. We assume that the agent will have the same

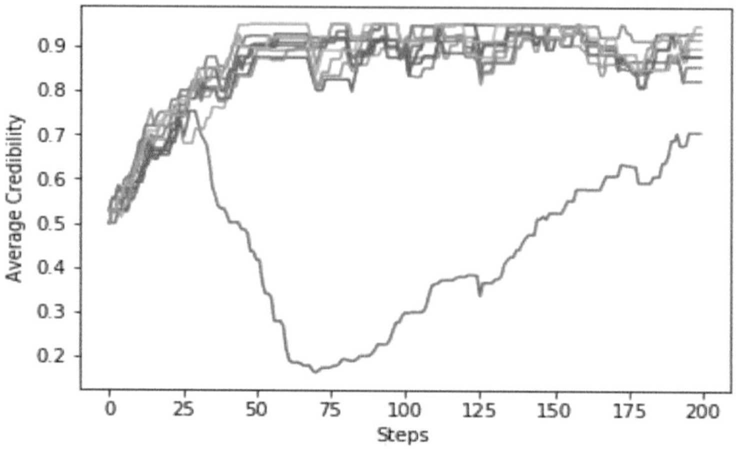

Figure 8: Change in average credibility for each information sharing agent.

attitude towards all agents. That is, if an agent is honest, it will treat all agents honestly. In addition, we compare our methods separately with previous literature. Some examples of actual implementation may also be helpful. In future work, we will improve this aspect and apply the trust model to cloud manufacturing.

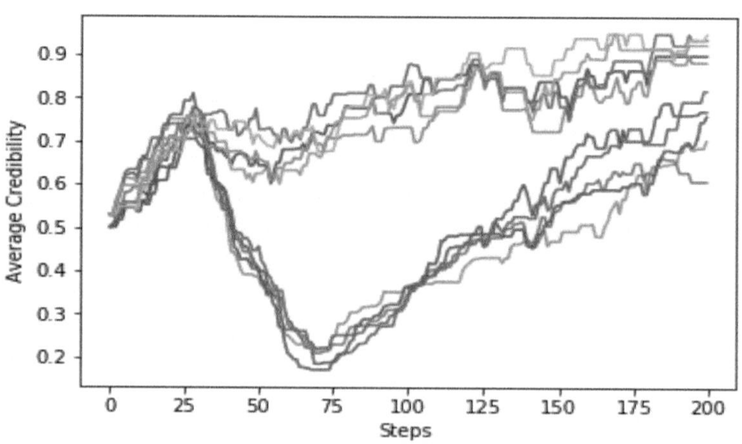

Figure 9: Change in average credibility for each information sharing agent when five agents change their performance.

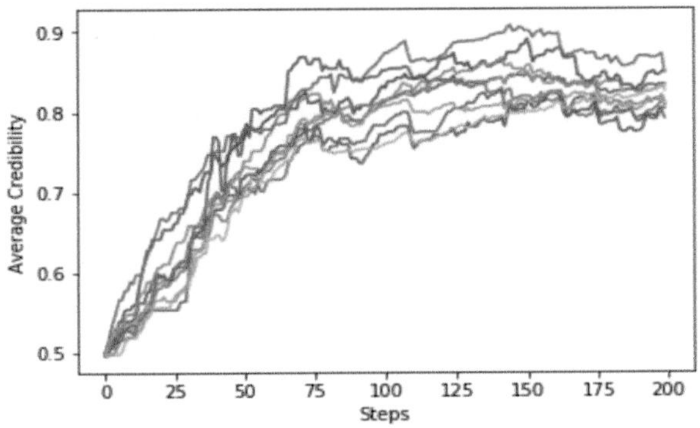

Figure 10: Change in average credibility for each information sharing agent when one agent changes its performance; the change process is as defined in[29].

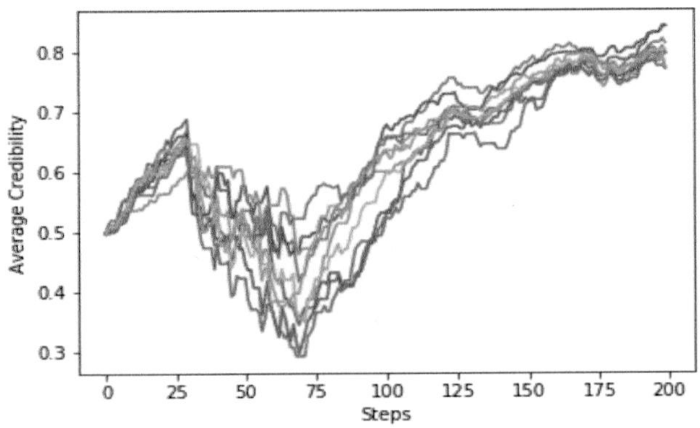

Figure 11: Change in average credibility for each information sharing agent when five agents change their performances; the change process is as defined in[29].

References

[1] C.-J. Liau, Belief, information acquisition, and trust in multi-agent systems—a modal logic formulation, Artificial Intelligence 149 (2003) 31–60.

[2] A. Josang, R. Ismail, The beta reputation system, in: Proceedings of the 15th bled electronic commerce conference, volume 5, Citeseer, pp. 2502–2511.

[3] C. Burnett, N. Oren, Sub-delegation and trust, in: Proceedings of the 11th International Conference on Autonomous Agents and Multiagent Systems - Volume 3, AAMAS '12, International Foundation for Autonomous Agents and Multiagent Systems, Richland, SC, 2012, p. 1359–1360.

[4] H. Yu, Z. Shen, C. Leung, C. Miao, V. R. Lesser, A survey of multi-agent trust management systems, IEEE Access 1 (2013) 35–50.

[5] A. Jøsang, R. Ismail, C. Boyd, A survey of trust and reputation systems for online service provision, Decision Support Systems 43 (2007) 618–644. Emerging Issues in Collaborative Commerce.

[6] R. Falcone, G. Pezzulo, C. Castelfranchi, G. Calvi, Why a cognitive trustier performs better: Simulating trust-based contract nets, in: Proceedings of the Third International Joint Conference on Autonomous Agents and Multiagent Systems - Volume 3, AAMAS '04, IEEE Computer Society, USA, 2004, p. 1394–1395.
[7] M. Peng, Z. Xu, S. Pan, R. Li, T. Mao, Agenttms: A mas trust model based on agent social relationship., J. Comput. 7 (2012) 1535–1542.
[8] A. Ferrario, M. Loi, E. Viganò, In ai we trust incrementally: a multi-layer model of trust to analyze human-artificial intelligence interactions, Philosophy & Technology 33 (2019) 523 – 539.
[9] W. T. L. Teacy, J. Patel, N. R. Jennings, M. Luck, Travos: Trust and reputation in the context of inaccurate information sources, Autonomous Agents and Multi-Agent Systems 12 (2006) 183–198.
[10] S. Jiang, J. Zhang, Y.-S. Ong, An evolutionary model for constructing robust trust networks, in: Proceedings of the 2013 International Conference on Autonomous Agents and Multi-Agent Systems, AAMAS '13, International Foundation for Autonomous Agents and Multiagent Systems, Richland, SC, 2013, p. 813–820.
[11] W. L. Teacy, M. Luck, A. Rogers, N. R. Jennings, An efficient and versatile approach to trust and reputation using hierarchical bayesian modelling, Artificial Intelligence 193 (2012) 149–185.
[12] E. Parhizkar, M. H. Nikravan, S. Zilles, Indirect trust is simple to establish, in: Proceedings of the 28th International Joint Conference on Artificial Intelligence, IJCAI'19, AAAI Press, 2019, p. 3216–3222.
[13] S. P. Marsh, Formalising trust as a computational concept (1994).
[14] J. M. Reagle, Trust in a cryptographic economy and digital security deposits: Protocols and policies, Ph.D. thesis, Massachusetts Institute of Technology, 1996.
[15] G. S. Basheer, M. S. Ahmad, A. Y. Tang, S. Graf, Certainty, trust and evidence: Towards an integrative model of confidence in multi-agent systems, Computers in Human Behavior 45 (2015) 307–315.
[16] G. Muller, L. Vercouter, O. Boissier, Towards a general definition of trust and its application to openness in mas, in: Proc. of the AAMAS-2003 Workshop on Deception, Fraud and Trust.
[17] R. Feng, X. Xu, X. Zhou, J. Wan, A trust evaluation algorithm for wireless sensor networks based on node behaviors and ds evidence theory, Sensors 11 (2011) 1345–1360.
[18] R. Urena, G. Kou, Y. Dong, F. Chiclana, E. Herrera-Viedma, A review on trust propagation and opinion dynamics in social networks and group decision making frameworks, Information Sciences 478 (2019) 461–475.
[19] M. Cheng, C. Yin, J. Zhang, S. Nazarian, J. Deshmukh, P. Bogdan, A general trust framework for multi-agent systems, in: Proceedings of the 20th International Conference on Autonomous Agents and MultiAgent Systems, pp. 332–340.

[20] Y. Wang, M. P. Singh, Formal trust model for multiagent systems., in: IJCAI, volume 7, pp. 1551–1556.
[21] C. J. Fung, J. Zhang, I. Aib, R. Boutaba, Dirichlet-based trust management for effective collaborative intrusion detection networks, IEEE Transactions on Network and Service Management 8 (2011) 79–91.
[22] E. Parhizkar, M. H. Nikravan, R. C. Holte, S. Zilles, Combining direct trust and indirect trust in multi-agent systems., in: IJCAI, pp. 311–317.
[23] G. Fortino, L. Fotia, F. Messina, D. Rosaci, G. M. Sarné, Trust and reputation in the internet of things: State-of-the-art and research challenges, IEEE Access 8 (2020) 60117–60125.
[24] D. Shehada, C. Y. Yeun, M. J. Zemerly, M. Al-Qutayri, Y. Al-Hammadi, J. Hu, A new adaptive trust and reputation model for mobile agent systems, Journal of Network and Computer Applications 124 (2018) 33–43.
[25] H. R. Berenji, Treatment of uncertainty in artificial intelligence (1988).
[26] K. S. Barber, K. Fullam, J. Kim, Challenges for trust, fraud and deception research in multi-agent systems, in: Proceedings of the 2002 International Conference on Trust, Reputation, and Security: Theories and Practice, AAMAS'02, Springer-Verlag, Berlin, Heidelberg, 2002, p. 8–14.
[27] B. Yu, M. P. Singh, Distributed reputation management for electronic commerce, Computational Intelligence 18 (2002) 535–549.
[28] B. Yu, M. P. Singh, Detecting deception in reputation management, in: Proceedings of the second international joint conference on Autonomous agents and multiagent systems, pp. 73–80.
[29] B. Yu, S. Kallurkar, R. Flo, A demspter-shafer approach to provenance-aware trust assessment, in: 2008 International Symposium on Collaborative Technologies and Systems, IEEE, pp. 383–390.
[30] Y. Zuo, J. Liu, A reputation-based model for mobile agent migration for information search and retrieval, International Journal of Information Management 37 (2017) 357–366.
[31] S. Ramchurn, C. Sierra, L. Godó, N. R. Jennings, A computational trust model for multi-agent interactions based on confidence and reputation (2003).
[32] A. Das, M. M. Islam, Securedtrust: A dynamic trust computation model for secured communication in multiagent systems, IEEE transactions on dependable and secure computing 9 (2011) 261–274.
[33] A. Bilgin, J. Dooley, L. Whittington, H. Hagras, M. Henson, C. Wagner, A. Malibari, A. Al-Ghamdi, M. J. Alhaddad, D. Alghazzawi, Dynamic profile-selection for zslices based type-2 fuzzy agents controlling multi-user ambient intelligent environments, in: 2012 IEEE International Conference on Fuzzy Systems, IEEE, pp. 1–8.
[34] M. Fraser, How to measure anything: Finding the value of" intangibles" in business, People & Strategy 34 (2011) 58–60.

[35] T. D. Huynh, N. R. Jennings, N. Shadbolt, Developing an integrated trust and reputation model for open multi-agent systems (2004).
[36] Z. Noorian, M. Ulieru, The state of the art in trust and reputation systems: a framework for comparison, Journal of theoretical and applied electronic commerce research 5 (2010) 97–117.
[37] H. Yu, Z. Shen, C. Miao, B. An, C. Leung, Filtering trust opinions through reinforcement learning, Decision Support Systems 66 (2014) 102–113.
[38] G. Rishwaraj, S. Ponnambalam, L. C. Kiong, An efficient trust estimation model for multi-agent systems using temporal difference learning, Neural Computing and Applications 28 (2017) 461–474.
[39] A. Dempster, Upper and lower probabilities induced by a multivalued mapping, The Annals of Mathematical Statistics 38 (1967) 325–339.
[40] G. Shafer, A mathematical theory of evidence, volume 42, Princeton university press, 1976.
[41] L. Sun, R. P. Srivastava, T. J. Mock, An information systems security risk assessment model under the dempster-shafer theory of belief functions, Journal of Management Information Systems 22 (2006) 109–142.
[42] Z. Li, G. Wen, N. Xie, An approach to fuzzy soft sets in decision making based on grey relational analysis and dempster–shafer theory of evidence: An application in medical diagnosis, Artificial intelligence in medicine 64 (2015) 161–171.
[43] M. Liu, S. Chen, Sar target configuration recognition based on the dempster–shafer theory and sparse representation using a new classification criterion, International Journal of Remote Sensing 40 (2019) 4604–4622.
[44] K. Wang, A new multi-sensor target recognition framework based on dempster-shafer evidence theory, International Journal of Performability Engineering 14 (2018) 1224.
[45] A.-L. Jousselme, D. Grenier, É. Bossé, A new distance between two bodies of evidence, Information fusion 2 (2001) 91–101.
[46] C. Wen, Y. Wang, X. Xu, Fuzzy information fusion algorithm of fault diagnosis based on similarity measure of evidence, in: Advances in Neural Networks-ISNN 2008: 5th International Symposium on Neural Networks, ISNN 2008, Beijing, China, September 24-28, 2008, Proceedings, Part II 5, Springer, pp. 506–515.
[47] B. Ristic, P. Smets, The tbm global distance measure for the association of uncertain combat id declarations, Information fusion 7 (2006) 276–284.
[48] F. Cuzzolin, A geometric approach to the theory of evidence, IEEE Transactions on Systems, Man, and Cybernetics, Part C (Applications and Reviews) 38 (2008) 522–534.
[49] D. Harmanec, G. J. Klir, Measuring total uncertainty in dempster-shafer theory: A novel approach, International journal of general system 22 (1994) 405–419.

[50] K. Yao, H. Ke, Entropy operator for membership function of uncertain set, Applied Mathematics and Computation 242 (2014) 898–906.
[51] A. Lesne, Shannon entropy: a rigorous notion at the crossroads between probability, information theory, dynamical systems and statistical physics, Mathematical Structures in Computer Science 24 (2014) e240311.
[52] U. Höhle, Entropy with respect to plausibility measures, in: Proc. of 12th IEEE Int. Symp. on Multiple Valued Logic, Paris, 1982.
[53] P. Smets, Information content of an evidence, International Journal of Man-Machine Studies 19 (1983) 33–43.
[54] R. R. Yager, Entropy and specificity in a mathematical theory of evidence, Classic works of the Dempster-Shafer theory of belief functions (2008) 291–310.
[55] H. T. Nguyen, On entropy of random sets and possibility distributions, The Analysis of Fuzzy Information 1 (1987) 145–156.
[56] D. Dubois, H. Prade, Properties of measures of information in evidence and possibility theories, Fuzzy sets and systems 24 (1987) 161–182.
[57] Y. Deng, Deng entropy, Chaos, Solitons & Fractals 91 (2016) 549–553.
[58] K. Özkan, Comparing shannon entropy with deng entropy and improved deng entropy for measuring biodiversity when a priori data is not clear, Forestist 68 (2018) 136–140.
[59] H. Cui, Q. Liu, J. Zhang, B. Kang, An improved deng entropy and its application in pattern recognition, IEEE Access 7 (2019) 18284–18292.
[60] R. Jiroušek, P. P. Shenoy, A new definition of entropy of belief functions in the dempster–shafer theory, International Journal of Approximate Reasoning 92 (2018) 49–65.
[61] C. E. Shannon, A mathematical theory of communication, The Bell system technical journal 27 (1948) 379–423.
[62] B. R. Cobb, P. P. Shenoy, On the plausibility transformation method for translating belief function models to probability models, International journal of approximate reasoning 41 (2006) 314–330.
[63] B. Yu, M. P. Singh, K. Sycara, Developing trust in large-scale peer-to-peer systems, in: IEEE First Symposium onMulti-Agent Security and Survivability, 2004, IEEE, pp. 1–10.
[64] Y. Deng, A threat assessment model under uncertain environment, Mathematical Problems in Engineering 2015 (2015).
[65] A. Jøsang, A subjective metric of authentication, in: Computer Security—ESORICS 98: 5th European Symposium on Research in Computer Security Louvain-la-Neuve, Belgium September 16–18, 1998 Proceedings 5, Springer, pp. 329–344.
[66] Y. Wang, M. P. Singh, Trust via evidence combination: A mathematical approach based on certainty, Technical Report, North Carolina State University. Dept. of Computer Science, 2006.

12. Optimization Model of Raw Material Selection for Construction Material Manufacturing

Xiuli He, Rongyuan Wang, Mengna Hong and Zhenglei He*

1 Introduction

Industrial production is a series of physical or chemical processes in which different types of raw materials are crushed, mixed, reacted, separated and heated[1]. Raw material selection is a necessary process before production, which aims to bring the quality indicators of raw material products and ingredient products up to technological standards. An accurate and efficient raw material selection process not only helps to improve the quality of the finished product and production efficiency, but also has important implications for energy saving and consumption reduction in the subsequent production process. For example, the ratio of long and short fiber in the papermaking process has an important effect on the power control of the refiner, the amount of chemicals added and the quality of the finished paper[2]. In the process of Portland cement production, the ratio of lime and clay has an important effect on the energy consumption of the rotary kiln[3]. In the process of thermal power generation, the ratio of single coal has an important effect on improving boiler combustion efficiency and reducing pollutant emission[4]. However, the quality of different batches of raw materials is uneven due to different sources, mining and processing levels, which negatively affects the raw material selection process. In production, the differences in raw materials are eliminated through the process of batch combination and formula calculation, so that the quality indicators of raw material products and ingredient products are as close as possible to the requirements. Operators determine the final raw mate-rial selection results through repeated testing, optimization and identification

State Key Laboratory of Pulp and Paper Engineering, South, China University of Technology, Guangzhou 510640, China.
*Corresponding author: hezhenglei@scut.edu.cn

steps, a tedious and time-consuming process that often leads to unsatisfactory results. Therefore, how to resolve the contradiction between natural raw materials and production requirements and improve the efficiency of the raw material selection process has become an urgent problem to be addressed in industrial production.

Existing literature reports on the optimization of raw material selection processes, mainly focusing on the food and construction sectors. On the premise of meeting the production constraints,[5] used the variable scale ant colony algorithm to optimize the beer formula, and achieved the lowest total cost of raw materials.[6] used the principle of fuzzy mathematics to abstract and quantify the sensory quality of corn wine, and established the regression model between formula and sensory quality.[7] analyzed the main nutrients of 15 kinds of extrudated reconstituted rice with rice flour, wheat flour and oat flour in different proportions. The results showed that when the ratio of rice flour: wheat flour: oat flour = 4.5:4.5:1, the sensory quality of rice was the best.[8] used projection pursuit regression model to predict the relationship between mortar ratio and compressive strength and hydration heat of LHP cement, and determined the mortar ratio under the best performance.[9] optimized the mixture ratio of raw materials for asphalt production through drainage test and Cantabro test, and the results showed that the optimal mixture ratio of PAC-10 and PAC-16 was 4.7% and 4.1-4.4%, respectively.[10] proposed a MOO method based on machine learning and meta-heuristic algorithm to optimize the concrete mix ratio.[11] established an artificial neural network model to analyze the influence of heavy clay composition on the important properties of clay brick. The results have shown that the clay brick with heavy clay oxide content of 53-66% SiO_2, 4.6-7.5% Fe_2O_3, 12.5-18.2% Al_2O_3, 0.9-8.8% CaO and 1.2-3.6% MgO has the best properties. To improve the mechanical properties of magnesium oxychloride cement (MOC),[12] used response surface methodology to optimize the ratio of raw materials. The results show that when $MgO{:}MgCl_2{:}H_2O = 7.56{:}1{:}13.51$, the compressive strength of MOC cement is the highest.[13] tested the combustion performance of the finished product of tobacco pole-corn stalk biomass pellet fuel in different proportions, and found that when the mass ratio was 7:3, the comprehensive combustion performance of the finished product of tobacco pole-corn stalk biomass pellet fuel was the best.[14] studied the mechanism rule of ceramist and used it to guide the optimization of the formula for preparing ultra-light ceramist from bauxite tailings, and obtained ultra-light ceramist products with excellent performance.[15] studied the effect of chemical composition on the soft magnetic properties and structure of amorphous $Fe_a.Ni_b.Nb_c.Zr_d$ alloy. The experimental results show that the soft magnetic properties of the alloy are the best when the atomic percentages of Fe, Ni, Nb and Zr are 31.3, 48.7, 0 and 20 respectively.

The above studies are mainly based on experimental or model simulation approaches to optimize the raw material selection process, but industrial production still faces the following problems:

(1) Raw material products are mixed from different batches of raw materials. When choosing a batch, the operator should consider not only the current combination, but also the rationality of the subsequent combinations. Otherwise, as production progresses, combinations that satisfy the requirements can no longer be selected. Due to the large number of inventory batches, the total number of combinations is large and its computational complexity can far exceed the computational capacity of human knowledge expression.

(2) The ingredient is composed by a variety of raw materials. In production, the quality parameters of the constituent products are required to float within a very small range of standard values. Due to the high dimensionality of the parameters with high coupling relations, manual calculations are nearly impossible and hence the feed weights of the raw material products in the present formulation are always unreasonable. It significantly retarded the production efficiency of the production process.

In this work, a batch combination model and a formula computation model are developed to solve the optimization problem of the raw material selection process in industrial production, respectively. The batch combination model is constrained by the quality indicators of the raw material products, and the logic combination algorithm is designed to solve the rational combination under the current inventory. The formula calculation model is constrained by the quality indicators of ingredient product, and solved based on the improved sequential quadratic programming (SQP). Finally, raw material data from the glass fabrication process is used to test the two models. From the point of view of optimizing batch composition and formulation, the model developed in this work can enable the intellectualization of the raw material selection process and provide support for improving production efficiency and product quality, as well as energy conservation and emission reduction in the production process.

2 Methodology

Due to the different sources, mining and processing levels, the quality indicators of different batches of raw materials are different. In production, batch combination is used to make the quality indicators of raw material product meet the requirements. Therefore, the batch combination model solves a discrete combination problem with multiple constraints. In industrial production, ingredient product is made from a mixture of various raw material products and used for subsequent production. The formula calculation model needs to solve the reasonable feeding weight of each raw material product according to the requirements of all the quality indicators of ingredient product. Therefore, formula calculation model solves a complex set of inequalities.

2.1 Raw material selection process in industrial production

Most of the raw materials used in industrial production are natural, used after mining and processing. For example, the raw materials of paper industry are mainly trees, straw and bamboo[16]. The raw materials of glass industry are silica sand, dolomite and mirabilite, among others.[17]. Figure 1 is a schematic diagram of the raw material selection process in industrial production. Firstly, according to the requirements of quality indicators, a number of batches of raw materials are selected from the inventory and mixed into raw material products. After each raw material product meets the requirements, the reasonable weight of each raw material product is calculated according to the quality indicators requirements of the ingredient product, and the raw material product is mixed to get the ingredient product.

2.2 Technical route

In this work, the batch combination and formula calculation models are established respectively to solve the optimization problem of the raw material selection process in industrial production. Fig 2 shows the overall technical route. The establishment of the two models is based on the quality parameters of raw materials and the constraints of product quality indicators. Therefore, relevant data should be collected before modeling. The quality indicators of ingredient product are calculated by the quality indicators of raw material products through certain calculation methods, so it is necessary to have a clear understanding of the calculation logic of the process. Then, according to the collected data and process calculation logic, the batch combination model and formula calculation model are established successively. Finally, the collected data are used to test the two models.

3 Modeling

3.1 The batch combination model

Raw material quality indicators vary slightly from batch to batch. In production, to simplify the assembly process, different batches of raw materials are fed the same weight. Therefore, there is no need to consider the feed weight of each batch of raw materials, but only the quality indicators when establishing

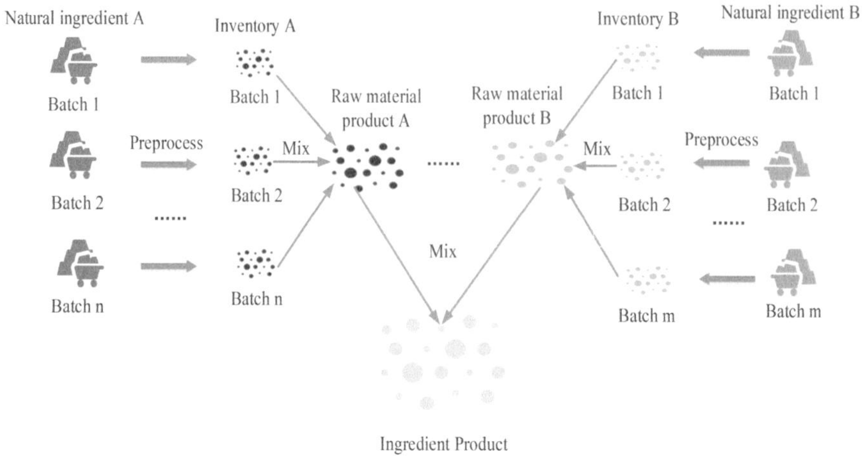

Figure 1: Raw material selection process of industrial production

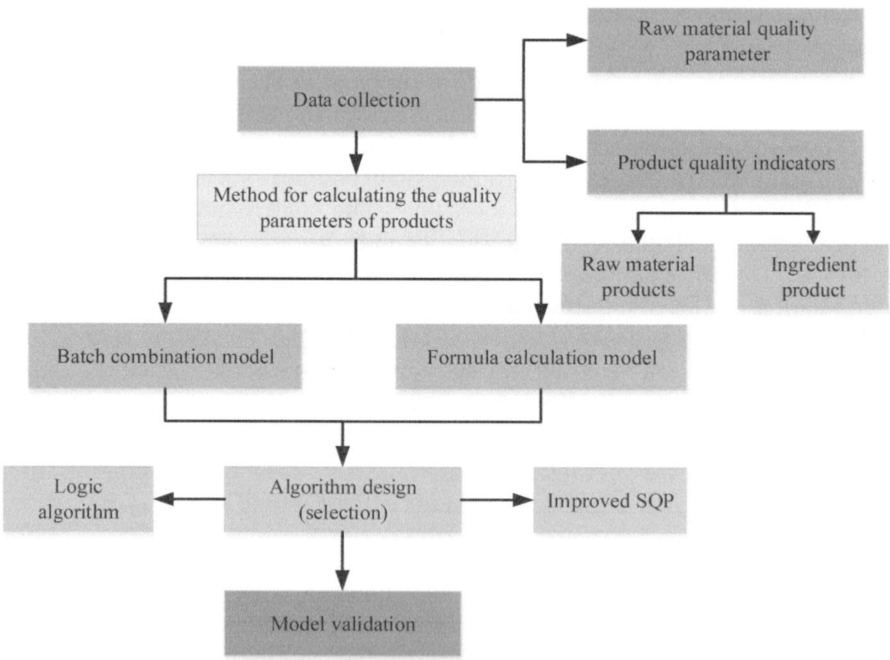

Figure 2: Overall technical route

the batch combination model. The calculation method of quality indicators of raw material product is as follows:

$$\bar{J} = \frac{1}{n}\sum_{i=1}^{n} J_i \tag{1}$$

Where, $\bar{J}$ represents the content of component J of raw material product, %; J_i represents the content of component J in batch i, %; n represents the number of batches contained in a combination. $\bar{J}$ needs to satisfy the following constraint:

$$\bar{J} = \frac{1}{n}\sum_{i=1}^{n} J_i J_{\text{stad}} \times (1 - \text{error}_J) \leq \bar{J} \leq J_{\text{stad}} \times (1 + \text{error}_J) \tag{2}$$

where, J_{stad} represents the standard value of $\bar{J}$, %; $error_J$ represents the maximum permissible percentage fluctuation of $\bar{J}$ allowed in production, %.

3.2 The formula calculation model

Ingredient products are composed of several raw material products, and the production has strict requirements on the quality indicators (such as total weight, and component content) The calculation method of the total weight of ingredient product is as follows:

$$M_{\text{sum}} = \sum M_q \tag{3}$$

where, M_{sum} represents the total weight of ingredient product, kg; M_q represents the feed weight of raw material product q, kg.

The calculation method of the content of each component in the ingredient product is as follows:

$$\text{pro}_k = \frac{ec_p \cdot m_k}{m_{sum}} \times 100\% \tag{4}$$

$$m_k = \sum M_q \cdot \text{con}_{qk} \tag{5}$$

where, pro_k represents the content of component k in ingredient product, %; ec_p represents the empirical coefficient and is a constant; m_{sum} represents the total weight of all components in the ingredient product, kg; m_k represents the weight of component k in the ingredient product, kg; con_{qk} represents the content of component k in the raw material product q, %.

In production, the above quality indicators are required to fluctuate within a certain range above and below the standard values, as shown below:

$$M_{stad} - M_{al} \leq M_{sum} \leq M_{stad} + M_{al} \tag{6}$$

$$pro_{kstad} - pro_{kal} \leq pro_k \leq pro_{kstad} + pro_{kal} \tag{7}$$

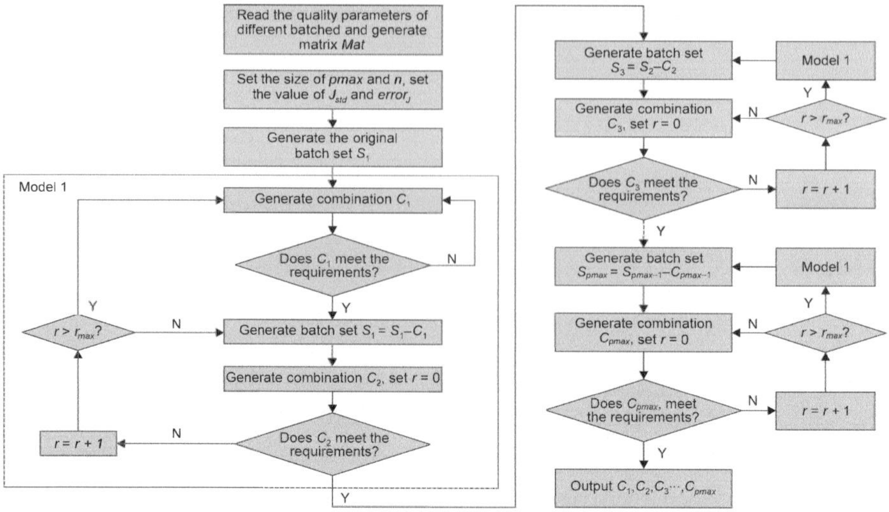

Figure 3: Flow chart of logic combination algorithm

where, M_{stad} represents the standard value of the total feed weight of ingredient product, kg; M_{al} represents the maximum allowable fluctuation value of M_{sum}, kg; pro_{kstad} represents the standard value of component k of the ingredient product, %. pro_{kal} represents the maximum allowable fluctuation value of pro_k, %.

3.3 Algorithm description

3.3.1 Logical combination algorithm

In this work, a logical combination algorithm is designed to solve the discrete combination problem in the batch combination model. The designed logic combination algorithm uses the convenience of a computer to generate random numbers and powerful calculation ability to solve all the combinations required for production. The algorithm flow chart is as follows (Fig3):

(1) Read the quality parameter data of each batch of raw materials in inventory and generate the matrix Mat, where each row of Mat represents a batch and each column represents the content of a certain component in this batch.

(2) Set the model parameters $pmax$, n, J_{stad} and $error_J$ where, $pmax$ represents the number of combinations required for production.
(3)Get the total number of batches in the current inventory sumbch and generate the original batch set S_1:

$$S_1 = \{1, 2, 3, \cdots, sumbch\} \tag{8}$$

(4) To solve the first combination C_1 in S_1, C_1 needs to meet the constraints shown in Equation (2). To avoid the enumeration of a huge number of combinations, this work uses a randomly generated method to solve the combinations. Randomly select n elements in S_1, calculate $\bar{J}$ and determine whether the constraints are satisfied, if satisfied, C_1 is solved successfully; if not, continue random generation until C_1 that satisfies the constraints is found. After C_1 is successfully solved, continue to solve C_2 in the same way. The calculation method of S_2 is as follows:

$$S_2 = S_1 - C_1 \tag{9}$$

When C_1 is unreasonable, C_2 may have no solution. This work defines the number of solutions r and the maximum number of solutions r_{max}. When r is greater than r_{max} and C_2 is still not successfully solved, C_1 will be resolved. Solve C_3 in S_3, when r is greater than r_{max} and C_3 is still not solved successfully, C_1 will be resolved. The calculation method of S_3 is shown in the following equation:

$$S_3 = S_2 - C_2 \tag{10}$$

and so on, until C_{pmax} is solved. Finally, the model outputs all the reasonable combinations.

3.3.2 Improved SQP algorithm

In production, M_{sum} is required to float within a certain range of the standard value M_{stad}, and the closer it is to M_{stad}, the better. The objective function of the formula calculation model is as follows:

$$min(M_{sum} - M_{stad}) \tag{11}$$

Therefore, the solving problem of the inequality set shown in Equations (6) and (7) is transformed into a nonlinear programming problem with multiple constraints in this work. However, the purpose of solving the model has not changed, that is, to solve for M_q that meets all production constraints.

Improved SQP is one of the most effective algorithms for solving nonlinear optimization problems. It transforms complex nonlinear optimization problems into a series of relatively simple quadratic programming problems for solving. Compared with other algorithms, the most prominent advantages of the algorithm are good convergence, high computational efficiency and strong boundary search ability, which has been widely paid attention to.

For example,[18] solve the patient appointment scheduling problem in outpatient chemotherapy clinics.[19] minimize time machining through Mixed-Integer Nonlinear Programming. As another example,[10] employed SQP to solve the optimization problem of the energy system for the tissue papermaking process in comparison with a designed optimization algorithm based on energy efficiency analysis. These literatures show the great potential of SQP with regard to complex nonlinear optimization problems in the industry.

The main description of the improved SQP algorithm is as follows: (1)Use the second-order Taylor formula to expand the original objective function at each iteration point X^k:

$$f(X) \approx f(X^k) + (X - X^k)^T \nabla \mathrm{f}(X^k) + \frac{1}{2}(X - X^k)^T H^k (X - X^k) \tag{12}$$

where, $\nabla \mathrm{f}(X^k)$ and H^k respectively represent the Jacobi matrix and Hessian matrix at X^k of the objective function. This step approximates the solution of the original problem into the solution of the quadratic programming (QP) problem as shown in Equation (12)[20].

(2)Further differentiate the above equation and let X=X^{k+1}, then the following equation can be obtained:

$$X^{k+1} \approx X^k + \left(H^k\right)^{-1} \left[\nabla f\left(X^{k+1}\right) - \nabla f\left(X^k\right)\right] = X^k + S^k \tag{13}$$

Equation (13) shows the iterative form of searching for the optimal solution X^* of the original problem, where S^k is called the iterative direction. Therefore, the solution of the above QP problem is the solution of S^k. This method of solving the QP problem is called Newton's method[21].

(3)In the process of iteration, a smaller step size will lead to a longer optimization time, and a larger step size will lead to the oscillation and non-convergence of the optimization process. To make the iteration of each step reasonable, it is necessary to determine the iteration step size α^k of each step, and the calculation method uses the backtracking line search algorithm[22].

(4)As can be seen from Equation (13), Newton's method needs to calculate H^k and its inverse matrix in addition to $\nabla \mathrm{f}(X^k)$, and the whole iteration process needs the support of extremely powerful computation and storage power provided by hardware. In addition, when H^k is a non-positive definite matrix, the iterative process will not converge, and Newton's method is completely invalid. To overcome the defects in the above two aspects, the quasi-Newton method is proposed, which uses a formula called BFGS to construct H^k of each iteration and then solves the QP problem. As shown in Equation (14) to (16), this solution method is called improved SQP[23].

$$\begin{aligned} H^{k+1} = H^k + \frac{1}{\left[\Delta X^k\right]^T \Delta q^k} \Bigg\{ & \Delta X^k \left[\Delta X^k\right]^T + \frac{\Delta X^k \left[\Delta X^k\right]^T \left[\Delta q^k\right]^T H^k \Delta q^k}{\left[\Delta X^k\right]^T \Delta q^k} \\ & - H^k \Delta q^k \left[\Delta q^k\right]^T - \Delta X^k \left[\Delta q^k\right]^T H^k \Bigg\} \end{aligned} \tag{14}$$

$$\Delta X^k = X^{k+1} - X^k \tag{15}$$

$$\Delta q^k = \nabla f\left(X^{k+1}\right) - \nabla f\left(X^k\right) \tag{16}$$

4 Case study and model validation

The glass industry is a basic material industry in the construction of the national economy[24]. China is the world's largest producer of flat glass. In 2019, the country produced 9.3 billion weight boxes of flat glass, accounting for more than 60 percent of the global total[25]. The glass production process is complex, from the processing of raw materials, melting, clarification, molding, annealing, cutting, packaging and transportation were each step requires energy consumption[26]. Among them, melting accounts for more than 80% of the total energy consumption in the production process, which is the key energy consumption link[27]. The emission in glass production mainly comes from the flue gas emitted after burning heavy oil or natural gas during the melting process of liquid glass. For example, when natural gas is used as fuel, the NO_x emission concentration is about 1800mg/Nm[28]. Relevant studies show that in 2018, the total annual CO_2 emission from China's flat glass industry was about 1.7 million tons, accounting for about 1.7% of China's total carbon emissions[29]. In this work, raw material data from a glass factory is used to verify the batch and formula calculation models. The production line mainly produces ultra-thin electronic glass.

4.1 Data acquisition and quality parameters calculation method

When the raw materials enter the factory, the quality parameters of different batches of raw materials are inspected and recorded into the digital operation platform. Dolomite is the most widely sourced raw material in glass production, and a large number of batches are often stored in the warehouse, so this work selects dolomite to verify the batch combination model. Combined with the production requirements on the quality indicators of the dolomite raw material product, this work collected the SiO_2, CaO, MgO, Fe_2O_3, Al_2O_3 component content data of 40 batches of dolomite from the digital operation platform. SiO_2 is the main component of glass, which gives glass a series of excellent properties, such as transparency, mechanical strength, chemical stability and thermal stability[30]. Al_2O_3 can reduce the crystallization tendency of glass and improve the chemical stability and mechanical strength[31]. CaO can reduce the high temperature viscosity of the glass solution and promote its melting and clarification[32]. MgO acts similar to CaO, and Fe_2O_3 is a harmful ingredient that can stain glass and reduce its transparency, ultraviolet radiation, thermal diathermy and mechanical strength[33]. In production, the above quality indicators are should fluctuate within a small range above and below

Table 1: Quality indicators' requirements of dolomite raw material product

J	J_{stad} (%)	$error_J$ (%)
SiO_2	2.24	6
CaO	29.81	0.1
MgO	21.09	0.2
Fe_2O_3	0.07	2
Al_2O_3	0.13	20

the standard values. The factory production requires four batches per combination. Therefore, $J \in \{CaO, MgO, \mathrm{Fe_2}O_3, \mathrm{SiO_2}, \mathrm{Al_2}O_3\}$, $pmax$=40, n=4. The standard values and allowable deviations of the above 5 quality indicators are shown in Table 1.

In addition to dolomite, the raw material products used in the production of the factory also include silica sand, alumina, soda ash, mirabilite, iron powder, carbon powder and potassium carbonate. Silica sand is the main raw material to provide SiO_2 [30]. Soda ash can provide the necessary Na^+, and the CO_2 gas released by thermal decomposition in the furnace agitates the glass melt [34]. The main component of mirabilite is Na_2SO_4, which is mainly used as a clarifying agent for liquid glass in production [35]. Carbon powder can adjust the redox atmosphere of the liquid glass, so as to adjust its melting quality, for example, to reduce the bubbles generated in the melting process [36]. Potassium carbonate can improve the transparency, strength and refractive coefficient of glass [37]. The following quality indicators of ingredient products are the most concerned in the production: total feed weight, oxide content (including SiO_2, CaO, MgO, Fe_2O_3, Al_2O_3, K_2O, Na_2O, MgO and CaO content sum), mirabilite content, carbon content, and glass density. The factory uses two different kinds of silica sand for production, namely silica sand 1 and silica sand 2. Therefore, the formula calculation model needs to solve the reasonable feed weight of the above 9 raw material products. The quality indicators of these 9 kinds of raw material products are tested and recorded into the digital operation platform. The specific test items are shown in Table 2. The standard values and allowable deviations of the quality indicators of ingredient products are shown in Table 3.

The calculation method of carbon content in ingredient products is as follows:

$$\mathrm{pro}_C = \frac{M_C \cdot \mathrm{con}_{CC}}{M_{\mathrm{mira}} \cdot \mathrm{con}_{Na_2SO_4}} \tag{17}$$

where, pro_C represents the carbon content in ingredient products, %; M_C and M_{mira} represents the feed weight of carbon powder and mirabilite, kg; con_{CC} represents the carbon content in carbon powder, %; $con_{Na_2SO_4}$ is the content of Na_2SO_4 in mirabilite, %.

Table 2: Test items of raw material product

q	SiO_2	Fe_2O_3	Al_2O_3	CaO	MgO	K_2O	Na_2O	Na_2CO_3	Na_2SO_4	C
Silica sand 1	*	*	*	*	*	*	*	-	-	-
Silica sand 2	*	*	*	*	*	*	*	-	-	-
Dolomite	*	*	*	*	*	-	-	-	-	-
Alumina	*	*	*	-	-	-	*	-	-	-
Soda ash	-	-	-	-	-	-	-	*	-	-
Mirabilite	-	-	-	-	-	-	-	-	*	-
Carbon powder	-	-	-	-	-	-	-	-	-	*
Iron powder	-	*	-	-	-	-	-	-	-	-
Potassium carbonate	-	-	-	-	-	*	-	-	-	-

Note: "*" represents the test items, and "-" represents the non- test items
t

Table 3: Quality indicators requirements of ingredient product

Quality parameters (Unit)	Standard value	Maximum allowable error
M_{sum}(kg)	1440	0.1
pro_{SiO_2}(%)	69.408	0.2
$pro_{Fe_2O_3}$(%)	0.064	0.0005
$pro_{Al_2O_3}$(%)	4.649	0.002
pro_{CaO}(%)	5.700	0.1
pro_{MgO}(%)	4.000	0.1
pro_{K_2O}(%)	1.019	0.001
pro_{Na_2O}(%)	14.697	0.05
pro_{mira}(%)	1.962	0.05
pro_C(%)	1.859	0.5
pro_{CaMg}(%)	9.700	0.04
$dens$	2.491	0.01

The calculation method of mirabilite content in ingredient product is shown in Equation (18):

$$\mathrm{pro}_{\mathrm{mira}} = \frac{ec_m \cdot M_{\mathrm{mira}} \cdot \mathrm{Con}_{\mathrm{Na_2SO_4}}}{ec_m \cdot M_{\mathrm{mira}} \cdot \mathrm{Con}_{\mathrm{Na_2SO_4}} + ec_i \cdot M_{\mathrm{Sa}} \cdot \mathrm{Con}_{\mathrm{Na_2CO_3}}} \tag{18}$$

where, pro_{mira} represents the mirabilite content in the ingredient product, %; ec_m and ec_i represent the empirical coefficients, which are constants; M_{sa} is the feed weight of soda ash, kg; $con_{Na_2CO_3}$ represents the content of Na_2CO_3 in soda ash, %.

Density is an important index in glass manufacturing. The calculation method is as follows:

$$\mathrm{dens} = \frac{1}{\sum ec_k \cdot \mathrm{pro}_k} \tag{19}$$

where, dens represents the density; ec_k represents empirical coefficients of component k in ingredient product, which are constants.

In production, the total feed weight, density and content of each component of the ingredient product are constrained, and these quality parameters should fluctuate within a certain range of the standard values, as shown in the following equations:

$$dens_{stad} - dens_{al} \leq dens \leq dens_{stad} + dens_{al} \tag{20}$$

$$pro_{Cstad} - pro_{Cal} \leq pro_C \leq pro_{kstad} + pro_{Cal} \tag{21}$$

$$pro_{mirastad} - pro_{miraal} \leq pro_{mira} \leq pro_{mirastad} + pro_{miraal} \tag{22}$$

$$pro_{CaMgstad} - pro_{CgMgal} \leq pro_{CgMg} \leq pro_{CaMgstad} + pro_{CgMgal} \tag{23}$$

$$pro_{CaMgstad} = pro_{CaOstad} + pro_{MgOstad} \tag{24}$$

where, $dens_{stad}$ and $dens_{al}$ represent the standard value and maximum allowable deviation of density respectively; pro_{Cstad} and pro_{Cal} represent the standard value and maximum allowable deviation of carbon content respectively, %; $pro_{mirastad}$ and pro_{miraal} represent the standard value and the maximum allowable deviation of mirabilite content respectively, %; $pro_{CaMgstad}$ and pro_{CgMgal} represent the standard value and the maximum allowable deviation of the sum of CaO content and MgO content respectively, %.

4.2 Results and discussion

4.2.1 Batch combination model

To compare with the results calculated manually, the quality index deviation mre_J and the standard deviation std_J between the combinations are selected as the measurement indicators[38]. The calculation methods are as follows:

$$\mathrm{mre}_J = \frac{1}{p\max} \sum_{i=1}^{p\max} \left| \frac{\bar{J}_i - J_{\mathrm{stad}}}{J_{\mathrm{stad}}} \right| \times 100\% \tag{25}$$

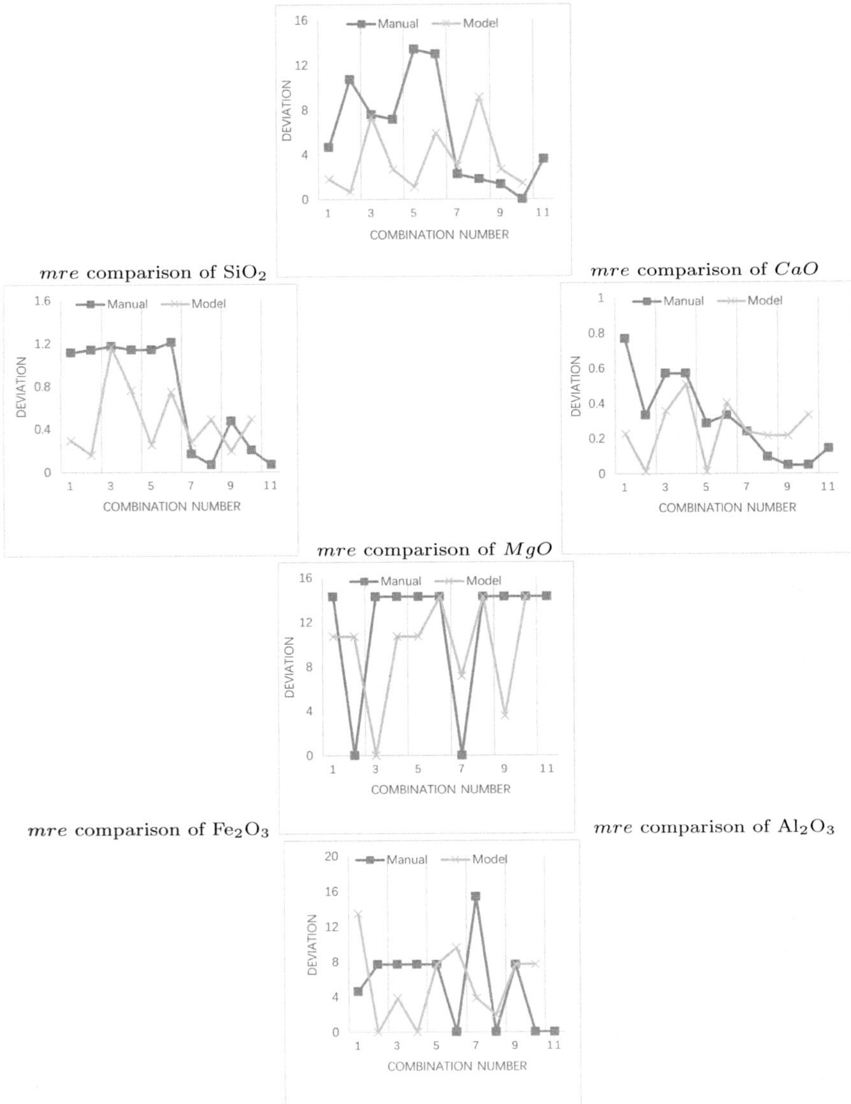

Figure 4: Comparison of manual and model deviations of quality indicators

Table 4: Comparison of *mre* and *std*

J	mre(%) Manual	mre(%) Model	std(%) Manual	std(%) Model
SiO_2	5.942	3.583	12.5	7.4
CaO	0.717	0.482	14.5	9.1
MgO	0.311	0.251	6.7	6.0
Fe_2O_3	11.688	9.643	0.4	0.3
Al_2O_3	5.315	5.577	0.91	0.90

$$std_J = \sqrt{\left((\bar{J}_1 - l)^2 + (\bar{J}_2 - l)^2 + \ldots + (\bar{J}_{p\max} - l)^2\right)/(p\max - 1)} \times 100\% \tag{26}$$

where, $\bar{J}_i$ represents the content of quality index J in combination i, %; l represents the mean of $\bar{J}_i$ of all combinations.

The batch combination model is validated using 40 batches of dolomite from the factory inventory. The 40 batches of dolomite had previously been used in production, and the combination results were derived from manual calculations. The original convention in production was $pmax$=10,n=4. Due to the limited computational capacity of the human brain, only 1-3 combinations are usually discharged for production. Therefore, the convention cannot be strictly observed in the subsequent combinations, and $pmax$ and n may change. According to the historical record of the digital operation platform, the actual $pmax$=11 was obtained, in which the combination of n=5 appeared once, the combination of n=3 appeared 5 times, and the combination of n=4 appeared 5 times. The batch combination model is used to solve the problem strictly in accordance with the production convention, that is, the combination results of $pmax$=10,n=4 are solved. Figure 4 (a)-(e) respectively show the comparison between actual deviations and model deviations of each combination 's quality indicators. Table 4 shows the comparison between *mre* and *std* of each index. The following conclusions can be drawn:

(1)It can be seen from the Figs that the index deviation of each combination solved by the model is not completely lower than the manual deviation. However, in all the combinations solved by the model, except Al_2O_3, the *mre* of the other four indicators are significantly reduced, with an average reduction of 0.89%. It shows that the calculation results of the model make the quality indicators of dolomite raw material products closer to the standard value, which indirectly improves the quality of dolomite raw material products.

(2)The *stds* of the five quality indicators are all reduced to different degrees, with an average decrease of 3%. This indicates that the quality difference between different combinations is reduced, which lays a foundation for the stability of the finished product quality.

Table 5: Comparison of e_q

q	e_q(Actual)	e_q(Model)	Improve (times)
M_{sum}(kg)	0.045	0.045	1
pro_{SiO_2}(%)	0.11	0.10	1.1
$pro_{Fe_2O_3}$(%)	0.00043	0.00043	1
$pro_{Al_2O_3}$(%)	0.0018	0.000018	100
pro_{CaO}(%)	0.03	0.008	3.75
pro_{MgO}(%)	0.02	0.008	2.5
pro_{K_2O}(%)	0.0001	0.0001	1
pro_{Na_2O}(%)	0.044	0.044	1
pro_{mira}(%)	0.049	0.000005	7000
pro_C(%)	0.42	0.42	1
pro_{CaMg}(%)	0.04	0.000003	13333
$dens$	0.0029	0.00071	4.08

4.2.2 Formula calculation model

In production, the operators adjust the feed weight of raw material products according to their quality parameters and the requirements of production indicators. Because of the high precision and mutual coupling, the manual calculation process is very time-consuming and often does not meet the requirements. In this work, 7 groups of historical samples from the factory are selected to test the formula calculation model, and the manual calculation results are compared with the model calculation results. The deviation of quality indicators and the standard deviation of deviations is selected as the measurement indicator. The calculation method is as follows:

$$e_q = \frac{1}{m}\sum_{i=1}^{m}\left|\frac{q_i - q_{stad}}{q_{stad}}\right| \times 100\% \tag{27}$$

$$\sigma_q = \sqrt{\left((e_{q1} - Q)^2 + (e_{q2} - Q)^2 + \cdots + (e_{qm} - Q)^2\right)/(m-1)} \tag{28}$$

where, e_q represents the average deviation of all samples with respect to the quality index q, %; m represents the number of samples, m=7; q_i represents the calculated value of the quality index q of the i-th sample, %, q_{stad} represents the standard value of the quality index q, %, σ_q represents the standard deviation of the deviation of all samples with respect to the quality index q, and Q represents the mean value of the deviation of all samples with respect to the quality index q.

As shown in Table 3, there are 12 types of quality indicators. The results of manual and model calculation are shown in Table 5 and Table 6. By contrast, the following conclusions can be drawn:
(1)In terms of many quality parameters, e_q generated by model calculation is

Table 6: Comparison of σ_q

q	σ_q(Actual)	σ_q(Model)	Improve (times)
M_{sum}(kg)	0.02	6.3×10^{-8}	308413
pro_{SiO_2}(%)	2.5×10^{-5}	8.8×10^{-8}	281
$pro_{Fe_2O_3}$(%)	2.5×10^{-6}	8.6×10^{-11}	28723
$pro_{Al_2O_3}$(%)	1.3×10^{-5}	8.5×10^{-9}	1580
pro_{CaO}(%)	6.3×10^{-4}	9.4×10^{-7}	670
pro_{MgO}(%)	5.0×10^{-4}	9.5×10^{-7}	530
pro_{K_2O}(%)	4.1×10^{-6}	2.1×10^{-9}	2015
pro_{Na_2O}(%)	1.0×10^{-3}	5.6×10^{-8}	18682
pro_{mira}(%)	7.9×10^{-4}	2.4×10^{-11}	32512493
pro_C(%)	7.4×10^{-3}	9.2×10^{-10}	8095730
pro_{CaMg}(%)	1.7×10^{-5}	9.0×10^{-9}	1907
$dens$	3.9×10^{-5}	4.7×10^{-7}	82

significantly improved compared that generated by manual calculation, such as pro_{CaMg}, pro_{mira}, which makes the quality indicators of the ingredient products closer to the standard value.
(2)The σ_q generated by model calculation is far less than that generated by manual calculation, with an improvement of 7-8 orders of magnitude for individual indicators, which significantly stabilize the quality of the ingredient products, thus making the quality of finished glass more stable.

5 Conclusions

In industrial production, it is of paramount importance to resolve the contradiction between the uneven quality of natural raw materials and the stringent production requirements. Instead of exploring new raw materials or developing novel technologies, this work builds an optimization model for the material selection process by optimizing batch combinations and formulations. The two models are tested by using the raw material selection process of glass manufacturing as a case study. The results show that the batch combination model can reduce the bias of the quality indicators of raw material products by an average of 0.89% and a standard deviation of 3% compared to the results of manual calculation. The formula calculation model can reduce the deviation of each quality index of ingredient products by 1-7000 times and the standard deviation by different orders of magnitude, among which the standard deviation of carbon content and mirabilite content can be reduced by 7-8 orders of magnitude. The optimization model for the raw material selection process developed in this work for complex industries can improve production efficiency and product quality. In addition, it can provide support for optimizing subsequent production processes and promoting energy conservation and emission reduction.

6 Declarations

Ethics approval and consent to participate: Not applicable
Consent for publication: Not applicable
Data availabilty: Data will be made available on reasonable request.
Conflict of Interest: The authors declare that there's no conflict of interests that could affect the work reported in this paper.
Funding and support: There is no funding or support for this paper.
Authors' contributions: Xiuli He carried out investigation, writing, review and editing, visualization. Rongyuan Wang carried out investigation, methodology, software, visualization. Mengna Hong took care of project administration and software. Zhenglei He carried out conceptualization, review, project administration, supervision.

References

[1] N. Sinaga, P. Paryanto, S. A. Widyanto, R. Rusnaldy, A. Hetzner, J. Franke, An analysis of the effect of gravitational load on the energy consumption of industrial robots, Procedia CIRP 78 (2018) 8–12.

[2] A. Balea, N. Merayo, E. De La Fuente, C. Negro, Á. Blanco, Assessing the influence of refining, bleaching and tempo-mediated oxidation on the production of more sustainable cellulose nanofibers and their application as paper additives, Industrial crops and products 97 (2017) 374–387.

[3] B. Li, X. Ling, X. Liu, Q. Li, W. Chen, Hydration of portland cements in solutions containing high concentration of borate ions: Effects of lioh, Cement and Concrete Composites 102 (2019) 94–104.

[4] Ž. Štancar, L. Snoj, An improved thermal power calibration method at the triga mark ii research reactor, Nuclear Engineering and Design 325 (2017) 78–89.

[5] A. D. Rodman, D. I. Gerogiorgis, Dynamic optimization of beer fermentation: Sensitivity analysis of attainable performance vs. product flavour constraints, Computers & Chemical Engineering 106 (2017) 582–595.

[6] B. Simonato, F. Mainente, E. Selvatico, M. Violoni, G. Pasini, Assessment of the fining efficiency of zeins extracted from commercial corn gluten and sensory analysis of the treated wine, LWT-Food Science and Technology 54 (2013) 549–556.

[7] H. Cetin-Karaca, M. C. Newman, Antimicrobial efficacy of phytochemicals against bacillus cereus in reconstituted infant rice cereal, Food microbiology 69 (2018) 189–195.

[8] J. Gong, C. Jiang, X. Tang, Z. Zheng, L. Yang, Optimization of mixture proportions in ternary low-heat portland cement-based cementitious systems with mortar blends based on projection pursuit regression, Construction and Building Materials 238 (2020) 117666.
[9] J. Hu, T. Ma, Y. Zhu, X. Huang, J. Xu, L. Chen, High-viscosity modified asphalt mixtures for double-layer porous asphalt pavement: Design optimization and evaluation metrics, Construction and Building Materials 271 (2021) 121893.
[10] Y. Zhang, M. Hong, J. Li, J. Ren, Y. Man, Energy system optimization model for tissue papermaking process, Computers & Chemical Engineering 146 (2021) 107220.
[11] M. Vasić, L. Pezo, Z. Radojević, Optimization of adobe clay bricks based on the raw material properties (mathematical analysis), Construction and Building Materials 244 (2020) 118342.
[12] K. Li, Y. Wang, X. Zhang, X. Wang, A. Zhang, Raw material ratio optimisation of magnesium oxychloride cement using response surface method, Construction and Building Materials 272 (2021) 121648.
[13] Y. Xu, M. Zhai, Y. Zhang, Q. Shen, G. Kumar, P. Dong, A. M. Parvez, Effect of the ash melting behavior of a corn straw pellet on its heat and mass transfer characteristics and combustion rate, Fuel 286 (2021) 119483.
[14] Y. Qi, B. Dai, S. He, S. Wu, J. Huang, F. Xi, Y. Ma, M. Meng, Effect of chemical constituents of oxytetracycline mycelia residue and dredged sediments on characteristics of ultra-lightweight ceramsite, Journal of the Taiwan institute of chemical engineers 65 (2016) 225–232.
[15] M. H. K. Feizabad, E. Sarvestani, G. R. Khayati, Modeling and optimization of chemical composition of nano/amorphous fea. nib. nbc. zrd alloy prepared via high-energy ball milling with enhanced soft magnetic properties; a mixture design approach, Journal of Alloys and Compounds 841 (2020) 155646.
[16] Y. Man, J. Li, M. Hong, Y. Han, Energy transition for the low-carbon pulp and paper industry in china, Renewable and Sustainable Energy Reviews 131 (2020) 109998.
[17] D. Peters, J. Drallmeier, D. A. Bristow, R. G. Landers, E. Kinzel, Sensing and control in glass additive manufacturing, Mechatronics 56 (2018) 188–197.
[18] M. Heshmat, K. Nakata, A. Eltawil, Solving the patient appointment scheduling problem in outpatient chemotherapy clinics using clustering and mathematical programming, Computers & Industrial Engineering 124 (2018) 347–358.
[19] S. Cafieri, F. Monies, M. Mongeau, C. Bes, Plunge milling time optimization via mixed-integer nonlinear programming, Computers & Industrial Engineering 98 (2016) 434–445.
[20] S. Khalilpourazari, S. H. R. Pasandideh, S. T. A. Niaki, Optimization of multi-product economic production quantity model with partial back-

ordering and physical constraints: Sqp, sfs, sa, and wca, Applied Soft Computing 49 (2016) 770–791.

[21] J.-P. Auclair, J.-F. Lemieux, L. B. Tremblay, H. Ritchie, Implementation of newton's method with an analytical jacobian to solve the 1d sea ice momentum equation, Journal of Computational Physics 340 (2017) 69–84.

[22] İ. Sabuncuoğlu, Y. Gocgun, E. Erel, Backtracking and exchange of information: Methods to enhance a beam search algorithm for assembly line scheduling, European Journal of Operational Research 186 (2008) 915–930.

[23] Q. Tong, G. Liang, X. Cai, C. Zhu, J. Bi, Asynchronous parallel stochastic quasi-newton methods, Parallel computing 101 (2021) 102721.

[24] K.-Y. Lin, User experience-based product design for smart production to empower industry 4.0 in the glass recycling circular economy, Computers & Industrial Engineering 125 (2018) 729–738.

[25] B. Huang, X. Gao, X. Xu, J. Song, Y. Geng, J. Sarkis, T. Fishman, H. Kua, J. Nakatani, A life cycle thinking framework to mitigate the environmental impact of building materials, One Earth 3 (2020) 564–573.

[26] Y. D. Ko, I. Noh, H. Hwang, Cost benefits from standardization of the packaging glass bottles, Computers & Industrial Engineering 62 (2012) 693–702.

[27] M. B. Durmusoglu, A. Nomak, Gt cells design and implementation in a glass mould production system, Computers & industrial engineering 48 (2005) 525–536.

[28] G. E. Alvarez, M. G. Marcovecchio, P. A. Aguirre, Optimization of the integration among traditional fossil fuels, clean energies, renewable sources, and energy storages: An milp model for the coupled electric power, hydraulic, and natural gas systems, Computers & Industrial Engineering 139 (2020) 106141.

[29] C.-L. Chang, T.-K. Mai, M. McAleer, Establishing national carbon emission prices for china, Renewable and Sustainable Energy Reviews 106 (2019) 1–16.

[30] P. Bandyopadhyay, A. Dey, A. K. Mandal, N. Dey, A. K. Mukhopadhyay, New observations on scratch deformations of soda lime silica glass, Journal of non-crystalline solids 358 (2012) 1897–1907.

[31] S. Uttiya, C. Bernini, M. Vignolo, I. Pallecchi, D. Marré, A. S. Siri, L. Pellegrino, Inkjet printing of conducting silver patterns on alumina and insulating ceramic-glass by saline precursors, Thin Solid Films 642 (2017) 370–376.

[32] S. Das, S. Gautam, C. Gautam, Mechanical and tribological characterization of lead calcium titanate borosilicate glass ceramic doped with ferric oxide, Materials Today: Proceedings 5 (2018) 17746–17752.

[33] A. Michelin, E. Leroy, D. Neff, J. Dynes, P. Dillmann, S. Gin, Archeological slag from glinet: an example of silicate glass altered in an anoxic iron-rich environment, Chemical Geology 413 (2015) 28–43.
[34] M. C. Sow, J.-P. Blondeau, N. Sagot, N. Ollier, T. Tite, Nanosecond (ns) laser transfer of silver nanoparticles from silver-exchanged soda-lime glass to transparent soda-lime glass and shock waves formation, Applied Surface Science 336 (2015) 255–261.
[35] A. Vavouraki, P. Koutsoukos, Kinetics of crystal growth of mirabilite in aqueous supersaturated solutions, Journal of crystal growth 338 (2012) 189–194.
[36] M. Xing, Z. Fu, Y. Wang, J. Wang, Z. Zhang, Lead recovery and high silica glass powder synthesis from waste crt funnel glasses through carbon thermal reduction enhanced glass phase separation process, Journal of hazardous materials 322 (2017) 479–487.
[37] F. Isa, H. Zabiri, N. K. S. Ng, A. M. Shariff, Co2 removal via promoted potassium carbonate: A review on modeling and simulation techniques, International Journal of Greenhouse Gas Control 76 (2018) 236–265.
[38] T. P. Carvalho, F. A. Soares, R. Vita, R. d. P. Francisco, J. P. Basto, S. G. Alcalá, A systematic literature review of machine learning methods applied to predictive maintenance, Computers & Industrial Engineering 137 (2019) 106024.

13. Research on Fault Diagnosis of Paper-making Industry based on Knowledge Graph

Qingyuan Zhang,[1] Zhenglei He,[1,*] Yi Man[1,2] Jigeng Li[1] and Mengna Hong[1]

1 Introduction

Industry 5.0 represents an intelligent era that drives industrial transformation by leveraging information technology, causing a revolutionary change in traditional manufacturing and value creation. However, with the rapid development of the industry, the problem of malfunctions is becoming increasingly difficult to solve, resulting in a series of severe losses. In 2005, the explosion at the Biphenyls unit of Jilin Petrochemical Company, which is affiliated with China National Petroleum Corporation, resulted in eight deaths and sixty in-juries, causing a direct economic loss of 69.08 million yuan. Over a hundred tons of benzene-type pollutants flowed into the Songhua River, leading to serious pollution of the water body and caused damage to the fishing industry. Similarly, on January 7th, 2010, the Lanzhou Petrochemical Company experienced an explosion and burning accident due to the failure to detect tank leakage in a timely manner, causing five people to go missing, one person to be seriously injured, and five people to be slightly injured[1].

The papermaking industry is one of the traditional industries, and it is the industrial sector that produces various types of paper and paper products. Al-though there is no separate assessment of losses due to malfunctions in the papermaking industry, the papermaking process itself is a production process with frequent malfunctions, and the coupling between its malfunctions is also

[1] State Key Laboratory of Pulp and Paper Engineering, South, China University of Technology, Guangzhou.
[2] Guangdong Artificial intelligence and Digital Economy, Laboratory (Guangzhou) (Pazhou Lab), Guangzhou.
*Corresponding author: hezhenglei@scut.edu.cn

relatively high. Moreover, with the continuous upgrading of technology in papermaking enterprises, their production systems are becoming increasingly large-scale and complex. Therefore, traditional experience-based methods of workshop maintenance workers and some handheld fault diagnosis equipment are difficult to use in the high-speed papermaking process. Existing intelligent fault diagnosis methods also face many problems, such as the complexity of industrial processes making it difficult to establish accurate mathematical models[2], poor interpretability of data-driven methods[3], making it difficult to truly diagnose faults, and expert systems with poor generalization abilities that require a lot of manual effort[4]. The emergence of knowledge graphs has brought new development directions for fault diagnosis in the papermaking industry. At present most knowledge graph applications emerge in fields such as healthcare[5;6], education[7;8]and finance[9;10]where knowledge is well-structured and readily available. Knowledge graphs have powerful knowledge integration capabilities and the ability to represent unstructured relationships[11], which can address data storage problems caused by the continuous large-scale development of the papermaking industry. Additionally, knowledge graphs can store various types of knowledge, and once constructed, they can greatly reduce the knowledge requirements for diagnostic personnel. Moreover, knowledge graphs have search and analysis capabilities[12], which can reason and trace the source of faults in complex environments, find the transmission path of faults, and determine the true root cause.

Although knowledge graphs possess the powerful functionalities mentioned above, they contain a vast amount of textual knowledge that is difficult for computers to directly recognize and process. This impedes the quantification of relationships between entities. Therefore, this study proposes to utilize word embedding, a common natural language processing technique, to convert the textual entities within the knowledge graph into vector representations and situate them in the vector space. The computer can then further process the entity's word vectors. To summarize, this study aims to develop a knowledge graph construction framework based on the characteristics of the paper-making industry, construct a knowledge graph for the paper-making industry's drying section using the aforementioned framework, convert the entities in the completed knowledge graph into vector form using a word embedding model that the computer can recognize, and ultimately achieve a complete fault diagnosis process using a computer.

Table 1: Difficulties in constructing process industry knowledge graphs

Difficulties	Description
complexity of knowledge	1) There are many types of equipment and processes, and different equipment and processes can be used for the same process, making it difficult to establish relationships uniformly. 2) Equipment information is scattered and disorganized. The knowledge of various equipment is stored in different databases, online and offline documents, and the level of structuring is low.
the lack of relevant data	1) Constructing knowledge graphs requires a large amount of labeled datasets. There are many types of process industries, each with a lot of data but low utilization. 2) Data interaction is poor. Most factory data is only stored in the factory and rarely used.
difficulty of knowledge management	1) Experts are required to manage knowledge graphs. Each situation needs its own experts, which makes the operation and management very costly. 2) Equipment changes and improves rapidly. The knowledge graph needs a mechanism to update its knowledge dynamically.

2 Construction of Knowledge Graph of Paper-making Industry

2.1 Characteristics of Knowledge Graph Construction in Paper-making Industry

The paper-making industry involves a multitude of phenomena, such as heat transfer, mass transfer, and momentum transfer, which alter the original characteristics of raw materials through energy input and lead to changes in their state, form, physical and chemical properties, etc., to meet the demands of the final product. The industry is also characterized by large-scale and continuous production, where a specific process flow must be followed to ensure product quality and production efficiency[13].

Therefore, unlike knowledge graph construction in other fields, the construction of a knowledge graph in the paper-making industry is based on industrial raw materials, specifically paper sheets, and established based on the changes in material and energy flows. The knowledge graph mainly describes the heat and mass transfer phenomena during the process, which eventually manifest as changes in various properties of the raw materials and form the desired product.

The difficulties in constructing process industry knowledge graphs lie in the complexity of knowledge, the lack of relevant data and the difficulty of knowledge management, as shown in Table 1.

It is especially important to build a good ontology framework for these problems. A good ontology framework can greatly reduce the dependence on domain experts' knowledge. You can create a knowledge graph by adding knowledge instances under the framework with some basic knowledge. This also lowers the experts' workload and increases data utilization. Some existing methods for building ontologies are: (1) skeleton method[14]; (2) IDEF5 method[15]; (3) TOVE method[16]; (4) KACTUS engineering method[17]; (5) METHONTOLOGY method[18]; and (6) seven-step method[19]. The first four methods are common for business ontologies, while the last two are typical for domain ontologies. The most popular method for domain ontologies now is the seven-step method, which was developed by Stanford University in the US and first applied to medical ontologies. It is very practical and versatile.

2.2 Establishment of paper-making process ontology and its fault ontology

2.2.1 Establishment of paper-making process ontology

The seven-step procedure for ontology construction in this study is shown in Figure 1. The first step is to define the domain scope of the ontology, which is designed for representing the papermaking process. As a typical process industry, paper production involves mainly heat and mass transfer from pulp to paper. Most of the reactions are physical rather than chemical. Therefore, the knowledge graph of paper production requires tracing both material and energy flows along the entire process.

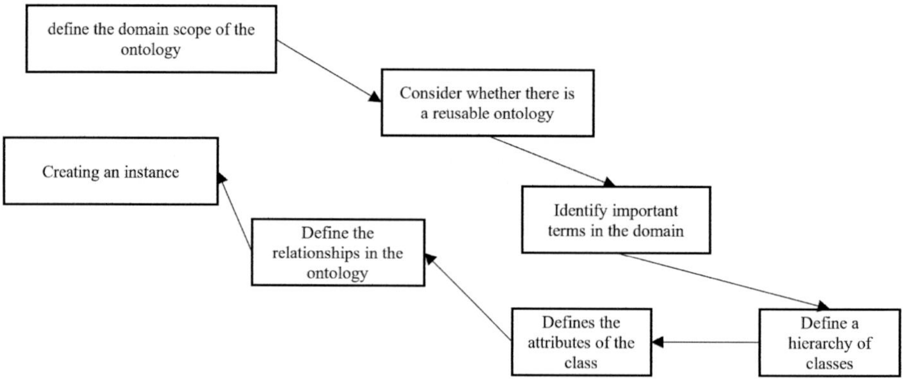

Figure 1: Specific steps of the seven-step ontology construction method

Considering the reuse of ontology, OntoCAPE is the result of nearly twenty years of research by a group led by Wolfgang Marquardt. It aims to establish a reusable chemical engineering process ontology for knowledge representation and information modeling in chemical engineering. This ontology covers almost all concepts in the chemical engineering domain and spans the entire life cycle of chemical process and plant design. It also provides good guidance for the construction of ontologies in other domains. The papermaking process differs significantly from the chemical process in terms of reaction types, but they have in common that they both focus on the material and energy flows of raw materials. Therefore, some descriptions of OntoCAPE for chemical processes, such as 'Heat exchange','Material flow', etc., can be reused in the construction of papermaking process ontology.

The next step is to identify the important terms in the domain and define them as classes, and then determine the hierarchy of classes. Some of the professional terms and hierarchies involved are shown in Figure 2, which describe the papermaking process from three aspects: the 'reactions' that occur, the 'phenomena' that form, and how to 'implement' them. From these three aspects, they are further divided until they reach a direct description of the material changes. The ontology includes reactions and parameters that are common to general physical and chemical processes, such as 'Vaporization','Liquidation','Temperature','Pressure', etc. There are also some descriptions unique to the papermaking process, such as 'Dryness','Strength','Bulk' and various papermaking process equipment.

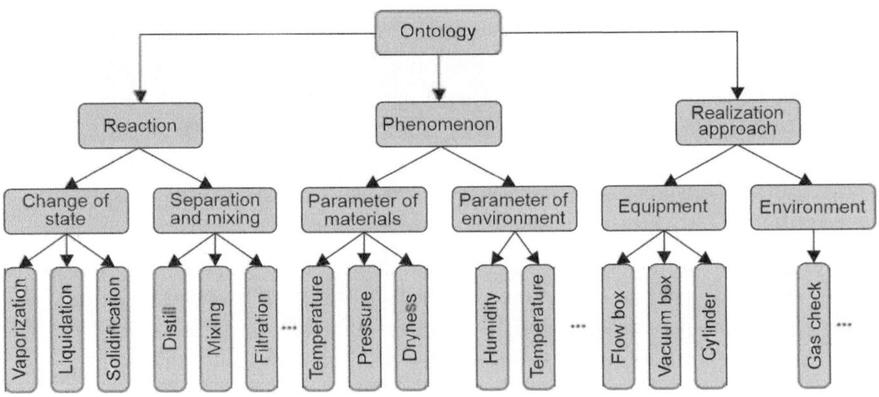

Figure 2: Specific steps of the seven-step ontology construction method

The last step is to define the attributes of the classes and the relationships between the ontologies. Figure 3 shows a schematic diagram of some of the relationships in the pulp drying process. The raw materials are 'Paper pulp' and 'Steam'. The flows into the 'Cylinder', and theis fed into the . The main components of the are 'Water' and 'Fabric', which correspond to two attributes: 'Moisture' and . The in the enters the and undergoes a process of . The phenomenon in terms of parameters is that thechanges. The raw material temperature and dryer cylinder temperature are both phenomena of this process, and their specific values are attributes of temperature. After all these steps, the papermaking process ontology is constructed.

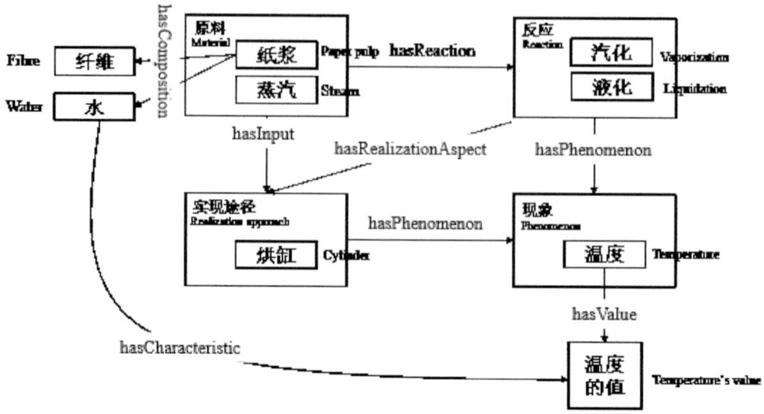

Figure 3: Specific steps of the seven-step ontology construction method

2.2.2 Establishment of fault ontology

For the process industry, normal operation is only part of the operating state, and there are many abnormal operating states. And because most of them are continuous production processes, that is, they operate continuously from inputting raw materials to outputting finished products, and any problem in any link will affect the production performance of the entire industrial chain. These abnormal operating states will greatly affect the economic benefits of the process industry and sometimes even affect safety. In order to describe the process more comprehensively and better play the role of knowledge graph, this paper establishes not only a knowledge graph of normal operating state, but also a fault knowledge graph of abnormal state.

The construction of fault ontology also chooses the seven-step method, and the fault domain terms are determined by combining FMEA 'Failure Mode and Effects Analysis' fault analysis method. Its analysis content mainly includes 'Fault mode', 'Fault occurrence location', 'Fault cause', 'Fault symptom', 'Fault impact' and 'Fault handling measures' , and these analysis contents are conceptualized into corresponding classes. The hierarchy of classes is determined according to the fault tree analysis method combined with the characteristics of papermaking faults, as shown in Figure 4.

The relationship between class attributes and ontology is shown in Table 2. Fault refer to 'referTo' , has warning 'hasWarning' , is related to 'isRelatedTo' , happen in 'happenIn' , has fault cause 'hasFaultCause' , lead to 'leadTo' , and has maintenance method 'hasMaintenanceMethod' after fault occur.

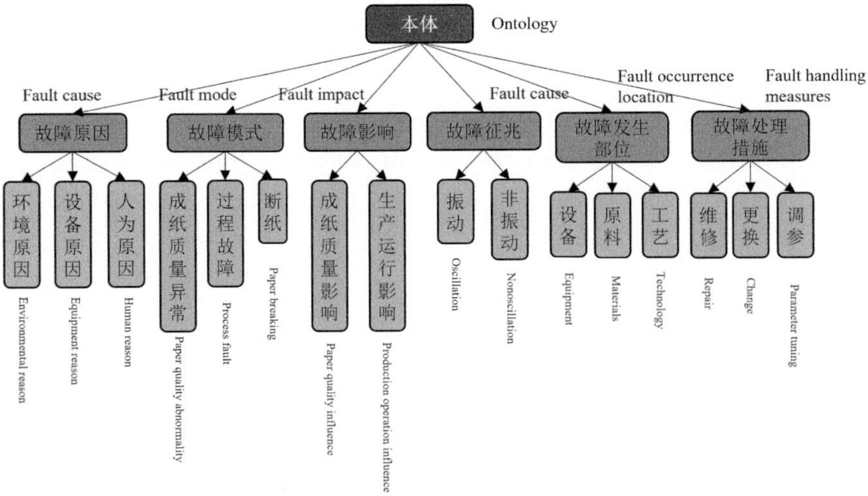

Figure 4: Specific steps of the seven-step ontology construction method

Table 2: The relationship between class attributes and ontology

Relation	Domain	Range
referTo		technological process
happenIn		equipment structure
hasFaultCause		fault cause
hasMaintenanceMethod	fault mode	fault handling measures
hasWarning		fault symptom
isRelatedTo		apparatus parameter
leadTo		fault impact

2.3 Knowledge extraction and import

The last step of the seven-step method for ontology construction is to create instances, which is actually the process of building the data layer of the knowledge graph. It extracts triples from structured, semi-structured and unstructured data, and then imports them into the framework of the ontology. The specific operation is to extract named entities from the data first. The extraction method chosen in this paper is a heuristic method, which is to construct a large number of entity recognition rules by experts in a specific domain and then match the rules with text characters to extract entities. This method can automatically extract entities more accurately in small-scale extraction, but the disadvantage is that constructing rules consumes a lot of manpower and has poor portability. This paper aims to construct a single-domain knowledge graph for papermaking process, which does not require high generality, so it chooses this method to establish a knowledge graph with higher accuracy.

The process of importing after entity extraction is shown in Figure 5. The entities are aligned with the ontology classes and imported into the ontology framework. Then, the relevant ontology relations are linked to construct the data layer of the knowledge graph.

float

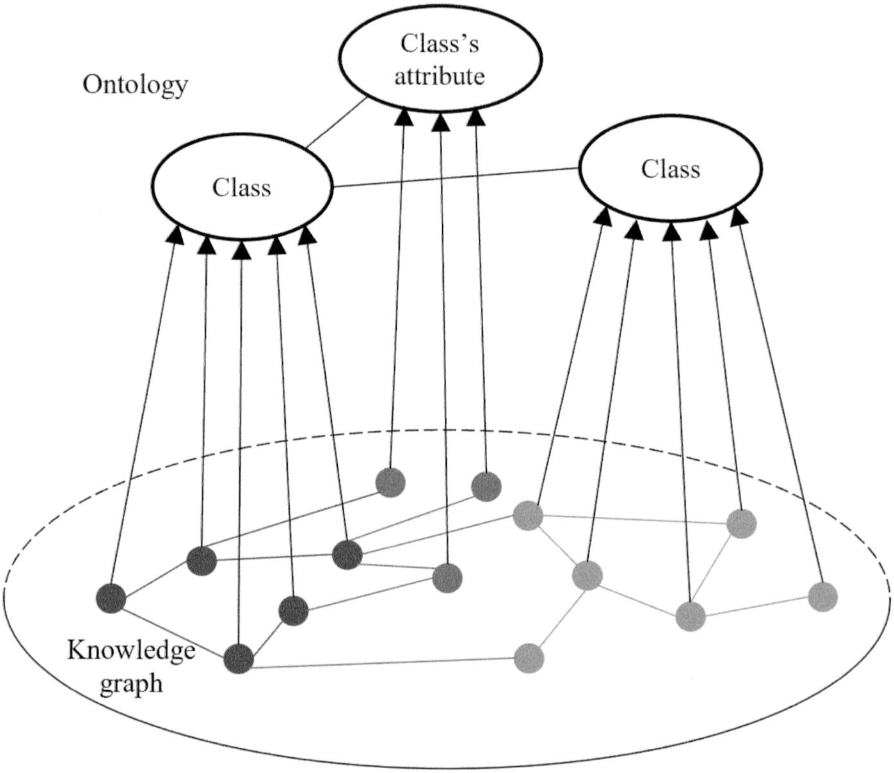

Figure 5: Schematic of the import process after entity extraction

3 Construction and Application of Word Vector Model

The knowledge graph constructed in the previous chapter is essentially a form of text data stored in a graph database. As computers cannot understand human language, text data cannot be directly processed by computers. Therefore, in order to further implement fault diagnosis through computers, the entities in these knowledge graphs need to be transformed into digital forms that are computationally processable. This chapter will use word embedding techniques to build a word vector model. The main process involves first collecting a large amount of text information about the papermaking process from various open channels, and then using the Skip-gram word vector model in the static word vector training method Word2vec to convert all vocabulary in the text into vectors and represent them in the word vector space.

3.1 Text Data Preprocessing

Data preprocessing mainly includes data cleaning and segmentation. All relevant knowledge about the papermaking process is obtained from various open channels, including factories, networks, data, articles, etc. The raw texts obtained from the papermaking process are all unstructured data. These unstructured data will contain a lot of noise, such as some pictures, meaningless particles, alphanumeric characters, etc. Therefore, data cleaning is necessary to remove irrelevant information and useless information such as pictures and formulas through manual discrimination, and to check and correct or remove incomplete parts of the text. In addition, since there is a natural segmentation between words in English corpus, but not in Chinese text, segmentation of Chinese text is necessary to meet the input conditions for building word vector models.

There are many pre-trained models and tools available online, including Jieba segmentation that supports traditional Chinese corpus and custom dictionary, PkuSeg that supports multiple domains, THULAC that has part-of-speech tagging function, etc. In this article, we used the open-source Chinese segmentation tool Jieba segmentation, and selected the precise segmentation mode.

Firstly, we removed all digits, letters, and some meaningless using regular expressions. Stop words like these can also be found in some publicly available stop word lists online. After removing stop words, we proceeded with text segmentation.

3.2 Word vector model

Word2Vec comprises two distinct models, namely the Continuous Bag of Words (CBOW) and the Skip-gram model[20]. In this paper, we employ the Skip-gram model, as depicted in Figure 6, which predicts the center word based on its surrounding context words. The specific steps for training the model are as follows: 1) Firstly, represent the text words as one-hot vectors and use the one-hot vectors of the context words as the model's input.

2) Next, multiply the one-hot vectors of all context words by the weight matrix W between the input and hidden layers.

3) Obtain new vectors by adding all resulting vectors and take their average as the hidden layer vector.

4) Multiply the hidden layer vector by the weight matrix W' between the hidden and output layers.

5) Apply the softmax function to the resulting vector to obtain its V-dimensional probability distribution, and take the vector with the highest probability as the predicted word vector.

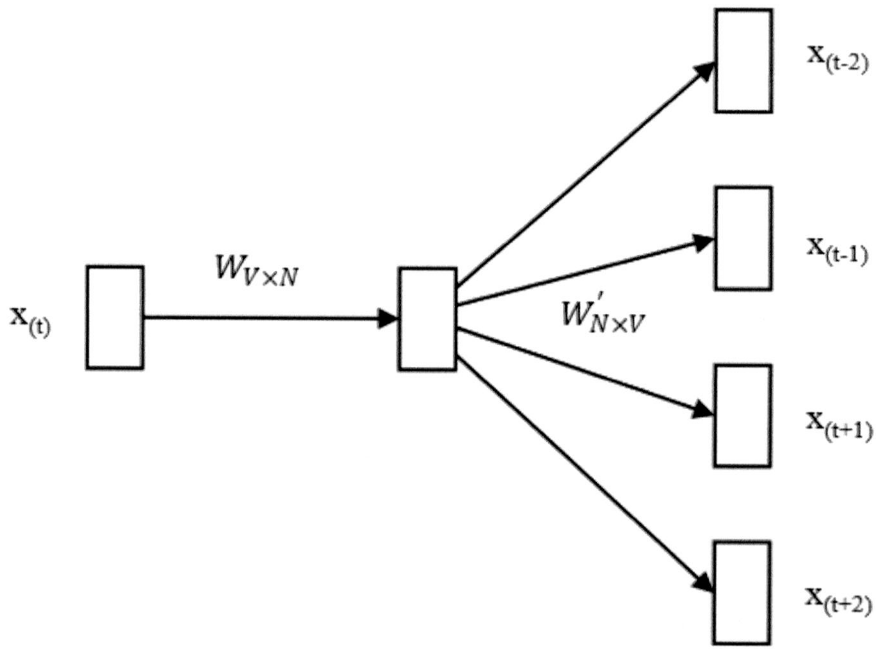

Figure 6: Schematic diagram of the Skip-gram model

3.3 Fault cause Analysis based on word vector distance

3.3.1 Gaussian distribution

In mathematics, a probability density function is used to describe the likelihood of a continuous random variable taking on a certain value at a specific point. If $f(x)$ is the probability density function, then $f(x = a)$ represents the probability of x taking on the value a. This is the theoretical basis for converting distances into probabilities.

The Gaussian distribution[21], also known as the normal distribution, is a very important distribution in probability theory and statistics. It has widespread applications in various fields such as natural sciences, social sciences, and engineering. Many phenomena in the real world can be described by the Gaussian distribution, including the occurrence of failures. The probability density function of the one-dimensional continuous Gaussian distribution is expressed as follows:

$$f(x) = \frac{1}{\sigma\sqrt{2\pi}} \exp\left(-\frac{(x-\mu)^2}{2\sigma^2}\right) = \frac{1}{\sigma\sqrt{2\pi}} e^{-\frac{(x-\mu)^2}{2\sigma^2}}$$

where σ is the standard deviation, μ is the mean, and the Gaussian Distribution with $\sigma = 1$ and $\mu = 0$ is called the standard normal distribution.

The main characteristic of the standard normal distribution, with $\sigma = 1$ and $\mu = 0$, is the 3σ rule as shown in figure 7 , which states that the probability of a value lying between $(\mu - \sigma, \mu + \sigma)$ is 0.6826 , between $(\mu - 2\sigma, \mu + 2\sigma)$ is 0.9544 , and between $(\mu - 3\sigma,\ \mu + 3\sigma)$ is 0.9974

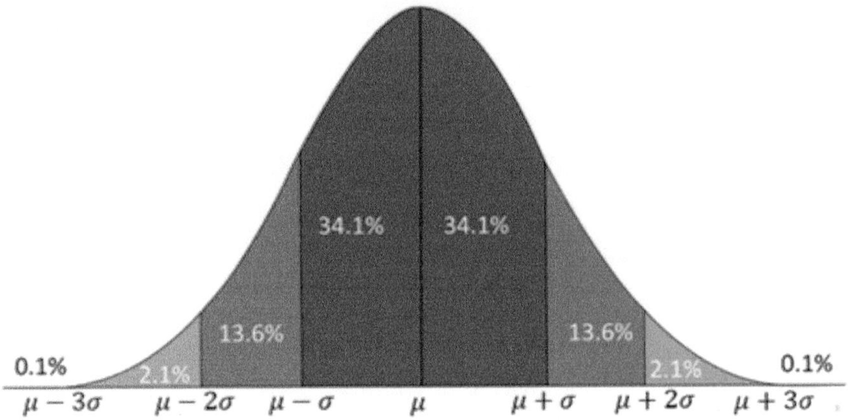

Figure 7: Characteristics of Gaussian distribution

3.3.2 Classification of fault causes based on Gaussian distribution

Based on the characteristics of failure occurrence and the Gaussian distribution, this study discretizes the Gaussian distribution as shown in:

$$P(x) = \frac{e^{-\frac{(x-\mu)^2}{2\sigma^2}}}{\sum e^{-\frac{(x_i-\mu)^2}{2\sigma^2}}}$$

By discretizing the Gaussian distribution and applying it to the distances between the vector representations of the faulty entity and other entities, we can obtain a discrete approximation of the Gaussian distribution with $\mu = 0$.

This allows us to calculate the probability by plugging in the distance between word vectors using the following formula.

$$P\left(X_{a,b}\right)=\frac{e^{-\frac{d_{a,b}^{2}}{2\sigma^{2}}}}{\sum e^{-\frac{d_{i,b}^{2}}{2\sigma^{2}}}}$$

The distance between the fault entity and the cause entity can be converted into the probability of the cause leading to the fault, thus quantifying the relationship between the fault entity and other entities. The discrete Gaussian distribution also conforms to the actual situation that in most cases, the fault is caused by the main reason. Finally, the causes of the fault are classified into three levels based on the 3σ rule. The first-level cause, called direct cause, is distributed within $(0, \sigma)$; the second-level cause, called potential cause, is distributed within $(\sigma, 2\sigma)$; and the third-level cause, called rare cause, is distributed within $(2\sigma, 3\sigma)$.

4 Case Study of Fault Diagnosis

4.1 Construction of Knowledge Graph of Drying Department

This paper mainly takes the drying process of papermaking as an example to illustrate the establishment process of the knowledge graph of papermaking process. Protégé is a free and open-source ontology editing tool based on java language. Protégé has a clear operation interface, simple ontology imports steps, and supports the construction of Chinese ontology. The storage file supports multiple formats, including Rdf, Turtle, Owl and other formats. Combined with the specific application scenario of this paper, Protégé is selected as the ontology construction tool. We import the previously built ontology framework into Protégé and show its schematic diagrams for both drying process and fault ontologies in Figure 8. We also import and define their relations with value domains and definition domains in protégé's 'Object properties' interface. Finally, we import instances in protégé's 'Individual by class' interface.

After adding instances, the ontology forms a knowledge graph, and the visualization display chooses the same open-source graph database software Neo4j based on Java implementation. Neo4j graph database is a kind of directed knowledge graph with attributes, which expresses and stores knowledge through nodes and relationships between nodes. The class in ontology model is the label in knowledge graph and the instance corresponds to the entity in knowledge graph. The relationship between ontologies is the relationship in the triple 'Entity-Relation-Entity', and the data attribute of

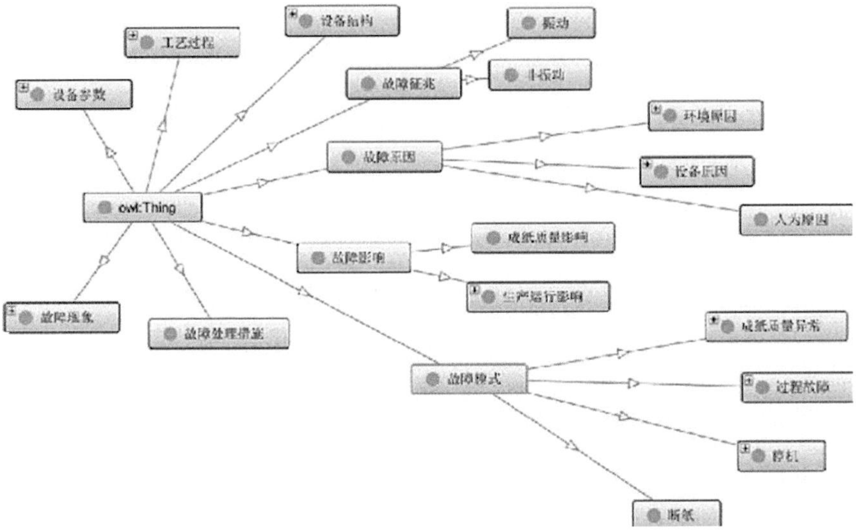

Figure 8: Schematic of the drying process ontology and the fault ontology

class is the attribute in the triple 'Entity-Attribute-Attribute value'. Through Py2Neo transformation, Protégé saved OWL file can be imported into Neo4j graph database. The knowledge graph of papermaking drying process contains 156 entities and 304 relationships.The fault knowledge graph uses the same method to import, which currently contains five faults of papermaking drying process.Including 'drop phenomenon', 'lack of air and air leakage in drying process', 'drying temperature is too high', 'drying wind speed is not uniform' and 'drying moisture is abnormal'.

4.2 Construction of word vector model for dry part text

Data Collection: The data used in this study were obtained from various public sources, including web pages, articles, and books containing text knowledge related to the paper drying process. These knowledge mainly include two parts. One is related to the operation process of the paper drying section, such as equipment description, process steps, reaction process, etc. It can also be the process of establishing models, such as energy consumption simulation of drying, cost optimization, etc. The other part is the collection of texts related to drying process failures, including descriptions of failures, causes of failures, and solutions to failures. These texts were mainly obtained from equipment manuals, plant operation logs, equipment maintenance records, and other channels. In the end, a total of approximately 200,000 Chinese characters were collected.

Data preprocessing: Data processing can be divided into two parts. The first part involves deleting useless information, images, tables, and other noise interference during the collection process. This part of the work requires manual identification. The second part involves Chinese word segmentation. As mentioned earlier, one of the biggest differences between Chinese and English natural language processing is that Chinese text needs to be segmented. The quality of Chinese word segmentation greatly affects the quality of the subsequent models. This article chooses to use the third-party Chinese word segmentation tool, jieba segmentation, with the precise mode. In addition, to obtain better word segmentation results, it is necessary to delete some stop words and input a custom dictionary in advance. Stop words are mainly used to delete some meaningless words, such as some exclamatory words, onomatopoeic words, prepositions, letters, and numbers. The role of a custom dictionary is mainly to define some domain-specific proper nouns in advance in the dictionary. During the word segmentation process, these words will be prioritized for segmentation, reducing some ambiguous situations. The custom dictionary in this article mainly consists of the entities in the knowledge graph established earlier, which are not only professional terms in the papermaking industry but also key vocabulary in this research. After segmentation, the effect is shown in Figure 9. The 200,000-word Chinese text is segmented into about 90,000 words after removing stop words and segmentation.

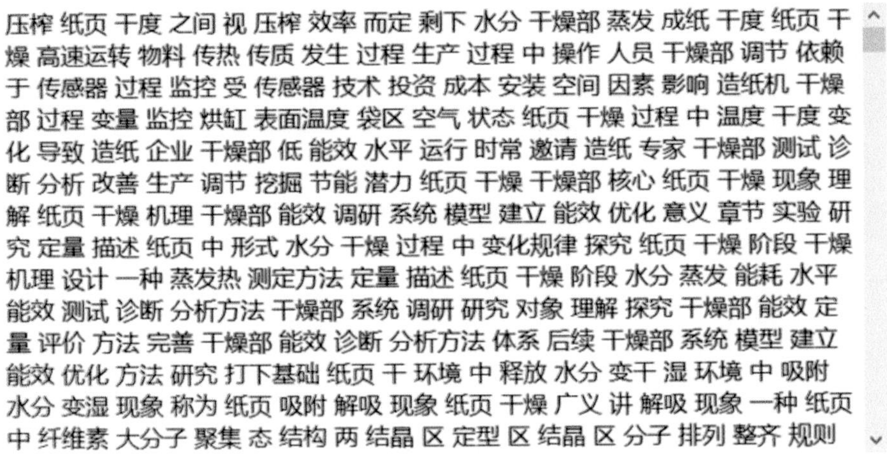
压榨 纸页 干度 之间 视 压榨 效率 而定 剩下 水分 干燥部 蒸发 成纸 干度 纸页 干
燥 高速运转 物料 传热 传质 发生 过程 生产 过程 中 操作 人员 干燥部 调节 依赖
于 传感器 过程 监控 受 传感器 技术 投资 成本 安装 空间 因素 影响 造纸机 干燥
部 过程 变量 监控 烘缸 表面温度 袋区 空气 状态 纸页 干燥 过程 中 温度 干度 变
化 导致 造纸 企业 干燥部 低 能效 水平 运行 时常 邀请 造纸 专家 干燥部 测试 诊
断 分析 改善 生产 调节 挖掘 节能 潜力 纸页 干燥 干燥部 核心 纸页 干燥 现象 理
解 纸页 干燥 机理 干燥部 能效 调研 系统 模型 建立 能效 优化 意义 章节 实验 研
究 定量 描述 纸页 中 形式 水分 干燥 过程 中 变化规律 探究 纸页 干燥 阶段 干燥
机理 设计 一种 蒸发热 测定方法 定量 描述 纸页 干燥 阶段 水分 蒸发 能耗 水平
能效 测试 诊断 分析方法 干燥部 系统 调研 研究 对象 理解 探究 干燥部 能效 定
量 评价 方法 完善 干燥部 能效 诊断 分析方法 体系 后续 干燥部 系统 模型 建立
能效 优化 方法 研究 打下基础 纸页 干 环境 中 释放 水分 变干 湿 环境 中 吸附
水分 变湿 现象 称为 纸页 吸附 解吸 现象 纸页 干燥 广义 讲 解吸 现象 一种 纸页
中 纤维素 大分子 聚集 态 结构 两 结晶 区 定型 区 结晶 区 分子 排列 整齐 规则

Figure 9: Schematic diagram of the effect of word segmentation

4.3 Fault diagnosis Example

Establishment of Word Embedding Model: The word embedding model is a crucial part of natural language processing, and it is the method that truly enables computers to understand human text. As previously mentioned, this research chooses the Skip-gram method in Word2vec to construct the word embedding model. The parameters are detailed in previous sections and will not be repeated here. The 'Window size' and 'Vector size' are chosen as 5 and 100, respectively.

Drop phenomenon is a frequently occurring malfunction in the paper drying process. According to previous studies[22 23], the failure mechanism is due to excessive humidity inside the hood of the drying section, causing water vapor in the air to condense and drop onto the paper. Based on actual production situations and a synthesis of existing research on drop phenomenon malfunctions, the main causes of dewing are five-fold.Including 'insufficient air supply temperature', 'insufficient power of air supply fan', 'insufficient power of exhaust fan', 'too fast drying rate', and 'insufficient air preheatin'. Insufficient air temperature supply leads to low temperatures inside the hood, causing water vapor to condense and form dew due to pre-cooling. Insufficient air supply power and exhaust power prevent the timely discharge of water vapor inside the hood, resulting in high humidity and dew formation. Too fast drying rate leads to excessive evaporation of water vapor and high humidity inside the hood, causing dew formation. Insufficient air preheating is an indirect factor that leads to drop phenomenon by affecting the air temperature supply.

The drop phenomenon can also cause other malfunctions, such as paper breakage and insufficient drying. Therefore, this section will diagnose and

Table 3: The top 10 entities with the closest distance to the word vector of 'Drop phenomenon'

Rank	Distance	Entity
1	0.4	Hot-air
2	0.512	Exhaust exhaust fan
3	0.523	Air supply fan
4	0.677	Waste heat recovery device
5	0.783	Steam pressure
6	0.856	Secondary steam
7	0.89	Water vapor
8	1.097	Paper
9	1.152	Moisture
10	1.156	Cylinder

analyze the drop phenomenon using the constructed fault diagnosis model. The diagnostic results of the fault diagnosis model in this study will be compared with the actual production situation and other existing research on drop phenomenon to verify its diagnostic capability.

According to the fault diagnosis method of this study, the drop phenomenon is diagnosed, and the specific process is as follows:

1) Calculate the distance between the word vector of 'drop phenomenon' and all other entity word vectors, and output the 10 closest entity word vectors to the word vector of 'drop phenomenon' in the knowledge graph, as shown in Table 3.

2) Calculate the standard deviation σ of all distances, which is 0.53 in this experiment.

3) By incorporating distance into probability formula , the probability can be calculated and plotted as a curve, as demonstrated in figure 10.

Comparing the fault diagnosis results of this study with those of other relevant studies regarding droplet phenomenon, the primary causes identified in this study, namely 'hot air', 'supply fan', and 'exhaust fan', corresponded to 'insufficient supply air temperature', 'insufficient supply fan power', and 'insufficient exhaust fan power', respectively. The secondary causes, 'steam pressure' and 'waste heat recovery device', corresponded to 'excessive drying rate' and 'insufficient air preheating', respectively. The fault diagnosis results obtained in this study encompassed all of the results from the other studies, thus demonstrating the comprehensiveness and accuracy of the fault diagnosis in this study.

The identified causes were searched for their fault propagation paths in the completed knowledge graph. Figure 11 illustrates the search results for

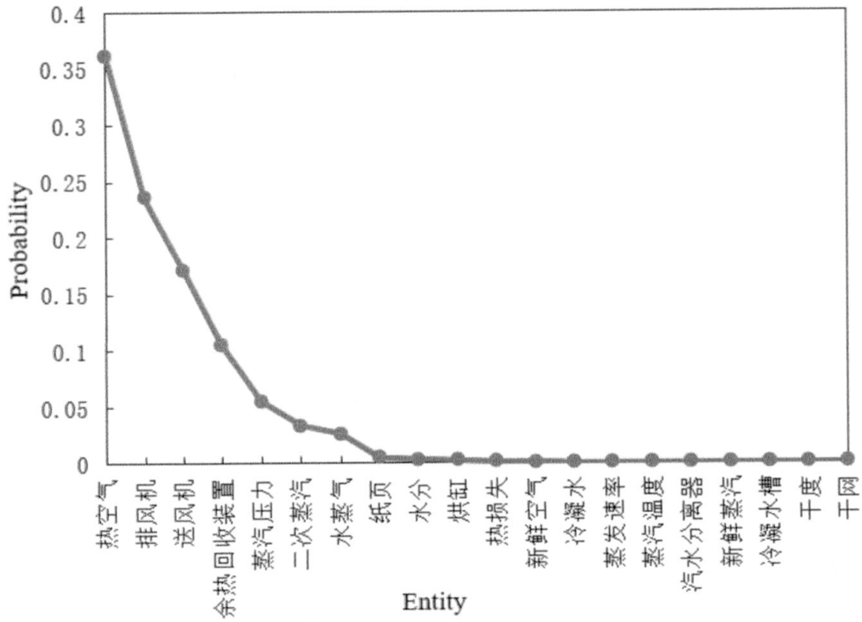

Figure 10: Probability distribution of the 20 entities with the closest distance to the entities of ‘Drop phenomenon’

the relationship between 'droplet phenomenon' and 'secondary steam'. 'Secondary steam' affected 'medium-temperature air' through the 'waste heat recovery device', which in turn affected the 'air heater' and ultimately led to the 'droplet phenomenon'. In actual processes, insufficient recovery of secondary steam results in lower waste heat recovery, leading to the insufficient heating of medium-temperature air and insufficient heat in the air heater during secondary heating. This results in lower supply air temperature and the occurrence of droplet phenomenon. However, this is a relatively low probability scenario, which is consistent with the earlier calculated probability of only 3% for secondary steam triggering droplet phenomenon. On the other hand, the primary fault causes, exhaust fan, hot air, and supply fan, are directly related to the droplet phenomenon. Any issues with the power of these two devices will likely cause droplet phenomenon. The relationship between hot air and droplet phenomenon is that the temperature of the hot air flowing into the supply fan is insufficient, leading to a low supply air temperature and ultimately causing droplet phenomenon.

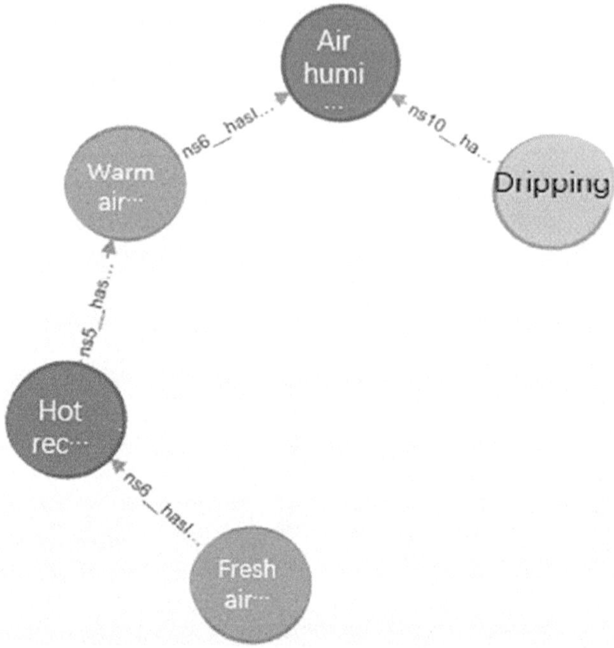

Figure 11: Schematic diagram of the path relationship search for 'Drop phenomenon' and 'secondary steam'

5 Summary and Prospect

Under the influence o f t he t rend t owards i ntelligent m anufacturing i n the global manufacturing industry, the production environment of the papermaking industry is becoming increasingly complex and uncertain. The probability of failures and the difficulty of di agnosis ar e al so in creasing, an d traditional fault diagnosis methods are gradually becoming inadequate to deal with the current papermaking production environment. Knowledge graphs have powerful knowledge representation and reasoning capabilities, and their emergence provides new avenues for addressing the above problems. Based on this, this paper has conducted research on the construction and application of knowledge graphs in the papermaking field to solve the problem of fault diagnosis in the papermaking industry, and has reached the following conclusions:

1) The knowledge graph construction method proposed in this article has good reusability for the paper industry, enabling the sharing of related domain knowledge and ensuring consistency in relevant terminology. It is innovative to divide the paper knowledge graph into a process knowledge graph for normal production status and a fault knowledge graph for abnormal production status, thereby encompassing richer knowledge and providing a sound data foundation for fault diagnosis research in subsequent sections. The process knowledge graph is further divided into three levels, which elaborately describe the material and energy flows of the entire process. This creates a more hierarchical and explicit relationship within the constructed knowledge graph.

2) Based on the characteristics of the data in this article, the Skip-gram method in Word2vec was chosen to establish a word embedding model. After collecting a large amount of text and preprocessing the data, the word embedding model was built to represent entities in the knowledge graph in the word embedding space. Then, the Euclidean distance between the fault entity's word vector and other entity word vectors was calculated, and the distance was discretized into probability using a Gaussian-like distribution, which can provide the probabilities of entities causing faults.

3) To illustrate the entire fault diagnosis process, the paper takes the paper drying section as an example. Firstly, the entire process of constructing the knowledge graph is implemented in open-source software Protégé and Neo4j, which enhances the visualization capability and facilitates editing and modification of the knowledge graph. The constructed drying process knowledge graph includes 178 entities, 259 relationships, and five fault knowledge graphs. Then, a drying section entity word vector model is built by collecting text, calculating the distance between entity vectors, and converting it into probability. Finally, the "drop phenomenon" is taken as an example to show the fault diagnosis result. The fault diagnosis results in this study include all the possible causes of the fault and achieve a comprehensive analysis of the fault by finding the fault transmission path through path searching in the knowledge graph.

This paper presents the construction of a knowledge graph for the paper drying process, as well as a knowledge graph for identifying and diagnosing faults within said process. The utilization of such knowledge graphs has resulted in significant progress towards the identification and resolution of certain drying process faults. However, this paper still contains some shortcomings that need to be addressed in future research:

1) The main focus of this paper is the analysis of the characteristics of the paper industry in order to construct an ontology for the papermaking process and related faults within said process, utilizing a selected software and

knowledge graph construction process. However, during the knowledge importation stage, an expert knowledge and heuristic approach was selected in order to maintain accuracy, which resulted in a low efficiency and high dependence on expert experience, thereby making it difficult to construct large-scale knowledge graphs. Currently, more and more scholars are utilizing various AI methods, such as named entity recognition and relation extraction, to increase efficiency. Although the accuracy of these methods is currently suboptimal, with the advancement of algorithms, these methods are expected to become more intelligent and even surpass the accuracy of manual recognition in the future. By utilizing these methods, large-scale knowledge graphs can be efficiently constructed and better applied in practical production.

2) Due to limited manpower and time, the knowledge graph constructed for the paper industry in this paper is not yet comprehensive. The current corpus is also insufficient, lacking actual data from factories, resulting in fault diagnosis results that do not meet expected requirements. As knowledge is collected and accumulated, the knowledge graph can be further improved through a completion mechanism to achieve better fault diagnosis results. Since the establishment of the knowledge graph in this study is mainly based on ontology construction, in the future, it may be possible to construct a knowledge graph specifically for a particular factory using an existing ontology framework. This will result in a more comprehensive and targeted knowledge graph, thereby improving the accuracy of fault diagnosis.

References

[1] D. GuangLi, Based on the statistical analysis and coping strategies of domestic production safety accidents from 2019 to 2021, Emergency Rescue in China (2023) 5.

[2] A. Ratni, D. Benazouz, Mathematical modelling of rolling element bearings fault for the diagnosis in the gearbox-induction machine, Mathematical Models in Engineering 6 (2020) 1–12.

[3] S. Taqvi, H. Zabiri, L. D. Tufa, F. Uddin, S. A. Fatima, A. S. Maulud, A review on data-driven learning approaches for fault detection and diagnosis in chemical processes, ChemBioEng Reviews (????).

[4] H. Borowczyk, J. Spychaa, On designing an expert diagnostic system of the aircraft turbine engine functional units, Journal of Konbin 50 (2020) 171–192.

[5] F. Zhang, Y. Zhang, T. Xu, A review of knowledge graph technology in the field of automatic question answering, in: 2020 International Signal Processing, Communications and Engineering Management Conference (ISPCEM).

[6] X. Wu, J. Duan, Y. Pan, M. Li, Medical knowledge graph: Data sources, construction, reasoning, and applications, Big Data Mining and Analytics 6 (2023) 201–217.

[7] WeiZheng, KaimanLi, YanrongWu, MingchenSun, ZhichunWang, WeiZheng, KaimanLi, YanrongWu, ZhichunWang, M. and, Building a large-scale knowledge graph for elementary education in china, Springer, Singapore (2019).

[8] Y. Lang, G. Wang, Personalized knowledge point recommendation system based on course knowledge graph, Journal of Physics Conference Series 1634 (2020) 012073.

[9] W. Wang, Y. Xu, C. Du, Y. Chen, H. Wen, Data set and evaluation of automated construction of financial knowledge graph, Data Intelligence (2021) 1–21.

[10] X. B. Tang, W. G. Fu, Y. Liu, Knowledge big graph fusing ontology with property graph: A case study of financial ownership network, Knowledge Organization: KO (2021) 48.

[11] Q. Qiu, K. Ma, H. Lv, L. Tao, Z. Xie, Construction and application of a knowledge graph for iron deposits using text mining analytics and a deep learning algorithm, Mathematical Geosciences 55 (2023) 423–456.

[12] S. Qi, Y. Wu, L. Peng, L. X. Dong, S. Hui, Mining summaries for knowledge graph search, IEEE Transactions on Knowledge and Data Engineering 30 (????).

[13] X. Zhu, Y. Ji, A digital twin-driven method for online quality control in process industry, The International Journal of Advanced Manufacturing Technology (2022) 119.

[14] J. Li, S. Liu, A. Liu, R. Huang, Knowledge graph construction for sofl formal specifications, International Journal of Software Engineering and Knowledge Engineering 32 (2022) 605–644.

[15] C. Kou, T. Liu, L. Ma, G. Jiang, Y. Wang, Y. Zhao, Construction and application research of knowledge graph in spacecraft launch, Journal of Physics: Conference Series 1754 (2021) 012180 (9pp).

[16] K. Messner, E. Srebotnik, Biopulping: An overview of developments in an environmentally safe paper-making technology, Fems Microbiology Reviews 13 (2010) 351–364.

[17] S. Karimi, I. Iordanova, D. St-Onge, Ontology-based approach to data exchanges for robot navigation on construction sites, Journal of Information Technology in Construction 26 (2021) 546–565.

[18] X. Dong, E. Gabrilovich, G. Heitz, W. Horn, N. Lao, K. Murphy, T. Strohmann, S. Sun, W. Zhang, Knowledge vault: a web-scale approach to probabilistic knowledge fusion, ACM (2014).

[19] A. Hur, N. Janjua, M. Ahmed, A survey on state-of-the-art techniques for knowledge graphs construction and challenges ahead (2021).

[20] F. Heimerl, M. Gleicher, Interactive analysis of word vector embeddings, Computer Graphics Forum 37 (2018) 253–265.

[21] J. C. S. Vasconcelos, G. M. Cordeiro, E. M. M. Ortega, G. O. Silva, R. Aykroyd, A random effect regression based on the odd log-logistic generalized inverse gaussian distribution (2023).

[22] W. Shen, Prevention and control of condensation drip in papermaking workshop, Air Conditioning HVAC Technology (2017) 5.
[23] Y. Wei, Analysis on modeling strategy of ventilation system in drying section of paper machine, Paper Science and Technology (2021).

14. Research on the Construction of Papermaking process Model based on Digital Twin

Yishui Zhang, Rongyuan Wang, Mengna Hong and Zhenglei He*

1 Introduction

The papermaking process involves a large number of complex physical and chemical reactions[1], accompanied with characteristics of multi-variable, strong coupling and nonlinear, etc., hindering the intelligent development pro-cess of the papermaking industry[2]. At present, the process models and control systems, established by paper-making enterprises based on a new generation of information technology, have not yet solved the problem of "data silos" and cannot integrate material flow and energy flow information to achieve real-time monitoring and control of paper production, which affects the sus-tainability of the process dramatically.

With the development of Industry 4.0 era[3;4], digital twin is gradually being studied and applied in the process industry[3]. In the steel industry, with the applications of technologies such as intelli-gent data sensing, multi-source heterogeneous data integration, efficient data transmission, digital twin creation, enhanced interaction, and conversion ap-plications, a production line that combines reality with virtuality can be estab-lished to realize the optimization of the production process[5]. In the machine building industry, simulation and optimization based on the digital twin's dynamic perception of the physical machine tool, it can effectively optimize machining conditions such as cutting parameters and reduce carbon emis-sions[6]. It is worth noting that, the establishment of a digital twin model of the process industry, can facilitate the simulation, analysis, monitoring and optimization of manufacturing processes in real time, assising dynamic man-agement without physical efforts[7].

State Key Laboratory of Pulp and Paper Engineering, South China University of Technology, Guangzhou 510640, China.
* Corresponding author: hezhenglei@scut.edu.cn

The papermaking industry is a typical process-oriented industry[3], and the information technologies such as big data and machine learning have been widely used in its energy-saving renovation[3;8]and process optimization[9] sectors. However, most of the previous studies only focus on a single process or a single equipment without considering the modeling and control of the whole process systematically, which tends to achieve local optimizations. Therefore, there is an urgent need to develop the digital twin model for the papermaking process that can perform online sensing, analysis, simulation, optimization, and decision-making of the process. The emergace of digital twin in the industrial domain provides feasible solution ideas[10;11;12]. As aforementioned, although a large number of scholars have conducted modeling studies on digital twins and achieved some results in related fields, there is a lack of specific digital twin frameworks and example studies for the chemical industry, especially for the papermaking industry.

2 Methods

2.1 A digital twin framework for the papermaking process

The papermaking process has the following characteristics: diverse raw materials with different nature and complex interrelationships in the process, and there are a large number of parameters that are difficult to be detected along with the process; various complicated physical and chemical reactions occur in its process[13].

The core of designing a digital twin of a papermaking process is to perform mechanistic or data-driven modeling of the process to rationally describe the production process[14;15;16]. Industrial data are first collected, pre-processed, and stored in a database; then data-driven methods are used to infer parameters that cannot be directly measured, find relationships in the production process, and simulate the process to enable assessment of the current state, diagnosis of past problems, prediction of future trends, provide more reasonable and reliable process parameters and optimize the process g process[17;18]. Figure 1 illustrates the digital twin framework for the papermaking process, and the individual modules are described below.

The physical layer mainly includes materials, equipment, production lines, environment and personnel. It mainly provides data on all production factors and production elements such as materials, equipment and processes in the papermaking process. Materials such as fibers, fillers, retention aids, dry strength agents, equipment such as pulper, headbox, wite part, core winder and cutter, production lines are also divided into different production lines such as newsprint, household paper, boxboard as well as the temperature and humidity of the environment and personnel operation.

The sensing layer is mainly responsible for data collection and transmission. The original data in the process is collected by the sensors on site to the distributed control system, and then uploaded to the enterprise's cloud service platform through the net gate it can be downloaded directly through the cloud service platform when used. The data collection also includes the data of offline inspection such as the quality index of raw paper.

The data layer is responsible for receiving the original data. Due to the different frequency of data collection, spatial location of sensors, etc., as well as the fact that some of the data are obtained by offline laboratory tests, and there are many uncertainties in the process, such as process abnormalities, equipment failures, or irregularities in operation, there are time differences and delays in the collected data. Therefore, after operations such as data matching, data filling, and outlier deletion, varied information needs to be stored in the database, which can later be used as input for the construction of the digital twin model in virtual space.

The model layer is the digital modeling of physical entities. The model layer includes a solver model for the process parameters, a prediction model, and a simulation model of the physical entity[19]. Based on the principle of mass and heat transfer, a parameter prediction model is built by analyzing the data generated in the process and simulating and calculating the production process to visualize the process parameters and process monitoring.

The application layer includes functions such as process optimization, parameter optimization, operation guidance, testing and diagnosis, and inventory calculation[20;21]. Finally, it is fed back to the physical entity to realize the control management and closed-loop optimization of the whole process. Based on the massive data, we can provide more reasonable and reliable process parameters and optimize the process through prediction and optimization models to achieve the evaluation of the current state, or the prediction of future trends.

2.2 Research on digital twin-based modeling techniques

The above research provides a method for constructing a digital twin framework for the papermaking process, but due to the complexity and variety of variables in the papermaking process, cost, and technical constraints that prevent access to certain key information on process variables, there is still a need to find suitable methods to address the problem that certain variables in the process are difficult to measure or cannot be measured directly by sensors.

Therefore, we propose a method for parameter solution and parameter prediction in which, when the process mechanism can be constructed, constraints are set on the unknown parameters, and the missing parameters are solved by combining the mechanism equation with a multi-objective optimization algorithm; in the case where the process mechanism is difficult to construct, the parameters are predicted by using a data-driven model by screening the characteristic variables through correlation analysis. Figure 2 shows the technical route of the method.

The raw data required for this paper is collected from the mill's distributed control system, which collects process data such as Vehicle speed, steam pressure, temperature, and humidity for four quantifications, as shown in Table 1.

The data collected had null values, missing values or data redundancy, and some of the data were collected at different frequencies, making modeling more complex[22]. To address the above characteristics, the originally collected data were matched in time series, with most of the data being sampled in 60s, 10min and 20min sets, and the collected data were matched in 5min sets to take the average value within the sampling interval. Missing values are filled by averaging and interpolation. Outliers are removed using box plots.

Paper production can be divided into working conditions based on basic weight, as shown in Table 2. In this paper, four basic weights are considered as four working conditions are modeled separately to compare the predictive effect of the model under different working conditions.

For processes where the mechanistic equations have been constructed, the solve function of the math module in Python is used to calculate the unknown solution. If only a partial solution can be found, the found parameters can be used together with the known parameters as known parameters, and the other parameters can be assigned values until the error in each mechanistic equation is minimised, and when the solution is stable, the value at this point can be considered to be the value of the sought parameter. Since

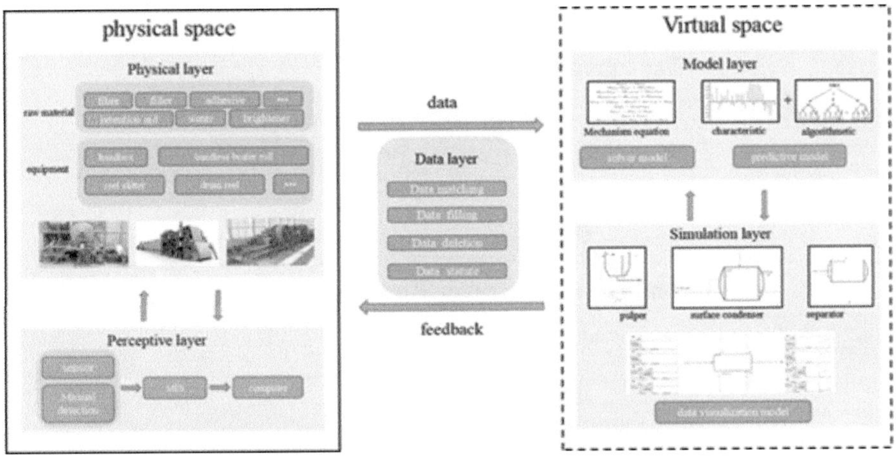

Figure 1: Digital twin framework of papermaking process

Table 1: Modeling basic data table

	Varibles	
R_s	Reel speed	$m{\cdot}min^{-1}$
P_s	Pressure of steam	kPa
P_h	Pressure of headbox	kPa
BW_s	Basis weight before drying sizing	$g{\cdot}m^{-2}$
M_{dz}	Moisture before drying sizing	%
M_p	Moisture content of paper	%
C_p	Pulp consistency of top wire	%
T_{ia}	Temperature of entering air	℃
H_{ia}	Humidity of entering air	%
T_{oa}	Temperature of exhaust air	℃
H_{oa}	Humidity of exhaust air	%
W_{oa}	Weight of exhaust air	$kgd{\cdot}a{\cdot}s^{-1}$
F_s	Flow of steam into cylinder	kg/s
P_{sc}	Pressure of steam into cylinder	kPa
T_s	Temperature of steam into cylinder	℃
BW	Basics weight	$g{\cdot}m^{-2}$
V_f	Former Vacuum	kg
V_{cr}	Couch Roll Vacuum	kpa
V_{pr}	Press Roll Vacuum	kpa

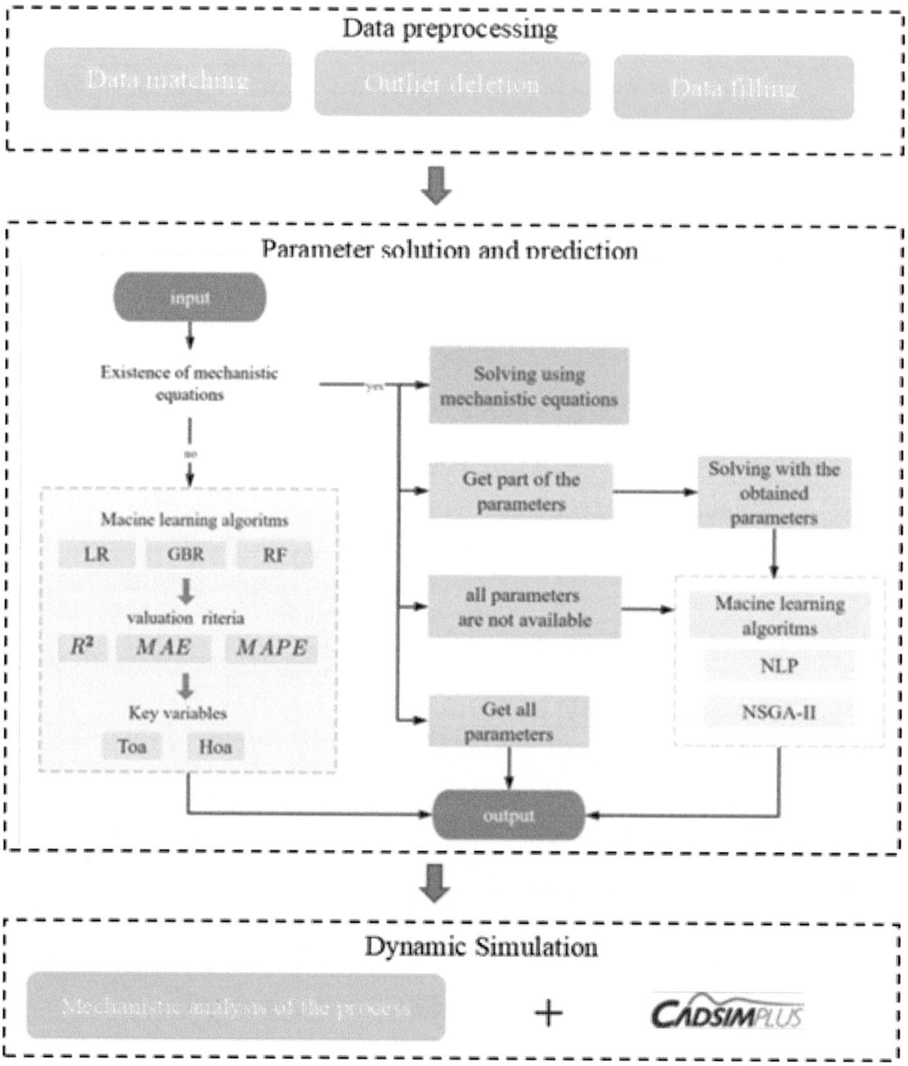

Figure 2: Technical route

Table 2: Working conditions and basic weight comparison

Condition	1	2	3	4
Basic weight ($g{\cdot}m^{-2}$)	115-130	105-115	80-95	65-80

there are multiple equations in which the computational error needs to be minimised simultaneously, this problem can be transformed into a multi-objective, multi-constraint optimization problem. Classical nonlinear programming and NSGA-II are used in this paper.

Sequential Least Squares Programming (SLSQP) is one of the nonlinear planning algorithms for constrained problems, which can solve iterative methods for nonlinear optimization problems with constraints, such as boundary equation and inequality constraints[23]. This method has the following characteristics: it can handle any degree of nonlinearity, including nonlinearity in constraints, and it must choose appropriate parameters such as initial solution, shift acceptance criterion, and step control strategy to ensure convergence and efficiency.

NSGA-II is an improvement on the first generation of non-dominated sorting genetic algorithm: it proposes a fast non-dominated sorting algorithm[24;25;26], which reduces the computational complexity on the one hand, and merges the parent population with the offspring population on the other hand, retaining all the best individuals; it introduces an elite strategy to ensure that some good population individuals will not be discarded in the evolutionary process, and improves the accuracy of the optimization results; it adopts the crowding degree and crowding degree comparison operators as the criteria for comparison among individuals in the quasi-Pareto domain to ensure that the diversity of individuals in the quasi-Pareto domain can be evenly extended to the whole Pareto domain. Crowding degree and crowding degree comparison operator are adopted as the comparison criteria among individuals in the population, so that the individuals in the quasi-Pareto domain can be evenly extended to the whole Pareto domain and ensure the diversity of the population.

When it is difficult to construct a mechanistic model to find the relationship between these parameters , a predictive model based on machine learning is established. Taking the ventilation system of the drying department as an example, the exhaust air temperature and humidity is an important index that reflects the operational status of the drying department and the rationality of the process, and the mechanism of the process is difficult to construct, so the prediction models for its exhaust air temperature and humidity are established. In this paper, we choose the more classical linear regression(LR), gradient boost regression(GBR) and random forest(RF) methods[27].

LR is a type of regression analysis that models the relationship between one or more independent and dependent variables using a least square function called a linear regression equation[28]. Linear regression models are very easy to understand and the results are very interpretable, which facilitates decision analysis, but for nonlinear data or polynomial regression with correlation between data features it is difficult to model and represent highly complex data well.

GBR uses a continuous approach to construct trees, with each tree trying to fix the errors of the previous tree[29]. By default, gradient boosted regression trees are not randomized, but use strong prepruning, and gradient boosting usually uses trees with very small depths. Such models have a small memory footprint and faster prediction speed[30]. The main idea of gradient boosting is to merge multiple simple models, where each tree makes good predictions for only part of the data, and the more the added trees, the more iterations can be made to improve performance.

RF is an algorithm that integrates multiple trees through the idea of integrated learning, based on which the samples are trained and predicted[31;32]. When used as a regression algorithm, the output value of the sample prediction is weighted by the regression values of the decision trees that make up the random forest as the output. It can handle both discrete and continuous data, and operates with high efficiency and accuracy.

2.3 Visualisation model based on digital twin

Most of the traditional simulation models of the papermaking process have only focused on a single section, without considering the whole process , or have only calculated a relatively small number of parameters; The visualisation model of the papermaking process is only based on a mechanistic model, but does not consider processes that cannot be modelled mechanistically, and does not incorporate advanced information technology; the built simulation process can only be calculated in a specific order. In view of the above problems, this paper uses the chemical simulation software CADSIM Plus to build the simulation process.

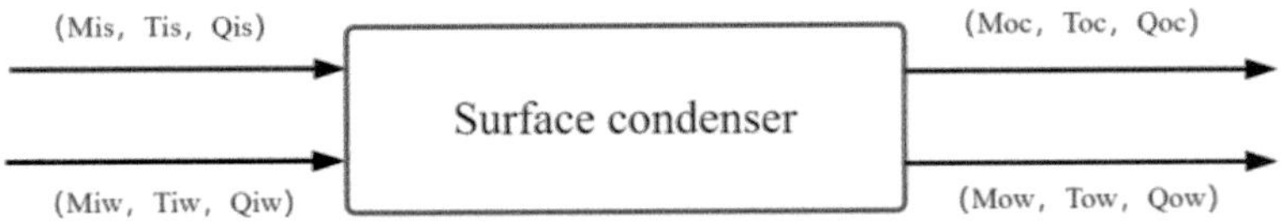

Figure 3: Surface condenser module

CADSIM Plus chemical process simulation software has a built-in physical property system database and operation unit module library of papermaking, including a lot of physical property methods and parameters, you can flexibly uses all the arrangement with and combinations of these methods to develop process simulation. At the same time, based on the previous work, we can use the mechanism model combined with process data to build modules of various parts in the software and store them in the model library. In the future, when building different processes, we can continue to reuse this module, which improves the simplicity of modeling and reduces the time required for modeling.

3 Results and Conclusions

3.1 Solver model

This paper takes the surface condenser of the drying process as the object of study. Firstly, the mechanistic process is analyzed: the secondary steam entering the surface condenser is discharged as condensate through heat exchange, and the incoming cold water is heated and discharged as hot water. Figure 3 is a schematic diagram of the surface condenser module, and Table 3 shows the input and output parameters of the surface condenser module.

In the ideal state, the mass balance and energy balance are equation (1-3):

$$M_{is} = M_{oc} \tag{1}$$

$$M_{ow} = M_{iw} \tag{2}$$

$$Q_{is} + Q_{iw} = Q_{ow} + Q_{oc} \tag{3}$$

Table 3: Input and output variables

Input variables		Output variables	
M_{iw} (kg)	Mass of cold water	M_{ow} (kg)	Mass of warm water
T_{iw} (°C)	Temperature of cold water	T_{ow} (°C)	Temperature of warm water
Q_{iw} (kj)	Heat of cold water	Q_{ow} (kj)	Heat of warm water
M_{is} (kg)	Mass of secondary steam	M_{oc} (kg)	Mass of condensated water
P_{is} (kpa)	Temperature of secondary steam	T_{oc} (°C)	Temperature of condensated water
Q_{is} (kj)	Heat secondary steam	Q_{oc} (kj)	Heat condensated water

Among them, secondary steam, cold water, hot water and condensate heat calculation equations (4-7), where C is the specific heat capacity of water:

$$Q_{is} = M_{is} \times (28.054 \times \ln(P_{is}) + 2583.6) \tag{4}$$

$$Q_{ow} = M_{ow} \times \ T_{ow} \times \ C \tag{5}$$

$$Q_{iw} = M_{iw} \times T_{iw} \times C \tag{6}$$

$$Q_{oc} = M_{oc} \times T_{oc} \times C \tag{7}$$

After constructing the mechanistic equations, the first assumption is made for a specific case and the two methods are compared with different numbers of known parameters for the solution. The number of known parameters is assumed to be 5, 4, 3, 2, and 1. Each parameter has different combinations and the number of known parameters is decremented in the order of Table 3 until the number of known parameters is one. The idea of the solution is to output the results directly if the known parameters can be calculated exactly by the mechanism equation. If the missing parameters do not give the full result, the solution continues in other ways. When substituting the unknown parameters into the mechanism equation, if the error and error sum of each equation is relatively small, the parameters can be considered as acceptable calculated values at this time.

As the relative error has the advantage of reflecting the magnitude and direction of the error and can more accurately reflect the level of confidence in the measurement, it is set in terms of relative error when setting the objective function over the error. The objective function is as follows: Equation (8).

$$\begin{cases} f_1 = (M_{is} - M_{oc})/M_{is} + (M_{is} - M_{oc})/M_{oc} \\ f_2 = (M_{ow} - M_{iw})/M_{iw} + (M_{ow} - M_{iw})/M_{ow} \\ f_3 = (Q_{is} - M_{is} \times (28.054 \times np.log(P_{is}) + 2583.6))/Q_{is} + \\ \quad (Q_{is} - M_{is} \times (28.054 \times np.log(P_{is}) + 2583.6)) \\ \quad (M_{is} \times (28.054 \times np.log(P_{is}) + 2583.6)) \\ f_4 = (Q_{ow} - M_{ow} \times T_{ow} \times C)/Q_{ow} + \\ \quad (Q_{ow} - M_{ow} \times T_{ow} \times C)/(M_{ow} \times T_{ow} \times C) \\ f_5 = (Q_{oc} - M_{oc} \times T_{oc} \times C)/Q_{oc} + \\ \quad (Q_{oc} - M_{oc} \times T_{oc} \times C)/(M_{oc} \times T_{oc} \times C) \\ f_6 = (Q_{iw} - M_{iw} \times T_{iw} \times C)/Q_{iw} + \\ \quad (Q_{iw} - M_{iw} \times T_{iw} \times C)/(M_{iw} \times T_{iw} \times C) \\ f_7 = (Q_{is} + Q_{iw} - Q_{ow} - Q_{oc})/(Q_{is} + Q_{iw}) + \\ \quad (Q_{is} + Q_{iw} - Q_{ow} - Q_{oc})/(Q_{ow} + Q_{oc}) \end{cases} \tag{8}$$

The constraints are the upper and lower limits of the range of each parameter in the actual case. To obtain better results, the limits of the error function are added to the constraints, as in equation (9):

$$\begin{cases} f_1 < 0.05 \\ f_2 < 0.05 \\ f_3 < 0.05 \\ f_4 < 0.05 \\ f_5 < 0.05 \\ f_6 < 0.05 \\ f_7 < 0.05 \end{cases} \tag{9}$$

Table 4 shows a comparison table of the errors in solving the parameters under different methods with the number of known parameters as 5, 4, 3, 2 and 1. The errors of the values in the table are in the form of percentages. From the table it can be seen that the solution result of nonlinear programming is better than that of NSGA-II. Nonlinear programming needs to be assigned reasonable initial values before running the model, so it is already closer to reasonable values in the first run, but it is difficult to provide feasible initial solutions in the face of difficult to determine upper and lower limits of

Table 4: Comparison of solution error results

	5		4		3		2		1	
	NLP (%)	NSGA-II(%)	NLP (%)	NSGA-II(%)	NLP (%)	NSGA-II(%)	NLP (%)	NSGA-II(%)	NLP (%)	NSGA-II(%)
M_{is}	-	-	-	-	-	-	-	-	-	-
P_{is}	-	-	-	-	-	-	-	-	369	891
M_{iw}	-	-	-	-	-	-	2	47	2	61
T_{iw}	-	-	-	-	10	22	8	15	7	27
M_{oc}	-	-	0	0	0	0	0	0	0	0
T_{oc}	9	11	9	12	18	24	18	1	18	3
T_{ow}	1	1	1	1	7	13	5	22	3	31
M_{ow}	0	0	0	0	0	0	2	47	2	61

parameters or when there are more parameters, so nonlinear programming is already closer to reasonable values in the first run, but when the upper and lower limits of parameters are not easy to determine or when there are more parameters to be assigned, it will not be possible to determine a suitable initial value. NSGA-II, on the other hand, uses random assignment of values in the first run, which can be applied to all cases. Normally, the more parameters are known, the more accurate the solution will be, whereas if there are only two or one known parameter values, the solution will no longer be credible. In this paper only five of these cases are listed, the rest are similar.

It can be concluded that the method can, to some extent, solve the problem of missing parameter values in the process of model calculation. Moreover, it is a relatively general method that can be used to solve the missing parameters in any process where the mechanistic equations can be established. When the range of parameters is known, the nonlinear programming solution is used; when the range of parameters is unknown and initial values are difficult to assign, NSAG-II is used to solve.

3.2 Prediction model

Considering that the production process is a relatively complex, non-linear, highly coupled and uncertain process, a correlation analysis between the parameters of the collected characteristics is necessary to eliminate irrelevant variables. The Pearson correlation coefficients of each collected variable with the exhaust air temperature and humidity are calculated, and a correlation coefficient generally considered with an absolute value greater than 0.4 representing a moderate correlation. In this study, the characteristic variables with correlation coefficients greater than 0.4 were selected as inputs to the model. Figure 4 shows the results of the correlation analysis.

The prediction results of the prediction model for the first twenty data sets under the four operating conditions are shown in Figure 5.

As can be seen from Figure 5, the predicted values of random forest and gradient boosting regression are closer to the actual values, and the prediction of linear regression is less effective. In principle, linear regression is suitable for situations where there are few variables and the relationships between the variables are not particularly complex. However, papermaking is a complex process involving mass and heat transfer, and the relationship between variables is complex, and in most cases the relationship between the parameters is non-linear, so linear regression is less effective in making regression predictions for non-linear data.

Both Random Forest and Gradient Boosting Decision Tree are integrated learning algorithms, they both consist of multiple decision trees, and the final result needs to be decided by all decision trees together. However, Random Forest and Gradient Boosting Decision Tree differ in their ideas. Random Forest adopts the idea of Bagging in machine learning, in which Bagging draws samples from the training set to train weak classifiers by uniform sampling with put-back, and the training sets of each classifier are independent of each other, and the training sets of decision trees are independent of each other, and the trees of Random Forest can be generated in parallel with each other. The gradient boosting decision tree uses the Boosting idea, where the training sets of each classifier are not independent of each other, and the composed trees need to be generated serially in order, and the training sets of each weak classifier are sampled from the results of the previous weak classifier.

From the above analysis it can be concluded that the prediction errors of both Random Forest and Gradient Boosting Regression are relatively small. From the values of the evaluation indices, the predicted values of random forest are better than those of gradient boosting regression, with better agreement with the actual value curve and a closer trend. It shows that the random forest model has better robustness and generalisation ability, and also shows more stable effect under different working conditions, so that the random forest can be applied to complex drying conditions to provide reference for index prediction; it also shows that the model is real-time and the prediction results can be used for online process monitoring.

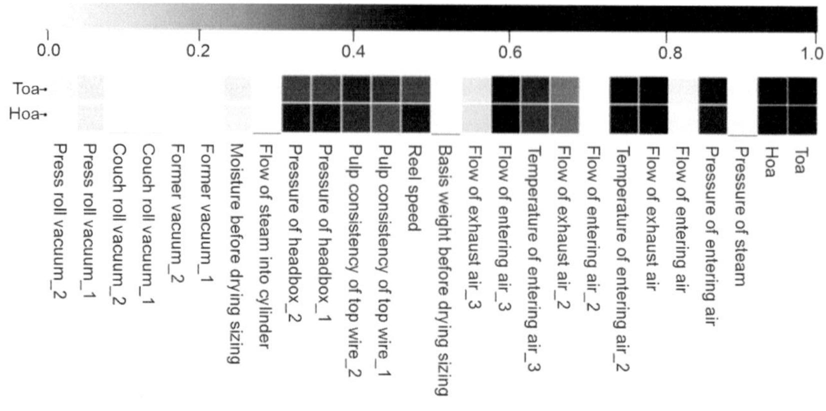

Figure 4: Correlation analysis results of T_{oa} and H_{oa}

Table 5: Comparison of the effects of exhaust temperature and humidity prediction models

Variable	Model	valuation riteria	1	2	3	4
		R^2	0.85	0.89	0.92	0.81
	RF	MAE	0.47	0.40	0.52	0.40
		MAPE	0.72	0.60	0.79	0.59
		R^2	0.82	0.88	0.92	0.78
T_{oa}	GBR	MAE	0.54	0.46	0.54	0.44
		MAPE	0.82	0.69	0.82	0.64
		R^2	0.65	0.80	0.85	0.59
	LR	MAE	0.79	0.61	0.80	0.62
		MAPE	1.19	0.92	1.22	0.90
		R^2	0.87	0.92	0.92	0.85
	RF	MAE	3.12	2.54	3.56	2.58
		MAPE	3.20	2.52	3.58	2.27
		R^2	0.84	0.92	0.91	0.83
H_{oa}	GBR	MAE	3.56	2.77	3.96	2.85
		MAPE	3.62	2.75	3.96	2.49
		R^2	0.76	0.85	0.84	0.58
	LR	MAE	4.65	4.08	5.44	4.37
		MAPE	4.68	4.03	5.45	3.82

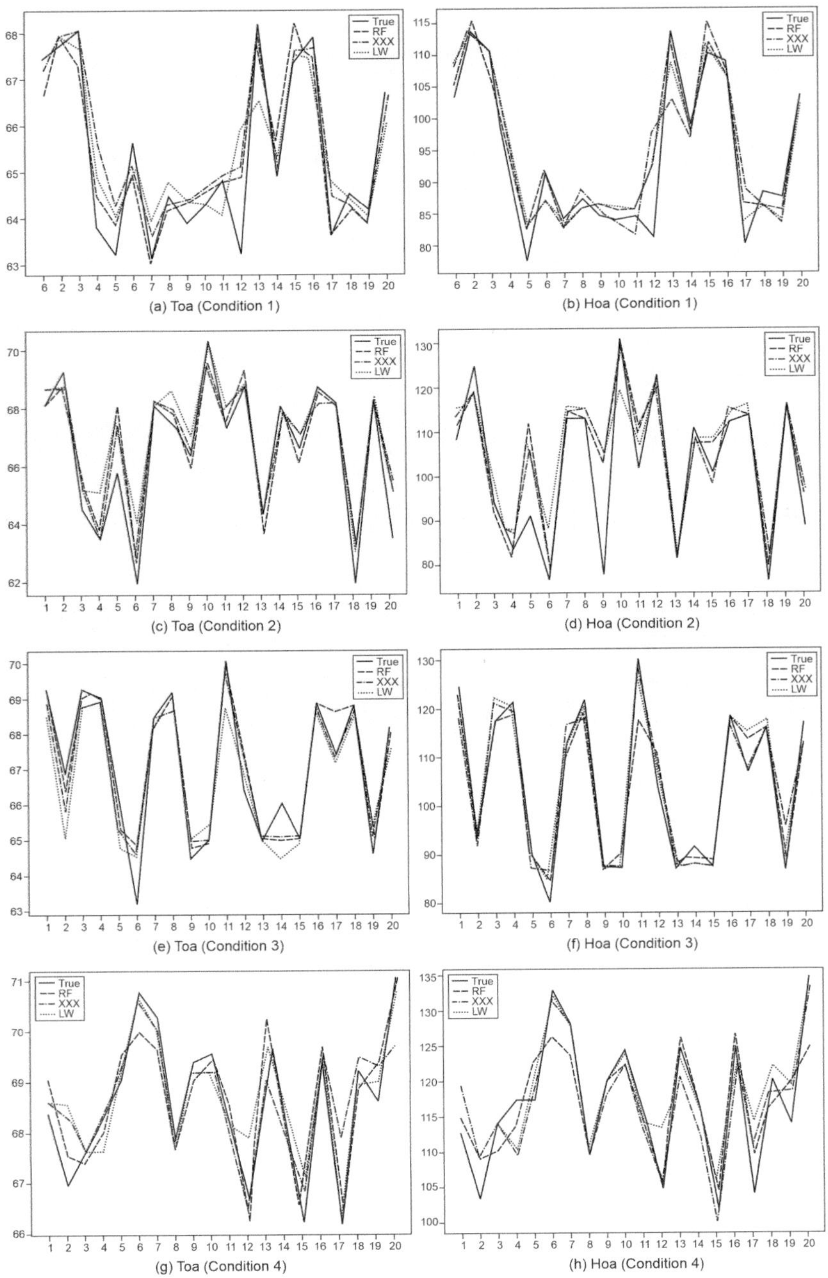

Figure 5: Prediction comparison of different models under four working conditions

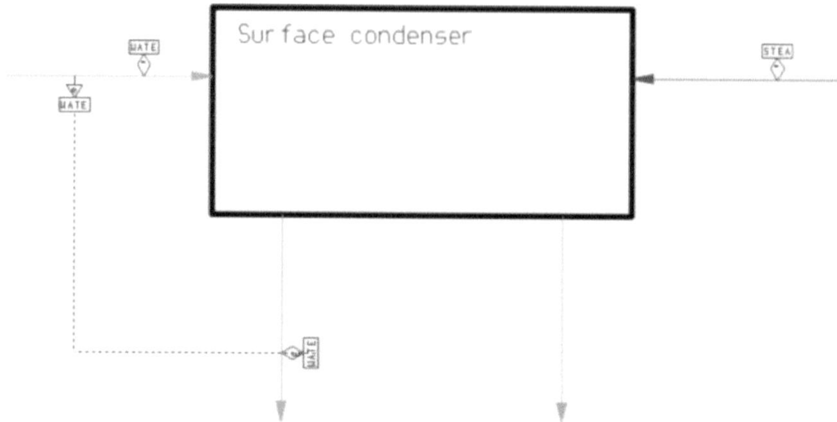

Figure 6: Interface display of surface condenser model

3.3 Visualisation model

The following simulation model is based on the example of a surface condenser, where the input and output processes of the surface condenser are analysed and the input and output streams are identified based on the mechanistic equations. A new module is created in CADSIM Plus as a "Polygon", which is a closed shape used to represent a unit of equipment and, when connected to the flow lines, forms the basis of the simulation model. Then to select the streams to be used and draw the flow lines, select "Create Stream Definition" to access the "Stream Definition Wizard" and select the streams to be used in the process Select the stream components to be used in the process, such as liquids, gases and solids, or some property values such as temperature and pressure. Select "Paper" for the "Stream Type", "STEAM" for the secondary stream and "Temperature" for the other three streams, depending on the mechanism equation. The other three streams are selected as "WATER" and temperature, and are constructed as shown in Figure 6.

CADSIM Plus communicates with the COM server through the COM-CLIENT module. External numerical data can be used in simulation calculations as part of the simulation model. CADSIM Plus can provide numerical data to an external application (e.g. Excel), which can perform its own cal-

Table 6: Interface display of surface condenser model

Heat less	M_{iw}	T_{iw}	M_{is}	T_{is}	M_{ow}	M_{oc}	T_{ow}
	0.51	11.42	100	30	0.51	11.42	55.87
	0.58	11	95	32	0.58	11	62.38
0%	0.66	11.55	99	35	0.66	11.55	68.00
	0.54	11.61	101	33	0.54	11.61	59.94
	0.67	11.32	102	32	0.67	11.32	66.29
	0.51	11.42	100	30	0.51	11.42	55.84
	0.58	11	95	32	0.58	11	62.35
3%	0.66	11.55	99	35	0.66	11.55	67.97
	0.54	11.61	101	33	0.54	11.61	59.91
	0.67	11.32	102	32	0.67	11.32	66.25

culations and then return the results to CADSIM Plus for further processing. Based on the pre-solved data, the accuracy of the simulation is verified by bidirectional data transfer via the software's COM function, and the operation is observed for different input parameters. It can be found that the simulation results are the same as the mechanistic calculations when the heat loss is set to zero, and the module can accurately simulate the surface condenser. The performance is affected when operating parameters such as the temperature and mass of the incoming secondary steam or the temperature and mass of the cold water are changed. Any change in these parameters requires a completely new simulation.

In real production processes, there is a certain amount of heat loss in the heat transfer process. In CADSIM Plus, it is possible to set the heat loss ratio for the equipment to obtain simulation results that are closer to the real situation. Table 6 shows that after adjusting the heat loss ratio for the equipment to 3%, the output term changes in a similar way to the real situation.

Figure 7 shows the interface of the surface condenser simulation run, the curves shown in the figure are the variation curves of condensate and temperature. In practical use, real-time data can be imported via EXCEL and the dynamic changes of output parameters can be observed after simulation calculations, and parameter variation curves can be constructed for each parameter in the model interface to provide reference for actual process decisions.

-0

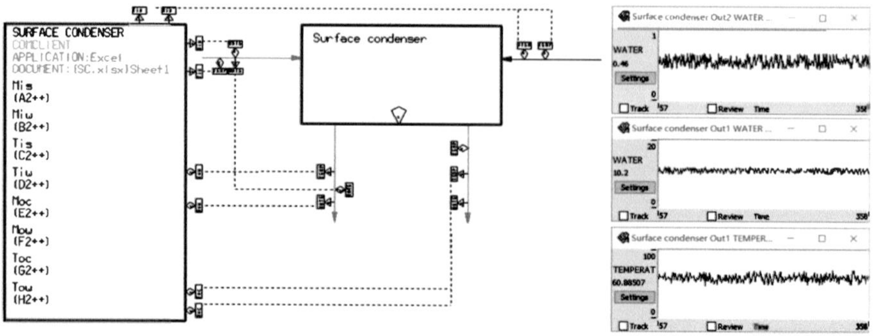

Figure 7: Real-time parameter monitoring interface of surface condenser model

At present, the papermaking industry lacks efficient means of whole life-cycle control. Therefore, this paper proposes a digital twin modelling framework for the paper manufacturing process, and develops two data completion methods for addressing the possible missing data in the modeling framework, including parameter solving based on mass and heat transfer mechanisms, and parameter prediction based on random forest.

The following conclusions are obtained: the mechanism-based parameter solution can be used as a general method to solve a variety of missing parameter problems and usually obtains better results; the random forest-based parameter prediction model is robust and has high accuracy, with the average value of R^2 above 0.9. Based on which, this the paper has implemented a visual modelling of the surface condenser in the dry section of papermaking process based on CADSIM Plus and the digital twin framework. The model runs well and is able to monitor the dynamic change of parameters in real time.

The digital twin-based visualization model proposed in this paper can reduce the coupling complexity of the physical entity modules, and has good scalability and generality, and can be subsequently extended to the whole process of paper production.

4 Future Works

In future studies, artificial intelligence and big data analysis technology can be further to be explored by online analysis on the basis of interaction with the production site thorough data and applying the established models to implement production process management for optimization and decision making, and so on. The integration of these techniques could significantly improve the production efficiency and reduce production costs, and ultimately achieve sustainable development of the process.

5 Declarations

Ethics approval and consent to participate: Not applicable
Consent for publication: Not applicable
Data availabilty: Data will be made available on reasonable request.
Conflict of Interest: The authors declare that there's no conflict of interests that could affect the work reported in this paper.
Funding and support: There is no funding or support for this paper.
Authors' contributions: Yishui Zhang carried out investigation, writing, review and editing, visualization. Rongyuan Wang carried out investigation, methodology, software, visualization. Mengna Hong took care of project administration and software. Zhenglei He carried out conceptualization, review, project administration, supervision.

References

[1] Y. Man, Y. Han, J. Li, M. Hong, Review of energy consumption research for papermaking industry based on life cycle analysis, Chinese Journal of Chemical Engineering 27 (2019) 1543–1553.

[2] F. Qian, D. Bogle, M. Wang, S. Pistikopoulos, J. Yan, Artificial intelligence for smart energy systems in process industries (2022).

[3] P. O. Antonino, R. Capilla, P. Pelliccione, F. Schnicke, D. Espen, T. Kuhn, K. Schmid, A quality 4.0 model for architecting industry 4.0 systems, Advanced Engineering Informatics 54 (2022) 101801.

[4] Y. Fan, C. Dai, S. Huang, P. Hu, X. Wang, M. Yan, A life-cycle digital-twin collaboration framework based on the industrial internet identifica-

tion and resolution, The International Journal of Advanced Manufacturing Technology (2022) 1–29.
[5] Y. Li, C. Yang, H. ZHANG, J. LI, Discussion on key technologies of digital twin in process industry, Acta Automat Sin 47 (2021) 501–514.
[6] L. Zhao, Y. Fang, P. Lou, J. Yan, A. Xiao, Cutting parameter optimization for reducing carbon emissions using digital twin, International Journal of Precision Engineering and Manufacturing 22 (2021) 933–949.
[7] T. A. Huynh, E. Zondervan, Process intensification and digital twin–the potential for the energy transition in process industries, Physical Sciences Reviews (2022).
[8] C. Lai, Y. Wang, K. Fan, Q. Cai, Q. Ye, H. Pang, X. Wu, An improved forecasting model of short-term electric load of papermaking enterprises for production line optimization, Energy 245 (2022) 123225.
[9] Y. Zhang, M. Hong, J. Li, J. Ren, Y. Man, Energy system optimization model for tissue papermaking process, Computers & Chemical Engineering 146 (2021) 107220.
[10] R. M. Soares, M. M. Câmara, T. Feital, J. C. Pinto, Digital twin for monitoring of industrial multi-effect evaporation, Processes 7 (2019) 537.
[11] M. Liu, S. Fang, H. Dong, C. Xu, Review of digital twin about concepts, technologies, and industrial applications, Journal of Manufacturing Systems 58 (2021) 346–361.
[12] G. N. Schroeder, C. Steinmetz, R. N. Rodrigues, R. V. B. Henriques, A. Rettberg, C. E. Pereira, A methodology for digital twin modeling and deployment for industry 4.0, Proceedings of the IEEE 109 (2020) 556–567.
[13] I. Shabbir, M. Mirzaeian, F. Sher, Energy efficiency improvement potentials through energy benchmarking in pulp and paper industry, Cleaner Chemical Engineering 3 (2022) 100058.
[14] J. Li, X. Tian, J. Liu, Dynamic data scheduling of a flexible industrial job shop based on digital twin technology, Discrete Dynamics in Nature and Society 2022 (2022).
[15] D. Bamunuarachchi, D. Georgakopoulos, A. Banerjee, P. P. Jayaraman, Digital twins supporting efficient digital industrial transformation, Sensors 21 (2021) 6829.
[16] C. Zhuang, T. Miao, J. Liu, H. Xiong, The connotation of digital twin, and the construction and application method of shop-floor digital twin, Robotics and Computer-Integrated Manufacturing 68 (2021) 102075.
[17] E. Negri, S. Berardi, L. Fumagalli, M. Macchi, Mes-integrated digital

twin frameworks, Journal of Manufacturing Systems 56 (2020) 58–71.
[18] F. Tao, F. Sui, A. Liu, Q. Qi, M. Zhang, B. Song, Z. Guo, S. C.-Y. Lu, A. Y. Nee, Digital twin-driven product design framework, International Journal of Production Research 57 (2019) 3935–3953.
[19] Y. Zhang, W. Wang, H. Zhang, H. Li, C. Liu, X. Du, Vibration monitoring and analysis of strip rolling mill based on the digital twin model, The International Journal of Advanced Manufacturing Technology 122 (2022) 3667–3681.
[20] G. Ding, S. Guo, X. Wu, Dynamic scheduling optimization of production workshops based on digital twin, Applied Sciences 12 (2022) 10451.
[21] T. Goodwin, J. Xu, N. Celik, C.-H. Chen, Real-time digital twin-based optimization with predictive simulation learning, Journal of Simulation (2022) 1–18.
[22] Y. Yin, J. Liu, Y. Wang, Y. Zhuo, Y. Meng, Modeling of ventilation's influence on energy consumption in multi-cylinder dryer section part1: Theoretical model, International Journal of Computational Intelligence Systems 15 (2022) 47.
[23] J. P. P. Marques, D. C. Cunha, L. M. Harada, L. N. Silva, I. D. Silva, A cost-effective trilateration-based radio localization algorithm using machine learning and sequential least-square programming optimization, Computer Communications 177 (2021) 1–9.
[24] Y. Liu, W. Shen, Y. Man, Z. Liu, P. Seferlis, Optimal scheduling ratio of recycling waste paper with nsgaii based on deinked-pulp properties prediction, Computers & Industrial Engineering 132 (2019) 74–83.
[25] A. Jadidi, R. Menezes, N. de Souza, A. C. de Castro Lima, Short-term electric power demand forecasting using nsga ii-anfis model, Energies 12 (2019) 1891.
[26] S. Verma, M. Pant, V. Snasel, A comprehensive review on nsga-ii for multi-objective combinatorial optimization problems, Ieee Access 9 (2021) 57757–57791.
[27] Y. Ao, H. Li, L. Zhu, S. Ali, Z. Yang, The linear random forest algorithm and its advantages in machine learning assisted logging regression modeling, Journal of Petroleum Science and Engineering 174 (2019) 776–789.
[28] G. Ciulla, A. D'Amico, Building energy performance forecasting: A multiple linear regression approach, Applied Energy 253 (2019) 113500.
[29] S. S. Rathore, An exploratory analysis of regression methods for predicting faults in software systems, Soft Computing 25 (2021) 14841–14872.
[30] D. A. Otchere, T. O. A. Ganat, J. O. Ojero, B. N. Tackie-Otoo, M. Y.

Taki, Application of gradient boosting regression model for the evaluation of feature selection techniques in improving reservoir characterisation predictions, Journal of Petroleum Science and Engineering 208 (2022) 109244.
[31] J. Cai, K. Xu, Y. Zhu, F. Hu, L. Li, Prediction and analysis of net ecosystem carbon exchange based on gradient boosting regression and random forest, Applied energy 262 (2020) 114566.
[32] L. Sun, Y. Ji, X. Zhu, T. Peng, Process knowledge-based random forest regression for model predictive control on a nonlinear production process with multiple working conditions, Advanced Engineering Informatics 52 (2022) 101561.

Index

D

E

Q

R

S

For Product Safety Concerns and Information please contact our
EU representative GPSR@taylorandfrancis.com Taylor & Francis
Verlag GmbH, Kaufingerstraße 24, 80331 München, Germany